How do peasants, producing mainly for themselves, become capitalist farmers, producing largely for sale? How far are peasants, anyway, involved in other pursuits besides subsistence farming, such as industry and dealing? What happens to farm sizes, farming practices, the farm family, farm workers, non-farming activities and the relationships between cultivators and other people, in the countryside and the town, in the process of this transition? How far does it vary from region to region? Is it inherent in the peasantry, or must it be instigated by landlords, towns, or the state?

These are some of the questions addressed by Göran Hoppe and John Langton in this study of regional change in Sweden. The authors have carefully combined theories about the transition from peasantry to capitalism with meticulous analysis of the abundant Swedish records. Through this 'new regional geography', they reveal the wide geographical variety and rich socio-economic complexity of the changes which occurred in the process of modernization in the nineteenth century.

The authors show that regional geography can be brought to bear on important questions about the way the world changes. They also show that explanations of economic and social change must get to grips with the wide variety of regional geographical experience if they are to be plausible.

T0269107

Cambridge Studies in Historical Geography 22

Peasantry to capitalism

Cambridge Studies in Historical Geography

Series editors:
ALAN R. H. BAKER, RICHARD DENNIS, DERYCK HOLDSWORTH

Cambridge Studies in Historical Geography encourages exploration of the philosophies, methodologies and techniques of historical geography and publishes the results of new research within all branches of the subject. It endeavours to secure the marriage of traditional scholarship with innovative approaches to problems and to sources, aiming in this way to provide a focus for the discipline and to contribute towards its development. The series is an international forum for publication in historical geography which also promotes contact with workers in cognate disciplines.

For a full list of titles in the series, please see end of book.

PEASANTRY TO CAPITALISM

Western Östergötland in the nineteenth century

GÖRAN HOPPE

Stockholm University

AND

JOHN LANGTON

University of Oxford

CAMBRIDGE
UNIVERSITY PRESS

CAMBRIDGE UNIVERSITY PRESS
Cambridge, New York, Melbourne, Madrid, Cape Town, Singapore, São Paulo

Cambridge University Press
The Edinburgh Building, Cambridge CB2 2RU, UK

Published in the United States of America by Cambridge University Press, New York

www.cambridge.org
Information on this title: www.cambridge.org/9780521259101

First published 1994
This digitally printed first paperback version 2006

A catalogue record for this publication is available from the British Library

Library of Congress Cataloguing in Publication data

Hoppe, Göran, 1946–
   Peasantry to capitalism: western Östergötland in the nineteenth
century / Göran Hoppe and John Langton.
      p.   cm. – (Cambridge Studies in Historical Geography: 22)
   Includes bibliographical references.
   ISBN 0 521 25910 X
   1. Agriculture–Economic aspects–Sweden–Östergötlands län–
History–19th century.   2. Peasantry–Sweden–Östergötlands län–
History–19th century.   3. Östergötlands län (Sweden)–Rural
conditions.   4. Land tenure–Sweden–Östergötlands län–
History–19th century.   I. Langton, John, 1942–   .   II. Title.
III. Series.
HD2020.O88H66   1994
305.5′633′09486–dc20   93-36927   CIP

ISBN-13 978-0-521-25910-1 hardback
ISBN-10 0-521-25910-X hardback

ISBN-13 978-0-521-02641-3 paperback
ISBN-10 0-521-02641-5 paperback

Contents

Illustrations

Plates

Figures

Tables

Preface

When we began the research for this book over a decade ago, we had a clear and limited end in view. It was to use ideas from time-geography to examine empirically, through the great wealth of data available for Sweden, what happened as capitalism developed in the rural areas which formed an overwhelming proportion of Europe when it occurred. The particular questions in which we were interested were those current when we began: how far and how quickly did farm size increase; what were the associated changes in capitalization and the labour process, and therefore in the mode of production? How were changes in agriculture linked to non-farming activities? Were rural changes supply driven, due to impoverishment from population growth, or a response to the transmission of new demands from external markets? How swift, uniform and comprehensive were changes across the rural economy and the farming population?

Much of our early work was concerned with collecting data from the almost overwhelmingly copious Swedish records.[1] The completion and analysis of data sheets took the first six or seven years of the time devoted to the book. Data and questions must be bound into a structure of argument to make them coherent. From the start, we were determined to hold together some of the categories which are often dealt with separately in analytically construed historical and geographical work.[2] First, we aimed to do this temporally: to get back to the early stages of modern development to examine how it emerged from – indeed, within – an indubitably 'pre-industrial' society. This (as well as the timing of the appearance of reliable and relatively standardized and comprehensive records) led us to begin our research in 1810 and end it in 1860, spanning the period when development was then thought to have been getting under way in Sweden.[3] Topically, we wanted to span the divides between rural and urban, agrarian and industrial, economic and social which obscure as much as they reveal of the past.[4]

These combinations operate in two dimensions. The first, 'vertical' one, is concerned with how different aspects of life in particular places affected each other: how economic changes were related to changes in social structure, and agrarian with industrial changes, for example. Secondly, they have a 'horizontal' dimension: how what occurred in one place affects what happened in another. As human geographers, we were intrigued by how these sets of integrations varied geographically within national boundaries. It is now common to argue that the process of development in early modern times must be sought at a regional scale,[5] so we decided at the outset to incorporate fundamentally different environments in our analysis. This is what led us to choose as the particular geographical arena of our research western Östergötland, where fertile plains and rugged shieldland abut sharply along a line followed by the Göta Canal after 1832 (see figure 3.1).

This is simply to say that we wanted to do regional geographical research, as now much advocated by aficionados.[6] However, because traditional regional geography treats regions integrally in order to characterize the processes at work within them, it ignores the second of the combinations just outlined: how interactions between neighbouring regions of differing character, and longer-distance contacts, affected what happened within each of them. This is why we included towns in our scheme of research. Interactions between individual places, our two subregions, and the region as a whole and places outside it must, at least to some extent, have been focussed through them.

It is easy to propose to study historical changes in topical and geographical integration at one and the same time; but it is another matter to devise a scheme through which this can be done. Lack of a firm conceptual structure and a rigorous methodology have always bedevilled regional geography.[7] Three aspects of modern work on peasant society attracted us to it as point of departure for our research. First, its relevance to the agrarian economy of much of today's Third World as well as to the European past, thus providing the widest possible comparative context for our findings. Secondly there is a close equivalence between the notion of peasant ecotype and the traditional regional geographical conception of *pays*. Thirdly, correspondences can readily be drawn between work on peasantry and research into the process of proto-industrialization.[8]

The concept of peasantry and the relationships between peasants and other sectors of society during economic development are the subject of chapter 1. Chapter 2 is an exploration of the (also currently much-touted) possibility of using the concepts and methods of time-geography to organize our enquiry into the dynamics of regional economic development in peasant society. Chapters 3–7 present the results of our empirical work. Three of them examine economic structures and the ways in which they were changing between 1810 and 1860, using modifed conceptions drawn from time-geography. The other two use ideas from time-geography to trace how flows of money became flows of capital and how the changing lifetime experiences of individual people had to change with structural economic change. The conclusion provides a summary of our empirical findings about the regional dynamics of the transition from pre-capitalist to capitalist modes of production and an evalution of the conceptual and methodological questions discussed in chapters 1 and 2.

We owe warm thanks to archivists in a number of Swedish Record Offices, particularly to Sven Malmberg and his colleagues in Vadstena. We are grateful, too, to the Swedish and British students who worked on inventories and the Göta Canal accounts in Vadstena in 1983. The maps and diagrams were drawn by Brita Hellichius-Pybus, Angela Newman, Peter Hayward and Ailsa Allen, and some of the text was typed by Marie MacAllister and Janet Bark. The book was conceived on an Anglo-Swedish Symposium in Historical Geography organized in 1978 by Alan Baker, to whom we offer warm thanks. We are very grateful, too, for the patient and skilful supervision by Richard Fisher of what proved to be a protracted and difficult delivery. Our archival research and numerous consultations about data analysis and writing were funded by the Swedish Research Council for the Social Sciences and Humanities, Carl Mannerfelt's Foundation, Hans and Lillemor Ahlmann's Foundation for Geographic Research, the University of Oxford Travel Fund, and St John's College, Oxford, to whom we also express our gratitude.

1
Introduction: peasants and economic development

Peasantry versus progress

Peasantry is a human ecological concept, concerned with how the economies and societies of human groups are related to the physical environments from which they satisfy their material needs. This makes it well suited to our task. However, its common use to demonstrate the nature of barriers to change in pre-capitalist farming systems suggests the opposite; indeed, it can be thought of as the obverse of a developed socio-economic system. Peasants operate in a moral economy, rather than according to market economic imperatives.[1] Their 'history is immobile',[2] running in long repetitive demographically driven cycles: there can be no long term material (or any other kind of) progress; '"historically speaking", the peasantry does not exist'.[3] If these things are axiomatic, the concept of peasantry cannot be articulated with modernity, but only set against it. It is often argued, too, that modern socio-economic systems and the values they embody developed in urban industrial societies,[4] whilst the survival of peasantry prevented the spread of capitalism into rural production.

To discover whether or not these things are necessarily so, we need to define some of the attributes of a market economic system of the kind whose emergence we are seeking.

The nature of capitalist farming

Capitalist market economic systems maximize wealth creation by directing productive effort to where the greatest relative scarcities are indicated by the highest prices. They require producers to act in accordance with the balance of costs and returns as defined at the economic margin and expressed in factor and product prices. This, in turn, means that the agents must benefit as individuals from operating in response to market signals. The resources transformed in doing so, and what is created in the process, belong to them. A

1

market economic system requires private rights in property to ensure behaviour which will maximize the monetary returns yielded by work.[5] It also needs 'free' pools of capital and labour which the owners of resources, each acting separately to maximize their private monetary returns from market exchange, can switch between different production projects in response to price changes: labour is proletarianized and land treated as a means of investment. The subdivision of labour in the performance of specialist tasks boosts the efficiency with which wealth is created and distributed further. This leads to economic and social differentiation between those who own the means of production and wage-workers, and within these two groups themselves. Subdivision of labour occurs spatially, too, with places as well as people concentrating on growing or making what can most efficiently be produced there according to comparative advantages. In these ways, a capitalist market system stimulates growth in aggregate wealth and population.

It has profound consequences for the structures of economy and society as well as their sizes. On the one hand, class divisions reflect differing relations of people to the means of production. A status society in which wealth follows the possession of power is superseded by a class society in which power follows wealth.[6] On the other hand, capitalism has little room for 'society' or 'community', but only individuals; everything is done by and belongs to individual people performing economically specialized roles, who band together in families, classes and states in response to biological and political necessity, but interact mainly through markets. Such social formations, unlike peasantries, actually *have* histories, driven by the aspiration of individuals to get more wealth to propel them upwards in and between the boundaries of the classes into which they were born.

In rural areas, the hallmarks of capitalist production are highly differentiated farm sizes,[7] labour proletarianization,[8] capital formation and investment,[9] the individuation of people as economic and social agents,[10] and regional differentiation.[11] Market exchange aimed at maximizing money returns to effort creates a highly stratified farming society. In it, large holdings are accumulated through investment on a land market and worked by hirelings acquired on a labour market to produce the bundle of goods which will yield the highest returns on commodity markets. Such behaviour implies that farmers are trying to maximize the accumulation of money wealth. If they are doing this in farming, they will see farming as only one of a range of possible options for the deployment of their resources to that primary end. Peasantry as a way of life must disappear as maximizing farmers become part of a bourgeoisie within which they are able to move out of farming.

The figures on table 1.1 show why competition between farms on commodity markets in environments where yields fluctuate from year to year inevitably means that 'the strong become stronger and the weak go under ... [which] leads to the conversion of independent producers into wage workers

Table 1.1. *The effects of harvest fluctuations on sizes of farm surpluses and deficits (in barrels)*

Farm size	Yield	Subsistence requirement	Crop surplus or *deficit*	Price ratio[a]	Cash surplus or *deficit* at £1 per barrel in a normal year (in pounds)
Normal harvest					
10 ha.	300	240	+ 60	1.00	60
20 ha.	600	240	+ 360	1.00	360
150% harvest					
10 ha.	450	240	+ 210	0.40	84
20 ha.	900	240	+ 660	0.40	264
50% harvest					
10 ha.	150	240	− 90	5.53	− 498
20 ha.	300	240	+ 60	5.53	332

Note:
[a] Calculated according to the formula of Bouniatian.[12]

and numerous small enterprises into [a few] big ones',[13] and why, once differentiation has occurred, the other characteristics of capitalist farming inevitably follow. The financial implications of variations in yields and inelastic demand imbue a commercialized farming system with a powerful tendency for farm size to rise to the point where surpluses can be produced in the worst possible years. Small farmers must suffer chronic indebtedness. They try to keep as far out of the market as possible and pray for good harvests. Large farmers wring every surplus grain they can from their land for sale and pray for low yields.[14] To reap maximal rewards, farms must grow beyond the size at which labour has to be hired from outside the family. The more it is possible to apply labour only when it is needed and to pay the same money wages per hour of work in bad years as in good ones, the more a farmer can keep for himself of the windfall gains brought by bad harvests. Farm labourers, like small farmers, suffer hardship in bad harvests. To run their large enterprises and benefit fully from specialization, farmers slough off many direct farming tasks and craft work and become professionalized.[15] Maximum benefit is gained from harvest fluctuations by growing the crop which thrives best in the environment where the farmer's land is situated, subject to relative prices. The long-term diversified activities of polyculture give way to monoculture on land which is considered simply as a disposable asset from which maximum returns must be made as quickly as possible.

Peasantry as an economic, social and cultural formation does appear to be the antithesis of the capitalist farming system we have just described. However, when followed beyond a superficial level, many of the logical consequences of the basic attributes of peasantry produce forms of behaviour and structures of farming which share many of the signs of a capitalist system. The empirical differentiation of a peasant from a capitalist farming system is far from straightforward.

According to the most succinct formulation, peasantries have four basic characteristics:[16] they comprise subsistence farmers, who live in kin-based household groups working family land, who are boorish and conservative compared with other social groups with whom they co-exist, and who are removed from the organs of political control over the territory they occupy. We will examine each characteristic in turn, and the mutual aid between neighbours, particular attitude towards money and exchange, distinctive demographic régime, and geographical differentiation to which the four primary characteristics give rise, to show why in theory they produce an immobile, unprogressive history, and why, paradoxically, peasantries exhibit many forms of social and economic behaviour which are, as empirical phenomena, indistinguishable from those of rural capitalism.

Peasantry as historical stasis

Peasants and subsistence farming
'Peasant life is devoted to survival. Perhaps this is the only characteristic fully shared by peasants everywhere'.[17] It was best guaranteed by self-reliance: dependence on others brought risks. As much as possible of what was required to sustain life must be produced on the peasant's own land, from animal and vegetable food and fibres to building materials and substances from which domestic utensils and tools could be made.[18] Household economies were generalized, diverse and similar: rather than different households occupying specialized economic niches in a subdivision of labour articulated through market exchange, different members of each household performed part-time as many as possible of the wide range of tasks necessary to provision a household completely.

Land on which to deploy this intricate set of activities was the *sine qua non* of this mode of survival. 'All in all, the fundamental aim of the peasant was not profit or the accumulation of capital. The aim, consciously or unconsciously pursued, was rather to ensure the continuity of the family and its lands from one generation to the next'.[19] Peasants worked to conserve the full range of resources their land provided.[20] Even so, too much wind or frost, or hailstorms, or too little rain or sunshine would bring disaster unless farming strategies aimed at producing enough in the worst possible conditions; at maximizing the minimum yield should the weather or pests do their worst; at minimizing risk, rather than maximizing output in one or more seasons.[21]

Because peasants had to pay taxes, and perhaps other dues, they must produce something for sale on a regular basis. These requirements for money were accommodated by producing a particular cash crop, animal, or marketable craft product specifically for the purpose.[22] When the environment was propitious, other surpluses would be produced inadvertently. Then, extra sales were made in order to accumulate a stock of money to insure against a future failure of subsistence, and lavish consumption occurred. Each farm family needed enough to keep themselves and their beasts alive, to exchange for the few necessary things that they could not produce for themselves, and to pay the levies imposed on them. But subsistence economy had no use for production beyond that. On the other hand, every family was absolutely required to produce all that was deemed to be necessary. The upshot was very little social differentiation within peasant societies, and very little variation between the produce and routines of different family farms.[23] Indeed, peasantries often had powerful mechanisms to eradicate inequalities that arose fortuitously:[24] conspicuously wealthy households had to do more to support the church and the poor, and to serve in the more onerous offices required for the community to function.[25]

The idea of marginal returns could not be formulated, let alone deployed. Labour inputs were not costed against a notional money return, and whether or not to expend them decided on that basis. People had to work as hard as necessary to produce their target output, however many hours it took. If environmental conditions worsened progressively, or if rents rose or prices fell, peasants worked harder, prone to 'self-exploitation'.[26] On the other hand, when conditions were good they seemed idle and feckless.[27] Normally, peasant families deployed their labour

in such a way that all opportunities that give a high payment may be used. Thanks to this, the peasant family, seeking the highest payment per labour unit, frequently leaves unused the land and means of production at its disposal, once other forms of labour provide it with more advantageous conditions.[28]

If upland grazing, fishing or part-time work on capitalist farms provided a more secure way to derive subsistence, effort would be directed outside farming, so that pastoral, fishing or worker-peasant communities developed.

The concept of investment was also absent. There was neither the opportunity nor the incentive to forego current consumption for greater future output; a peasant family must, in order to exist at all, already be achieving its target output, and the accumulation of surpluses to fund investment was prevented by the redistributive mechanisms described above. This was entirely rational for subsistence farmers. No one knew better than they that the fruits of the earth are finite and only yielded in return for almost incessant toil; if someone had more than their necessary share, it must have been gained at the expense of someone else. Peasant societies were permeated by 'the envy principle',[29] or 'the principle of the limited good':[30] someone could only gain progressively by causing others to lose. This acted to prevent both differentiation and the investment necessary to produce it.

Peasants and family

Holdings must be provided with considerable labour per unit area to produce the large and varied output that the subsistence imperative required. The be-all and end-all of secure survival meant that only labour which was completely beholden to the family unit through belonging to it would be used. There was symmetry in peasant societies between the size of holding that could support a family and that which could be worked by one, so that the terms peasant farm, family farm and labour farm are synonymous. This symmetry was maintained by choosing appropriate techniques, cropping strategies and labour intensities.

for the peasant farm, the task is to compose its field cultivation of such crops and proportions that the critical moments characteristic of their labor organization do not coincide, and the general labor intensity of the farm should be more or less uniform. In this respect the labor farm's tasks differ sharply from those of the capitalist farm which solves the critical moments of its labor organization with the help of temporary labour.[31]

This reinforced the tendency towards polyculture induced by the subsistence imperative, and meant that the social relations of production on it were completely imbricated with those of kinship. The family farm was not only the basic unit of production, consumption and biological reproduction, but also almost the sole agency of socialization, transmitting the ideology and training appropriate to peasants' economic and social roles, and providing support for young, aged, infirm and other relatives without membership of landed households of their own.

Resources were transmitted through successive generations of these kinship groups, and the composition of households was closely related to the inheritance system used.[32] In the Balkans joint family households formed *Zadruga*, in which groups of married kinsmen of the same generation lived together and jointly owned and worked family land, often with different nuclear sub-units specializing in different tasks.[33] Elsewhere in Europe it was more common for different families of the same cohort to live in separate households, with partible inheritance splitting the father's land between sons and movable property between daughters.[34] If parents survived beyond the age at which they could work, they distributed their property pre-mortem and lived in one of their children's households.[35] If there was only enough to allow marriage and family formation by one child, other children remained unmarried and lived in an extended family household with the married sibling. In these ways, a peasant family projected its survival by binding together successive generations which, at the extremes of youth and age, were unable to provide for themselves.

Extended and stem families change in composition through the life cycle of the central nuclear generation,[36] so that more or less labour became available, and more or less consumption had to be provided. But the two series do not run in parallel:[37] With many young children, consumption needs rise faster than labour power; as the children grow up, the opposite occurs, although consumption continues to rise. Symmetry between the consumption needs of a holding and the labour power available on it can only be maintained if either cropping systems with different labour intensity are used at different stages of the family life cycle, or if labour in excess of its own consumption needs is imported into the household at the start of the central nuclear family's life cycle, and labour (and consumption) exported, or more land acquired, at the end of it. Given the control exerted over cropping systems by the need to provide the fullest range of products and even-out labour demand through the year, the former course was *a priori* less likely than the latter. Peasant households therefore often contained members who did not belong to the stem family group, usually men and women in their late teens or early twenties who were surplus to the labour requirements of their own family farms. They were not like the wage workers of capitalist farming, but lived on the farm as family members. A snapshot cross-section would, therefore, produce the impression that the peasant farms of a village were highly differentiated, with some hiring labour. But this differentiation was cyclical rather than cumulative or permanent.

It could extend into other activities than food production for family consumption. Food and drink must be prepared and cooked; milk converted into butter and cheese; hides and skins, bones, feathers and bristles rendered into useful articles; yarn spun, cloth woven, clothes made and mended, tools and implements crafted and repaired, buildings and fences constructed and kept in good order. The combination of arable polyculture and animal rearing ensured that most of what was consumed could be supplied by the land and labour of a peasant family farm. This required a wide range of craft skills and an intricately organized subdivision of labour within the household. Generally, it was related to age and gender, usually under patriarchal direction, and varied as farming tasks varied from one physical environment to another. In arable farming regions of temperate latitudes men worked the fields, built, repaired and maintained farm buildings and fences, and did other work needing physical strength, such as slaughtering or carpentry. Women did most of the work in the house and homestead, such as spinning, weaving and tailoring, churning butter and cheese, cooking and brewing. Children learned their specialist adult roles from an early age, although there were some age-specific tasks, such as bird-scaring for boys and egg-collecting for boys and girls. Given their control over and operation in the homestead, where outsiders made contact with the farm, and their greater involvement in rendering produce into a consumable state, women also usually did any buying and selling from the homestead and, by extension, any marketing.

These specialist gender roles made marriage more of an economic alliance than a romantic attachment, and it was accompanied by the formation of a unitary conjugal property.[38] An imbalance of males and females in a household could be dealt with by exchanges between the farms of kinsmen. Or a family with numerous older girls could produce more bread, beer, cheese, butter or thread than the family itself required, exchanging them for goods in which the vagaries of demography had made the household deficient, and acquiring off the farm any extra raw materials that the greater than normal production of particular goods made necessary.

The mere existence of differentiated farm sizes and transactions in factors of production and products does not indicate that peasantry had ceased to exist: 'all these combined economic and social relationships ... did not run counter to the primary fact that every household had an economy of its own, the furthering of which was the basic interest of the peasant and his wife.'[39]

Peasants and money

Peasants needed money to pay dues and store surplus value after good harvests. They might also use it to measure relative value and as the medium of exchanging value in the transactions made necessary by the incompleteness of family subsistence and shifts in the ratio of household consumption and labour through the life cycle. However, the expression of the size of debits and credits in terms of money does not mean that money changed hands,[40] and even if it did this does not necessarily imply that the sum involved represented a value fixed at the margin by the intersection of supply and demand schedules. Prices paid per acre of land or per unit of labour could be conventional and unchanging or oscillate wildly[41] in peasant societies, and peasants might deliberately choose to sell at prices below the maximum available.[42] Their attitude towards money was paradoxical: on the one hand there was its miserly accumulation to no apparent purpose; on the other hand, they could not value their work in money terms, paid more to pedlars for buttons, ribbons and trinkets than they could buy them for in towns, and sold crops and animals to neighbours and dealers who came to their village for less than they would fetch in urban markets.

Peasants and community

Considerable interdependence between farms was necessary if each one was to maximize the probability of its own survival. Close and intricate bonds between neighbours were inevitable, in the form of shared beliefs and daily practices, and mutual obligations and favours owing from the past and predictable in the future. These bonds were numerous and 'multiplex: that is to say, ... not specialized to deal with a single activity', but inseparably intertwined 'the economic, political, ritual and spiritual, the functional and the evaluative, the expedient and the moral'.[43] Peasants operated in 'closed corporate communities' or 'moral communities' comprising all the house-holds near enough (or with close enough kin ties) to interact in these ways. Relationships with outsiders were usually characterized by a sharp concern to derive maximum immediate advantage. There was a deep division between 'us' and 'them' in peasant consciousness.

If the harness of a peasant's horse broke during farming, or a cartwheel collapsed bringing the harvest home, or a wife took sick, immediate help was needed. It was given unquestioningly by neighbours, in the full knowledge that there would be some future reciprocation. Most exchanges of this kind were unequal as well as widely spread through time and confined to neighbours. They took generations to be settled, if they could be fully settled at all. The use of money would have allowed each obligation to be terminated immediately,[44] and we have seen that money was available in peasant societies. However, the money value of each transaction was not calculable. Each particular transaction occurred so rarely that it was impossible to fix relative exchange values. The need for money in peasants' obligatory dealings with outsiders, whilst they had no use for it in their everyday activities, accounts for their apparently paradoxical attitude towards it. Pedlars were patronized because they visited the village regularly, marking cyclical rhythms, bringing news and stirring recollection. The sole purpose of buying from them was not, as in dealings with full outsiders, to maximize immediate material or monetary gain, but to bind them into the nexus of mutual obligations on which village life depended.[45]

The interdependence of neighbours, their common dependence on the land they occupied, the existence of a resource which could not be parcelled up into private property, and in feudal times their common duty to farm the lord's land interspersed with their own, gave rise to common property. With its adjunct of communal institutions and endeavour, this was also instrumental in best-guaranteeing survival for the individual peasant farm. They varied greatly in form and importance according to the types of resources they used.[46] It was impossible to split resources such as alpine pastures, extensive scrub grazing, or surface water and coastal fisheries into private parcels, even though they could be exploited by private herds, pumps or boats. Communal use brought economies of scale through a unitary institution such as a village council, or from some people performing tasks such as supervising herds, or harvesting hay or grapes, for all. Sharing rights to resources over a wide area also meant that risks due to spatial differences in fluctuations in the weather or pests were shared, and their effects on particular peasant families minimized.[47] The relationships between community, shared work, spread risks and common property reached their apotheosis in the plainland arable farming (*champion*) areas of Northern Europe, where there were common plough teams, herds and flocks; common wastes for grazing and fuel gathering; meadows grazed in common after the hay crop had been taken from privately allocated strips, and common open arable fields in which the holdings of each family were split into many scattered parcels for cropping, but combined into a single block of common grazing in fallow years. Communal effort was needed in ploughing, harvesting, and fencing to separate common pastures from arable fields and the land of neighbouring villages.[48] This was a risk-minimizing system *par excellence* on heavy clays under cloudy skies in Northern Europe. In return for its guarantee of survival, individual peasants forfeited the right to decide what crops to plant, when to plough for and harvest them, and how many and what types of beasts to keep. They vested decision-making in communal institutions such as manor courts and village meetings, which directed the nature of effort on individuals' strips and organized communal work routines as well as supervising, through its decisions and appointed officers, the exploitation of common property.

In another paradoxical way, then, the drive to guarantee the survival of the peasant family on the subsistence provided by family labour from family land gave rise to strong extra-familial bonds between neighbouring families and with outsiders who habitually visited their village. But there were few links of any kind between different peasant communes because physical distance precluded cooperative effort, shared resources, the necessary immediacy of aid, and the security with which recompense could be expected. Each peasant village was an integral 'little community'.[49]

Peasants and conservatism

Oral transmission of material and non-material culture in the household and village served to maintain customary practices and values across the generations.[50] And why should the practices which had ensured their forebears' survival be changed; and how could one family change its practices without all others in the community doing the same? Carefully regulated communal farming and common property completely prevented the adoption of maximizing strategies by individual peasants. Suggestions to change came from outside the peasant community and were 'therefore automatically categorized as dangerous and sinful . . . those villagers who [adopted] . . . new roles [ran] the risk of being marked as deviants and punished'.[51] 'Nobody is less opportunistic (taking the immediate opportunity regardless) than the peasant.'[52]

There was nothing innately 'irrational' or 'non-maximizing' about this conservatism in the face of the need for self-provisioning from limited resources, harsh nature and social oppression: it is chopping logic to argue that minimizing the risk to subsistence is different from maximizing the chance of survival. There was as powerful a logic to peasant behaviour as to that of capitalist farmers; their objectives simply differed.[53] But this 'backward referencing culture of survival'[54] induced ever more earnest claims that change must be forced by enlightened people with power over peasant land, or by the state during the nineteenth century.

Peasants and powerlessness

Everywhere at all times, 'peasant society is part society'.[55] Members of the other group, the patriciate, were always more powerful. They usually comprised about 10 per cent or so of the total population, rising to 15 per cent in Western Europe in 1300 and 20 per cent by 1500.[56] All the great civilizations of the world between the Roman Empire and the Industrial Revolution were established by patriciates,[57] and all were based on surpluses abstracted from peasantries. 'The clumsy lout who was deprived and mocked by court, noble and city', 'the stepchild of the age', was also 'the broad patient back who bore the weight of the entire social pyramid'.[58] The rights and privileges of the patriciate and the subservient status of the peasantry were based in law codes: power was defined by status at birth in pre-modern society, and enshrined in law. Almost without exception (although, as we shall see, Sweden was one) this state of affairs was based on government under a royal sovereign by three Estates of the Realm: nobility, clergy and townsmen. The 'fourth estate' of the peasantry was almost always unrepresented.[59]

Peasants rarely had ultimate power over the land on which they subsisted, and which supported the patriciate by abstractions from their crops. Least beholden to others were those with *dominium directum* as well as *dominium utile* over their land: that is, they owned it according to the modern conception of freehold tenure, with full power to alienate the land itself, as well as rights to use it and alienate its produce.[60] Such peasants paid no rents and could not be dispossessed, although they paid taxes to central government and tithes to the church and were usually subject to military conscription and had to quarter troops. This type of tenure was almost absent from Europe, except in its extreme northern and western fringes. Freehold over use rights (that is, the right to use land in perpetuity unless it was neglected) without *dominium directum*, was much more common, especially in the west. This tenure required the payment of dues to the lord who held *dominium directum* as well as taxes and tithes. They varied from being almost negligible to onerous exactions in the form of hunting rights, payments in kind on the transfer of land to heirs or by sale, threshing, carting and road repair duties, and the duty to grind corn only at the lord's mill.

At the other end of the scale from free tenure was serfdom, still common across Eastern Europe in the early nineteenth century.[61] A serf was the property of his lord and could be sold or bought like a farm animal. His duty was to provide labour on the lord's land. In return, he was supplied with land on which to grow subsistence for his family, but his rights to it were exiguous: he had *dominium utile* only at the lord's discretion.

Between serfdom and free tenure lay various forms of leasehold, in which a regular rent was paid (in addition to dues to the lord who held *dominium directum*) in return for *dominium utile* over land. Often lords followed a combined strategy, farming some of their domains directly and leasing the rest out in return for the labour needed to do it. In many parts of Western and Southern Europe most lords' domains were leased out; lords were rent-gatherers rather than cultivators, and share-cropping tenure was common. Further north fixed rents were more usual, paid in labour, cash or kind. Terms could be for fixed numbers of years, for the life of the tenant, or for his and a number of succeeding male lives. Fixed money rents over long terms were beneficial to tenants during inflation, and this is how free *dominium utile* tenures developed in France.[62] Labour services became more valuable to lords as produce prices rose, which is probably why (like serfdom) they were retained over Eastern Europe, where lords were more usually directly involved in farming on their own account than in Western Europe.[63] Rents were high where landlords kept control over their levels. In parts of Eastern Europe, for example, labour services of more than 365 man-days per year, plus the use of children at wages lower than the market rate, plus craft goods from the wife might be charged for a holding.[64]

The second Estate, the church, took its share of peasant production directly through tithes, paid in money or kind. The third Estate of the urban bourgeoisie also took its share directly where town corporations had either or both forms of *dominium* over farmland. More usually their share was abstracted indirectly by controlling the prices received and paid by peasants.

The total weight of dues and rent payments to which peasants were subject were alternatives, their total weight could be very heavy. They amounted to one-half of the total production of a typical Irish peasant farm in the early nineteenth century,[65] and over 40 per cent in eighteenth-century France. In parts of continental Europe, especially in the east, they could account for more than the value of the annual crops of a holding, forcing peasants to add value to their produce through craft production, or take additional work outside farming. Output from common property could ameliorate these swingeing ratios; indeed, access to commons might be the most important resource conveyed in a lease.

Peasants and population

The combined demographic effects of the aspects of peasantry we have discussed so far were homeostatic: a change of population in a particular direction, relative to available resources, was countered by the feedback mechanisms it automatically induced.[66] The relative autonomy of the peasant farm as a subsistence work unit in Western Europe threw the need for symmetry between the resources and labour power of family holdings into particularly acute relief.[67] This was coupled with strong discouragement for children to be born out of wedlock, and a social system in which the formation of a new family unit required access to sufficient resources to support it independently.[68] In consequence, the marriage of sons was delayed until the inheritance of parental land, or the pre-mortem transfer of a holding or its lease in return for support for parents when they were too old to work.[69] Marriage partners were therefore often in their thirties, although if parents died young marriage was early.

Other things being equal, a late age at marriage meant few children. Of course, the partner whose age is significant in this respect is the wife, not the land-holding husband. However, strong influences worked to ensure that men did not marry much younger women,[70] and marriage on a husband's inheritance, or the incapacity through age of his parents, meant that his wife had little more than a decade of fertile married life. This was still long enough to produce four or five children,[71] but breast-feeding lengthened birth-intervals, high infant mortality removed some of the children who were born,[72] and because at least one parent would usually have to depend on children in old age, they would ensure a progeny small enough to produce enough food and other necessities for the purpose from the family holding.

These demographic mechanisms would only operate if peasant family holdings were barely sufficient to produce their own subsistence plus the dues they must pay to others. However, if land were relatively abundant, peasant populations had the capacity to expand very rapidly by creating more household units through the marriage of all children at younger ages than would be fixed by their age at inheritance. In these circumstances parents would be tempted to relinquish their holding to a son long before they were unable to work it themselves to ensure that younger people were present on the holding when they ceased to be capable of farming. Thus, population growth would occur through larger family sizes as well as a higher rate of nuptiality and younger age at marriage if land were abundant.

What was being kept constant was the ratio between population numbers and the amount of sustenance available: a target income per head. Epidemics, environmental conditions so bad that even the strategies of peasants could not yield adequate harvests, and wars all produced land surplus to the needs of survivors, and all were followed by rapid population recovery.[73] The innovation of new techniques of production or food crops increased the subsistence capacity of a constant area of land, and allowed a higher population at the same target household income on smaller farms. The conservatism of peasants and their suspicion of innovations suggested by outsiders made sudden innovatory changes unlikely, but the continual elaboration of tried and tested techniques was commonplace. The gently rising trend in the troughs and peaks of population cycles in Western Europe owed something to cautious developmental changes of this kind. Sometimes innovation was rapid, if it seemed to offer a surer route to the peasants' target of independent and reliable subsistence on the basis of their own efforts on their own land. An acre of potatoes could support about five times as many people as an acre of oats.[74] Like grain, potatoes can be used as animal fodder and to distil alcohol, but in addition more of the crop is edible because the proportion which must be saved for seed is much smaller, and no equivalent of threshing, milling or other preparation outside the homestead is needed to make them edible. It was of no significance to subsistence farmers that, unlike grain, potatoes were expensive to transport and could not be stored beyond one season. Their cultivation led to rapid subdivision of peasant holdings, higher rates of nuptiality and larger family sizes. Given the apparent reliability of the crop, landlords also behaved predictably, allowing tenant farms to be split between heirs, and peasant populations grew explosively in consequence.[75] Because potatoes yield particularly heavily on erstwhile pasture land (especially on sandy soils and immediately after introduction) and can be used as animal fodder as well as human food, their cultivation allowed the extension of arable land into pastoral and forested areas and their colonization by denser populations. The rapid diffusion of potato cultivation amongst the European peasantry was associated with rapid population growth in the eighteenth and nineteenth centuries, *without* the cessation of their normal homeostatic demographic mechanism. Left to play itself out, it might simply have produced a long cycle of population increase which was steeper, but different in no other significant way from the long surges of demographic increase eventually halted by diminishing returns which had always existed in Western Europe.

Peasant ecotypes
An ecotype is the means adopted by a society for deriving energy for human consumption from the environment it occupies. Classification according to 'the degree of use of a given piece of land over time'[76] gives five distinctive peasant ecotypes, each broadly related to a different world climatic, biogeographical and edaphic zone, including the grain farming with annual fallow combined with herding of northern temperate regions. Within this broad category there were major variations reflecting physical geography, such as the Plainland, Atlantic fringe, alpine and Mediterranean variants.[77] At a smaller scale again, within the last, for example, mountains, plateaux, hills and foothills, and the plains, split between inland plains and plains of the sea each supported a distinctive ecotype.[78]

Particular means of harnessing energy from nature had distinctive effects on social relations, demography and beliefs about the natural and human worlds. Generally and on a large scale, the length of the fallow period was positively related to the amount of common property and communal endeavour.[79] As we have already seen, the heavy dependence on annually fallowed grain farming in European *champion* regions gave rise to communal work routines and a complex mixture of private and common property rights which, in turn, were generally associated with primogeniture inheritance systems and with a stolid and conformist temperament.[80] It was the existence of great variety within the environment, each village having its own 'vertical north' and sites for extensive grazing, hay meadows, forest, orchards, vineyards, and grain fields within a small area, which gave communities in alpine ecotypes their unusually high degree of economic, social and political autarky.[81] Further north, exiguous dependence upon grain harvests and reliance upon upland grazing in *bocage* regions, or fishing grounds on rugged coastlands, brought heavy dependence upon rights in common property which was exploited by private endeavour rather than communal effort, greater reliance upon kin than community, partible inheritance, anti-authoritarianism, and strong beliefs in the power of good and bad luck.[82]

Dissatisfaction with nationwide stereotypes of rural society and culture has been accompanied by greater attention to these relationships between habitat, economy and society in medieval and early modern Europe. But 'nothing will more surely guarantee the banality of ecological study ... than the insistence that cultural phenomena are ... a consequence of nature's unqualified cunning'.[83] It might also be argued that the reflection of environmental differences in farming practices will be less marked the more self-contained a peasant economy is; that 'a more precise, less coarse meshing of land use to technology and geography becomes possible' with commercial farming.[84] The geography of ethnic groups caused social structural and cultural variety with no relationship to physical geography, and 'within the range of our [one] subsistence base in hoofed herds and plowed-field grains, there are many ways of linking herds and herding to the grainfields, many diverse types of social organization on the subsistence base [so that] we are far removed from easy ecological determinants'.[85] Moreover, peasants, and other farmers, alter the physical environment in using it, and the actual resources which exist for exploitation by a peasantry are as much a product of their farming system as their farming system is a product of natural resources.[86] Self-contained causal chains did not link given characteristics of the natural environment to farming systems, and farming systems to social structure and culture in a simple deterministic way:[87] it is 'an erroneous supposition that it is possible to deduce the rest of the structure of a reproductive totality from the process of material production alone'.[88] Particular natural ecosystems contained quite different ecotypes,[89] and the further we move from farming system to culture, the more fully and subtly other variables intervene along the chain of contingencies which links nature to culture. Similarly, the smaller the scale, the less clearly such influences can be discerned: peasantries in similar environments in England had different farming systems and inheritance patterns, whilst those occupying different environments had similar ones;[90] adjacent alpine valleys had different inheritance practices and religious beliefs, and responded to identical outside stimuli to change in different ways.[91]

The most important intermediaries between physical environment and peasant ecomomy, society and culture came from its connections with other parts of society. Whether peasants had freehold legal title to land; whether serfdom was legal or not; whether peasants were represented in the state, and the power of their local village courts; whether particular inheritance systems or religious practices were prescribed by law; whether peasants were heavily or lightly taxed; whether markets were available for surplus agricultural and craft products, and the extent to which peasants could enter those markets freely and directly; whether paid work was available in farming or other activities nearby: all these things were controlled outside peasant society. They all affected the farming system adopted in an environment and impinged at other junctures along the line of influence from farming practices to social structure, demography and culture. The particular structure of domination to which a peasantry was subjected had a profound influence on how it adapted to a particular environment.[92] 'We have only to compare ecological adaptations in the same village community over a long span of time to find that although roughly the same resources are exploited in 1650, 1750 and 1850, the changing macro-structural conditions make the adaptations very different'; the concept of ecotype must be defined as 'a pattern of resource exploitation within a given macro-economic framework'.[93]

However, *within* any particular structure of domination and macro-economic framework, peasantry varied from place to place in ways which were related, *inter alia*, to variations in the resources provided by the physical environment. Although no more than caricatures when presented in list-like simplicity, we can draw for early modern Western Europe a set of intelligible relationships between plains, open field arable farming, communalism, primogeniture, lack of industrial by-employments, religious conformity and political quiescence, on the one hand, and between uplands, pastoral farming, individualism, craft by-employments, partible inheritance, religious non-conformity and political protest on the other.[94] Whilst it is true that to focus too narrowly on human interaction with the natural ecosystem in the analysis of human ecological relationships is 'to elevate a methodological emphasis ... to the status of an ontological reality',[95] to deny any significance to the influence of environment upon peasant economy, society and culture is at least equally perverse.[96]

But how firmly did ecotypes gain regional geographical expression? Given the dependence of the whole ensembles of relationships within ecotypes upon macro-economic structures and patterns of socio-political domination, they would only be patterned according to the natural environment if these variables were, too. Striking correlations did exist between natural conditions, peasant economies and social formations, and structures of political domination. But legal and administrative systems, religious affiliations and political allegiances could equally well vary in ways that were unrelated to the natural environment,[97] and some of the macro-economic structural variables would express themselves on the ground as punctiform markets or linear trade channels which would differentiate between areas within the same natural region. Although it is possible to predicate the existence of ecotypes within which local natural and technical, and wider political and economic influences interacted together in distinctive ways, the extent to which the latter would confound any tidy regional patterns incipient within the former is problematical.

Peasant exchange

Subsistence farmers best guaranteed their own survival by producing the fullest possible range of outputs from the complete range of micro-environments at their disposal. Land which was marginal for one crop but optimal for another, and nearby land which was different, could be used at different times of year, as in transhumance.[98] However, if distance prevented this or the labour requirements could not be combined into a single régime of seasonally meshed activities, each environment might support separate ecotypes which regularly exchanged products. This was common everywhere. In Central America, for example, certain villages produced all the straw goods, others all the earthenware, others all the output of particular food crops, and so on. But these marketing systems retained constant prices because the relative quantities of each good produced remained constant, and commodity markets existed without the emergence of markets in labour and land in the production of the goods which were bought and sold.[99]

For price-fixing through the supply and demand of commodities to occur and spread into the exchange of factors of production, 'network markets' must exist: that is, marketing systems with complex, overlapping and spatially wide-ranging exchanges, arbitrated by specialist middlemen who freely shifted goods around to where they fetched the highest prices. In network markets, peasants had to sell for whatever they could get, and buy at prices fixed by forces completely outside their control.[100] Prices in any one market also fluctuated, according to the outcomes of very complex interplays between ever-changing demand and supply schedules over a wide and generally ever-expanding area. In consequence, peasants could never predict what their surpluses would earn in a network market, nor what the things that they must buy would cost. The principles on which market exchange operated therefore seemed completely unintelligible to peasants,[101] and anyway their complex polycultural farming system and the intermeshing of social formations with systems of cultivation prevented a response to changing prices by changing crops. Similar conditions existed if labour rather than products were sold by peasant households into a free market economic system. Wages were fixed by the price of the crop on whose specialized production they were hired to work, which itself was fixed in the network market system; because of their own complex production system, peasants could not modulate the amount of labour released in response to variations in wages.

All over the world, and as far back as can be known, surpluses were extracted from peasants through the marketplaces which physically embodied the marketing system. Tolls and stallage were charged by the landowners or urban authorities who owned the marketplaces; middlemen came in the hope of buying cheap and selling dear; townsmen offered goods produced by monopolistic guilds in exchange for the output of individual unorganized peasants. To avoid these sources of disadvantage, peasants behaved in marketplaces according to their own moral economic standards, rather than in response to the price-fixing imperatives of the marketplace itself.[102] The surest means of escape was to depend as little as possible upon organized exchange in the marketplace, and as much as possible upon transactions with neighbours. Most villages had specialist craftsmen such as blacksmiths and tailors who usually also farmed and were fully imbricated in the peasant moral economy.[103] Another way was to draw particular traders into the moral economy of the village by refusing to deal with them in a narrowly specialized role, whether as a regular itinerant through the village or as a habitually visited trader whom they regarded as their agent and friend in town.[104]

A fourth means of avoiding formally instituted marketplaces was for peasants or their wives to sell surpluses directly to consumers.[105] Often, these exchanges were made once a year at local fairs (where the entertainment, jollification and reunion with old acquaintances which accompanied trading indicated an origin outside the imperatives of price-fixing exchange).[106] Or, setting out once a year in the slack farming season to peddle a huge variety of wares from door to door through the countryside or from fair to fair, they traversed well-worn and often surprisingly long circuits.[107] Seasonal surplus labour was sold in the same way. From medieval times, habitual journeys from pastoral or forested regions to the arable plains were common;[108] by the end of the seventeenth century they formed a cumulating cascade to the shores of the North Sea for the hay, madder, flax and grain harvests.[109] These regular periodic spells of migrant labour provided supplementary income rather than the mainstay of peasant households, and they carried all kinds of social and cultural as well as economic connotations. Like dealing with 'their own' merchants and selling their own products door-to-door on annually repeated circuits, they were a means of keeping out of the grip of formalized specialized marketing structures rather than an entry into them. Taking place outside marketing institutions, and therefore invisible to those who exercised political control, made records and abstracted peasant surpluses, these exchanges articulated a vast 'hidden economy' through which varied peasant ecotypes interacted.

Peasantry as process versus capitalist farming

Peasantries had complicated and varied economies, social formations and cultures; they were not slab-like formations demarcated by national boundaries. Nor were they static, but changed constantly in response to natural and induced environmental changes, demographic imperatives and outside influences. Peasantry is not a structure but a process.[110] Nonetheless, all peasantries had fundamental attributes in common, which remained constant as other aspects changed in homeostatic cycles rather than progressively, ensuring that they could have no real history.

The need for all peasant households to ensure their material survival and reproduction on family land with family labour,

determined both the relative stability of peasant economy – its ability to survive and revive itself in the most unfavourable circumstances – and its relative stagnation, its low level of labour productivity and its inability to broaden production. ... Market integration could only take place at the level of surplus production (rather than at a level allowed by the production of specialist crops deliberately for sale).[111]

Land and labour were not construed as factors of production which could yield money income, but as family resources to be used at the levels of intensity and endurance necessary to ensure social reproduction. Money was earned, but only to satisfy the impositions of those with power over them. Uncosted cooperation with neighbours in work and in the ownership, maintenance and administration of land and other property helped to keep peasantry as far outside the remit of formal market exchange as possible. They also prevented the full alienation of land and labour by an individual peasant. Neither, therefore, could be subjected to the economic disciplines of competitive allocation and price-fixing exchange. Commodity exchange between communities with different resources was integral to peasant economies, but it was firmly geared to achieving social reproduction, not maximum money returns. Family-formation practices were heavily homeostatic, ensuring as far as possible that social reproduction was maintained by producing enough but not too many members of successive cohorts. A holistic, introverted and backward-referencing world-view reflected these imperatives, reinforcing and validating the peasantry's secure survival as far as possible beyond the influence of other parts of society, where forces inimical to its own concerns constantly threatened.

Although 'this self-sufficient society was indestructible',[112] its stability was of a very particular kind: it was stability of the target income and lifestyle of individually reproducing families. This *required* short-term cyclical changes of personnel within peasant households, in their holdings of land and money, and in the goods they produced, in line with the cycles of biological reproduction. It also required long-term swings in overall population levels to accommodate exogenously generated mortality crises and any innovations which were obviously and immediately conducive to the better achievement of peasant family reproduction. Any cross-sectional study of peasantry is therefore likely to discover holdings of different sizes, the employment of non-family labour, specialist crop and craft production, and large holdings of money: features which might be taken to be symptoms of capitalist farming. If the cross-section were allowed to run for a short while through time, land would be seen to change hands for money, sales of crops, animals and craft produce would be registered. None of these features would, by itself or in combination, indicate that peasant society was in the process of transformation to capitalism. All were necessary to maintain the security of the peasant family on its own land.[113] Similarly, peasant ecotypes might develop, through exchange, a pattern of regional variety which is difficult to distinguish from the geographical outcome of commercial production across a varied physical environment.

To infer from these various criteria whether capitalist farming had emerged needs two further questions to be answered. First, the nature of the holdings in which they occurred has to be discovered: did they, through their temporal changes and geographical differences, maintain the character of diversified family farms producing a wide spectrum of crops, animals and craft goods on the basis of an intricate subdivision of mainly family labour? Secondly, the system must be observed over a long period of time, to discover whether changes connoted the cyclical movements necessary to maintain stability over time, or the beginnings of cumulative change. Did larger farms grow beyond the apex reached when the size of the peasant family household was at its maximum; did hired labourers maintain that status through the whole of their lifetimes, rather than hold it temporarily before inheriting their own land; was the production of particular kinds of crops or craft goods in increasing quantities accompanied by a reduced range of output and sustained over a long enough period to indicate specialized commodity production for the market rather than surpluses beyond self-consumption within families of changing composition or in different environments; was money systematically accumulated (or borrowed) in order to invest in equipment and buy more land upon which to deploy hired labour to produce commodities for sale, rather than hoarded against bad times or used to buy land to accommodate increases in family size? In short, the structural *and* longitudinal analysis of holdings, plus longitudinal analysis of the balance of craft and crop production on holdings *and* of individual people, are all necessary to reveal whether a farming system is best interpreted as surviving peasantry or incipiently capitalist.

Chapter 2 deals with how both longitudinal and structural analysis can be done. However, we must first examine the conceptual problem of how commodity production, the individuation of people and their operation as economic agents with specialist roles, private property, investment, aspiration after monetary gain, competition and openness to outside economic and cultural influences emerge in a system whose hallmarks are the antitheses of all of them.

From peasant economy to market exchange

These changes were transmitted into peasantries from outside: they were 'incorporated' or 'emancipated' into a capitalist economy developed within the 'Great Tradition' within which the myriad 'Little Traditions' of peasant ecotypes were set.[114] We will distinguish between the roles of landowners, clergy and bourgeoisie, plus the state institutions they created, to give some systematic coherence to models of change which are usually presented separately.

The role of landlords

The political power of landlords varied across Europe, and so did the degree of their dominance over the peasantry and the means through which they abstracted its surplus labour. In most of continental Europe landowners using land for their own purposes paid no taxes.[115] In Hungary they did not pay tariffs, tolls or tithes either, and in Poland they could fix the prices of goods on their domains. Generally, landowners had a monopoly over corn milling and making and selling alcoholic liquor on their domains; sole responsibility to administer justice to people resident on them, to hunt game over them, and to determine the conditions under which land was held by those who used it.[116] Peasant surplus produce was abstracted through the labour services of serfs, through the labour services, produce or money rents of tenants, and through tolls, taxes and other imposts levied on nominally freeholding peasants. The distinctions between serf, tenant and free peasant were not completely clear cut. The broadest definition of serfdom as the enforced transfer of surpluses to lords[117] categorizes all peasants who did not have *dominium directum* over their land as serfs. However, it is better for our purposes to distinguish between different forms and degrees of peasant subordination: between serfs *sensu strictu*, who belonged to their lords but, unlike slaves, had individual rights in law; tenants, who paid labour services or rents in kind or money and were free to leave their holdings, and free peasants, who had either *dominium directum* over their land or paid rents which were a fixed token of ownership rather than a significant impost under the control of the landlord.

Serfdom and the new feudalism[118]

After the end of the fifteenth century East European landowners pumped more and more grain and cattle out of their estates directly into world trade,[119] bypassing local urban merchants and their price-fixing markets. To do so they deepened the enserfment of their tenants by reducing their land holdings and increasing the labour services of men, women and children. Peasants owned neither their land nor their labour, and any propensity to produce surpluses beyond what was necessary to pay taxes and other dues was smartly negated by reducing the size of their holdings or increasing labour services or the supplies of linen, hemp, honey or tar required in return for their subsistence plots. Serfs were forced to remain in the natural economy by lords trying to wrest the maximum marketable surplus out of their labour. For each individual serf family, extra hands were an additional resource which, *in extremis*, would be kept alive by hand-outs from landlords. The servile community maximized its freedom from oppression by cleaving to common property rights (which could not be commandeered, but would be degraded through over-use with increasing dependence on them),[120] and to non-market exchange which could not be tapped by landowners. Peasant traits were propelled into extremity by a particular system of domination in a particular part of the world oecumene during the particular historical epoch of the second feudalism.

Neither did landowners operate in capitalist ways in this system. Their purpose was to maximize the annual money surplus which accrued to their operation which, with food and other goods produced on the estate, was used to support not only the landowner and his family, but also clients, retainers and acolytes in numbers ensuring political power, and to purchase expensive status-conferring consumer goods from Western Europe. To achieve this the feudal estate, like a peasant holding, produced everything possible towards its own consumption, often including fuel, cloth, metalware and glassware as well as grain, hay, flax, hemp and other crops, with uncosted serf labour and the estate's own land and other resources. As the terms of trade of Baltic grain against west European manufactures worsened from the seventeenth century, the tendency towards estate autarky increased in the trans-Baltic lands, which severely restricted the extent to which market exchange penetrated commodity production. Like peasants, feudal estate owners responded to changes in the prices of the goods they sold in the opposite way from capitalist farmers: as the relative value of what they produced for exchange fell they produced as many other goods for direct consumption as they could in order to maintain the size of their cash surplus, and increased the output of their cash crop. Both strategies meant reducing the size of peasant plots and imposing more onerous labour services, further reducing the possibility that peasants could enter market exchange.

In the new feudalism of Eastern and parts of Southern Europe a peasantry and a patriciate of particular kinds were locked together in mutually determining relationships of a particular kind in a system dependent on the form of its external economic relationships. Why the peasantry of Eastern Europe was inimical to the development of capitalist farming cannot be understood with reference to the peasantry alone.

Freedom and involution

Peasant freedom from feudal dominance was no more propitious for the development of capitalist behaviour, but tended to generate 'involution': that is, the support of ever greater numbers of people at a constant level of subsistence, without fundamental changes in technology or lifestyle, by ever more intricate elaborations of current practices.[121] 'Freedom' was anyway usually heavily qualified, signifying simply than the annual rents into which labour services had been commuted in the twelfth and thirteenth centuries had, through inflation, become valueless. This encouraged lords with *dominium directum* to increase tolls, monopoly milling and distilling rights, hunting rights, and rights over resources on wastes and commons.[122] It also prompted them to take land back into their own possession whenever possible. In many areas of free peasantry lords had a preferential right to buy peasant land offered for sale, and reversionary rights if peasant heirs failed. Estates were common and became more so through the seventeenth and eighteenth centuries.

Free peasantries generally practised partible inheritance, which was sometimes – as in Sweden and, later, areas subject to the Napoleonic Code – imposed by statute law. In bocage regions, where arable and meadow land were mainly important as means of gaining access to large tracts of upland commons, partible inheritance remained current throughout the early modern period even in England, where primogeniture developed early.[123] In longer and more densely settled open field regions it was usually succeeded by primogeniture as arable holdings shrank to the size where subsistence was only attainable if they were not split further. However, younger children did not lose all rights to family property with primogeniture, but only to its land, and elder sons were often burdened with debts incurred to provide younger children with portions and dowries transferred *inter vivos*. The availability of markets for cash crops outside the peasantry had the same effects on free peasants as new crops or techniques which increased yields per acre: cash proceeds were hoarded and used to sustain larger families. By funding the purchase of enough land for all sons to inherit, or an additional cash portion for the eldest son to buy out the younger ones if sufficient land was not available for all, cash hoards sustained partible inheritance. With primogeniture hoarded cash was used to purchase land for pre-mortem distribution to younger sons and for daughters' dowries. Greater fragmentation was inevitable, given that land adjacent to the peasant's own was unlikely to be available at the time when more was needed, which inevitably reduced the efficiency with which it could be utilized.

Free peasants with access to markets outside the peasantry for their products (or labour) simply drove up land prices, sucking capital out of the commercial economy into ever multiplying, ever shrinking, and ever less efficient primarily subsistence holdings. As yield on the capital invested in their purchase, the outputs of these holdings were far below commercial rates of interest.[124] The predisposition of peasants to provide portions for all children prevented the emergence of markets in land and labour.[125]

Free peasants with access to markets were also prone to the classic version of 'the tragedy of the commons'.[126] The gain from selling an animal accrued to its owner, but the cost of production was shared by all with common rights. This was also true of draught animals required for ploughing or to transport grain crops to market, so that greater arable output from private plots also led inevitably to pressure on common resources. In consequence, they were heavily degraded in peasant farming systems with access to outside markets, despite the introduction of complex rules to allocate common rights. The formulation and enforcement of these rules increased the total cost of production,[127] locked the peasant into even more constraining communal bonds and further impeded the penetration of market exchange into the allocation of land resources.

It was not, therefore, simply that most aspects of peasant farming could survive while free peasants produced commodities for the market,[128] but that they might be exacerbated by it. Again, we see that the characteristics of peasantry in early modern Europe were in the process of elaboration, not surviving archaisms, and that the exact nature of the process was dependent upon what was happening outside the peasantry itself: on the existence of an external market for goods and, in this case, on the ability of lords to prevent peasants getting title to their land.

Tenancy and agrarian capitalism
Tenancy was more amenable to the emergence of capitalist behaviour: 'with the peasants' failure to establish essentially freehold control over the land, the landlords were able to engross, consolidate and enclose, to create large farms and lease them to capitalist tenants who could afford to make large invest-ments'.[129] Landlords destroyed peasantry. Certainly, they had a vested interest in forcing onto their tenants the preference for primogeniture which developed in their own circles in the sixteenth century,[130] to prevent holdings from shrinking to a size where rents could not certainly be paid. In addition, capitalist transformation required that they were also able to ensure that high money rents could be charged, forcing tenants both to maximize the produc-tion of marketable surpluses and to transfer the bulk of it to them, rather than to children, and via that route into inflated land values and population growth.[131]

The debate over how this was achieved, and about why or whether it was confined to England, is far too complex to reproduce here.[132] In essence, the argument presents a sequence of changes leading from medieval serfdom to rack-rent-paying individuated capitalist tenants who, prised from communal practices and common resources though enclosure, worked large integral private holdings using rotations incorporating fodder crops, needing large hired labour forces.[133] In the process, the commoditization of land and labour, and the supercession of feudal lords and peasant communities by a class structure of landlord, tenant and wage labourer developed in English agriculture by early modern times.[134] But landlord power alone was not enough to bring these changes about. As we have already seen, in Eastern Europe an identical ambition of landowners with powerful control over the conditions of peasant tenure forced servile peasants even further into natural economy and communal behaviour. Their drive to maintain the productive potential of estates and prevent their despoliation[135] was a long way from the 'spirit of improvement' (that is, a drive to increase the productive potential of an estate through investment to yield higher money rents) which characterized English landowners by the eighteenth century.[136] Money rents under the control of landlords rather than labour services came to obtain where market relations were *already* extensive in economic organization: capitalist economic relations must exist somewhere else in patrician society before they could develop out of feudal agriculture.[137] And even where money rents were charged by landlords, peasantry could spiral alarmingly downward in involution: the pursuit of higher cash income through serving growing markets for grain and cattle in England after 1815 caused an expansion of directly farmed estates in Ireland, a concomitant collapse in the size of peasant holdings, and potato-dependent peasant misery.[138] Absentee landlords, also common in Southern Europe and most regions where large estates were dominant,[139] were much more prone to collude in involution than those who lived on the estates which provided their income. On the plains of Northern and Central France large landowners

had little interest in farming. Most behaved as *rentiers*, quite satisfied with the tendency for revenue from farms or sharecropping to increase as population pressure grew and competition for the privilege of becoming a tenant intensified ... Instead of encouraging innovation, leases tended to forbid the modification of established canons of good practice in case change would cause the impoverishment of the soil.[140]

And in France, unlike England and the Low Countries, tenantry was closely correlated geographically with partible inheritance rather than primogeniture.[141] Landlord power could be used to reinforce peasant conservatism as effectively as to destroy it.

Even in England, where resident landlords became more and more common in the seventeenth and eighteenth centuries,[142] the destruction of peasant farming required active collusion by the state as well as the capacity of landlords to fix money rents. Whilst English law was developing to encourage enclosure and the development of full private rights in landed property, governments elsewhere in Europe were expending equal energy in retaining the privilege of tax exemption on land directly farmed by landowners, thereby ensuring the continuance of labour services, and protecting the peasantry in other ways as the bases of taxation and military recruitment.[143] The development of a wage labour market would have been greatly retarded, if not completely blocked, without the existence of statutory poor relief to tide labourers over slack seasons and bad years in English commercial farming.[144]

The role of the church

Large areas of land were held by the church in medieval Europe.[145] Its estates were innovative in farming techniques, and heavily involved in producing wool and grain, according to the environment in which they were located, for the market.[146] In countries which became Protestant in the sixteenth century many well-run, market-orientated farms passed to the Crown, to be sold to landowners who borrowed in order to purchase them.[147] The notion of land as an investment and the institution of a land market received an enormous fillip from the Reformation.

The influence of the church reached all aspects of everybody's lives: 'medieval men and women possessed (or, perhaps more accurately, were possessed by) a galaxy of strong and interconnected concepts . . . [M]ost had been crafted for centuries by the Church'.[148] Roman Catholicism was at least partly responsible for the emergence of the West European family, through its powerful sanctions against illegitimacy and clear rules of property descent,[149] which was crucial to the demographic homeostasis of the peasantry. It also legitimated and allied with the power of landlords, and emphasized the role of oral rather than written communication and community rather than individuality. The view of cyclical time and past perfection propounded by the church militated against any notion of progress, and it represented the forests which provided the chance of escape from landowner and church oppression as a haven of demons and outcasts. This ideology and the hegemony of land and church 'proved remarkably durable, outlasting both Renaissance and Reformation, at least in Catholic Europe', where it survived until the nineteenth century.[150]

Protestantism sapped four foundations of traditional peasant non-material culture: it promoted individuality rather than community; confidence in the power of action rather than resignation; willingness to explore alternatives rather than unquestioning reliance on customary models from the past, and literacy rather than the oral transmission of culture. These traits also fragmented the unity of the church, its teachings and its authority. But in countries which seceded from the authority of Rome, the established Protestant sect was used to communicate national political, social and economic doctrines to the populace at large. Weaker communalism and freedom from outside control in ecotypes in upland regions of Central Europe and bocage and mountain regions of the north and west made them particularly receptive to Protestant beliefs. In turn, these traits were powerfully reinforced by Protestantism.[151]

The role of the bourgeoisie

Towns and the patriciate

Although the conversion of peasant surplus labour into cash did not need permanent urban markets and merchants, but simply annual fairs or peripatetic selling, the operation of the patriciate itself did. Temporal and spiritual administration required a permanent bureaucracy and other personnel, courts and other institutions. Servicing them with provisions, equipment, labour and specialist expertise multiplied out into other dependent ways of making a living. The resources left after providing for the administrative and spiritual needs of the peasantry and themselves was spent by the patriciate on goods and services which did not figure in peasant households: heavily wrought cloth and metalware; ornaments and utensils containing precious metals; exotic spices, wines and fabrics, and other specimens of art and artifice with which to adorn their castles, mansions and churches. They were provided by craftsmen and merchants in the towns: 'the intimate relations between the luxury consumption of the nobles [and churchmen] and the occupations of merchants and handicraftsmen should be noted. Towns often grew up in the shadow of a castle',[152] cathedral or abbey. At first, their relative size depended on the area over which these institutions exercised administrative power and drew in agricultural surpluses, ranging from the nationwide hinterlands of capital cities, through the smaller catchments of provincial centres and cathedral towns, to the church or noble estates scattered through the regions surrounding castles, monasteries and abbeys. The vast majority were in the last category: 'one can view the European economy of the late sixteenth century as consisting of hundreds of towns with hinterlands of 50 to 100 square miles. The great bulk of all agricultural output that left the farms was disposed of within these units'.[153] Nonetheless, towns were sharply marked off from the rural world by the ways in which the majority of townsmen made their livings, and by the institutions and legal status necessary to circulate, transform and capitalize agricultural surpluses. 'The neat separation of small market towns from the surrounding countryside ... was the mark of a predominantly agrarian economy with minimum industrialisation'.[154]

In Western Europe generally, the bourgeoisie formed by the craftsmen, merchants, administrators and other professional people of the towns was small, but it comprised the Third Estate of government, with commensurate economic privileges. Monetized exchange was restricted by law to regulated urban marketplaces. Only members of the appropriate urban craft guilds were permitted to buy materials, add value to them through manufacture, and sell the finished product. Other people could by law only work up in by-employments raw materials which they had themselves produced, and could only sell what they made directly to consumers or to urban dealers. Members of merchant guilds held a monopoly over buying goods at one price and selling them unaltered more expensively, either in local shops or through trade with a distant market.

The development of capitalist attitudes and modes of behaviour is generally ascribed to this bourgeoisie, especially the merchants within it. Unconditional rights in private property, laws of contract, individual personal freedom, and the hiring of wage labour, as well as the use of money to make more money rather than to fund consumption, came early to them, protected and legitimated in their own special urban courts.[155] Why landowners with supreme military and political power protected the distinctive activities of the small bourgeoisie is explained by Adam Smith's 'Bauble Thesis': they had an overweening desire, for the sake of prestige, for the rare and precious goods with which only long-distance merchants could supply them.[156] In most of Western Europe, unlike to the east, the bourgeoisie was able to defend its privileges over craft manufacturing and trade, and therefore an urban monopoly over them, until the second half of the nineteenth century.[157] Production in large factories, on the other hand, was usually reserved for royalty and its grantee monopolists, or the landowner Estate, especially in Eastern Europe where serfs provided free labour.[158] A necessary corollary of the reservation of particular economic activities for each Estate was that the peasantry was stuck fast by law, irrespective of any innate cultural inclination, into an economically subservient and almost purely agricultural role. Its functions were to provide taxes to finance governments, and manpower to fight wars to defend them; to produce crops for sale to urban dealers and craftsmen, and to labour so that landowners could do so, keeping itself fit for these roles through natural economy.

Capitalist economic development pulsed more and more strongly through the ever-thickening network market of the World Economic System subtended by the ports and capital cities of Western Europe in late medieval and early modern times.[159] It flowed fastest where factors of production were released most fully into market exchange, and therefore where the peasant economy was drawn furthest into it, despite the predilection of peasantry to avoid this and the formidable legal barriers preventing it. In parts of Western Europe, primarily England, this incorporation was achieved indirectly: where landowners had the means and inclination to maximize money rents, the peasantry were pushed into maximizing cash crop output, and therefore into greater specialization and greater dependence on the market for household provisioning as well as money income for rents. However, peasants could also be incorporated directly by the bourgeoisie itself, either through consuming 'Z goods' or via proto-industrialization.

'Z goods'

In England the proletarianized farm labourers and specialist petty craftsmen spawned by the early commercialization of farming were dependent on the market for their day-to-day requirements. Domestic workshops and shops supplying clothing, furnishings, bread, cheese, tea, sugar and so on were scattered thickly through the small towns and villages of arable southern England by 1759.[160] Although people with land were able to provide many more of their everyday needs for themselves,[161] technical improvements in farming and greater production for the market led to a marked increase in the size and quality of housing, the number and value of purchased household goods, and the amount of cash and precious metals possessed by English farmers as early as the sixteenth and seventeenth centuries.[162] The English government's seventeenth-century policy to introduce better quality manufacturing techniques from the continent to reduce the cost of of imports,[163] and its abandonment of town protection through prohibition of rural industrial activities,[164] both testify to the aspirations of English farmers after 'Z goods': that is, goods of a type and quality which they could not produce for themselves, but had to buy from skilled specialist producers or merchants.[165] By the mid-eighteenth century

The English ... have better Conveniences in their Houses and affect to have more in Quantity of clean, neat Furniture, and a great variety, such as Carpets, Screens, Windows, Curtains, Chamber Bells, polished Brass Locks, Fenders &c. ... than are to be found in any other country of *Europe, Holland* excepted ... [A]lmost the whole Body of the People of *Great Britain* may be considered either as the Customers *to*, or the Manufacturers *for* each other.[166]

To supply these commodities, specialist retail shops and itinerant traders multiplied prodigiously, forming a fully integrated system from large-scale city distributors down to part-time village chandlers, hawkers and chapmen by the late eighteenth century.[167]

Where peasantry and the system of Estates survived over the rest of Europe, self-provisioning peasantfarms produced surpluses to sell, but proceeds beyond the needs of tax, tithes and rents were hoarded or spent on land or larger families. Although dependence on specialist craft or dealing work in the countryside was usually illegal, peasant demand for goods and services from the bourgeoisie was low, and any increase in the price of cash crops simply resulted in lower production.[168] A stagnant urban system[169] was testimony to the weakness of peasant demand for urban goods and services compared with class-structured English rural society.[170] However, where urban manufactures or other goods perceived by the peasantry to be superior to those which they could produce for themselves were available, market penetration might occur in a similar way to that predicated by the 'bauble thesis' for the landowners, and like that amongst English farmers. These 'Z goods' would, in principle, bring profound changes in peasant economic behaviour.[171] To acquire them more time must be devoted to cash crop production, less to leisure and the production of craft goods for self-consumption. Land, labour and other resources of the farm would begin to be considered in terms of how much of the 'Z goods' they could be exchanged for if peasants voluntarily entered the market economy in order to acquire valued goods which they could not produce for themselves. This must, of course, have been an important impetus behind English commercial farmers' pursuit of ever higher outputs: the carrot to the landlords' stick of higher rents. In the seventeenth and early eighteenth centuries the prosperous dairy farming peasants of Friesland also 'avoided the trap of using their income to chase after parcels of land',[172] spending it on better house furnishings, glass, earthenware, clocks, books, paintings and gold and silver wares. They used the rising relative value of their butter and cheese in avidly assimilating urban cultural norms, at least in terms of patterns of consumption, buying mundane goods from rural craftsmen who increased prodigiously in number without an English-style proletarianization of farm labour, and precious commodities from urban specialists in preference to the substitutes they had previously made for themselves. In the Low Countries' towns growing fat on inter-national trade, craftsmen and merchants whose existence initially depended upon the service of landowners and other merchants were already in place as sources of 'Z goods' for the peasantry: a buoyant market for peasant output went hand in hand with a source of highly desirable products.

In early modern continental Europe the Low Countries were unique in the scale of their commercialization and urbanization,[173] and even there the 'Z goods' effect seems to have been confined to regions in the immediate vicinity of towns and coasts. Friesland was also Calvinist, with all that implied in material and non-material cultural terms (many of the books the peasants bought were bibles and other religious texts). Also as in England, its farmers were tenants, and thus prevented from subdividing their farms, and its towns were not only at the hub of a vast commercial overseas empire, but were also not so cocooned within craft guild protection and merchant monopolies as in most of Western Europe.[174] The 'Z goods' model of peasant incorporation is not commonly applicable to early modern continental Europe, but its appropriateness to Friesland as well as England shows, yet again, that what are usually considered to be irrefragable attributes of peasant economy actually existed only if particular conditions obtained in the other part of the society in which peasantry was embedded.

Proto-industrialization

Proto-industrialization was similar to the 'Z goods' model of change in that it involved peasants in devoting more time to one activity, this time craft work, in order to produce more goods for sale. It was diametrically opposed in that it culminated in the proletarianization of peasantry, rather than its embourgeoisement. It was as widespread as the consumption of 'Z Goods' was uncommon.[175]

The proto-industrialization model of development has a high degree of specificity to bocage regions and regions of even less agricultural promise: 'there can be no doubt of . . . the close correlation between adverse agricultural endowment and rural industry' in Europe generally,[176] and in England 'the industrial revolution started in the pasture farming regions'.[177] Although the maintenance of soil fertility and self-provisioning required crop cultivation, animal husbandry and craft work everywhere, the mix of these activities varied according to ecotype. Because grain farmers usually had to mill their crop on their lord's machinery and were often forbidden from distilling it for sale, and because they were locked into communal routines of farming and firmly under lordly control, their scope for producing worked-up goods for sale was usually circumscribed. This was not invariably so. Whilst the peasants of Friesland were specializing in the production of butter and cheese, those on the plains of Flanders were spending more time transforming flax crops into linen at the expense of food production and craft work for self-consumption.[178] Less fertile upland pastoral regions not only produced commodities which could be rendered less perishable and/or more valuable by working on them before sale, such as milk, hides or fleeces, but animals yielded more by-products than grain. Horn, antler, hoof, bone, hair and bristle could be made into cups, hafts, buttons, needles, toggles and brushes. Bocage and mountain regions had more timber, stone and other minerals, and more water for washing, scouring and retting and for driving mills, bellows, hammers, rollers and grinding wheels. Most of the exposed coalfields of Europe were, too, under bocage landscapes. As colonists pushed further into the common woodland and rough grazing of less hospitable upland regions, economic survival was only possible if agriculture were from the start heavily reinforced by, or even supplementary to, mining, textiles, charcoal, firewood or metal production.[179] As we have already seen, these bocage and mountain regions, like coastal ecotypes, were characterized by greater individualism and a social structure more focussed upon independent kin household groups than communal village institutions, and less trammelled by direct landlord exploitation stemming from demesne labour services. Tenure was characteristically freehold or copyhold, inheritance partible, and the independence of their close-knit family household groups was expressed in disrespect for outside authority and the controls it attempted to impose, and in religious nonconformity. This potent brew maintained complex peasant households into the eighteenth and nineteenth centuries even in England, where arable regions had long been dominated by commercial farming and full-time wage labour.[180] However, greater involvement in craft work, greater dependence upon commons, and fragmenting farm holdings was starkly involutionary, and the diffusion of the potato applied a further powerful ratchet in the eighteenth century.

The standard model of proto-industrialization posits a rapidly multiplying peasantry in bocage regions, depending more and more on earnings from craft goods marketed over longer and longer distances. Work eventually became minutely specialized in particular parts of the process of production, which therefore 'becomes more roundabout as it becomes more efficient'.[181] Freeing household economies from dependence on land and local markets allowed a younger age at marriage because the inheritance of land was no longer necessary for family formation, and children were a productive resource.[182] This cheapened labour further and made it even more dependent on craft income. Squeezed from its ties with the land, labour became a marketable commodity, which the domestic industrial family could produce for itself. It was commensurately less able to produce its own subsistence, or to provide security for loans to finance production on its own account. This brought a profound cultural change: as employees 'they think it no Crime to get as much Wages, and to do as little for it as they possibly can, to lie and cheat, and do any other bad Thing; provided only that it is against their Master, whom they look upon as their common Enemy'.[183] This, and the inordinate amount of fetching and carrying of semi-finished goods needed as production became more and more roundabout brought diseconomies of scale, which led to factory production as the machines on which the operatives worked were brought under one roof.[184] The application of inanimate power to groups of machines, first from flowing water then from steam generated from coal, followed in regions where resources were available.[185]

Where it was forbidden by law to practise craft work in rural areas except as a supplement to agriculture, this process was irresistibly braked; as it was if private or common rights to land were necessary for access to resources required in craft production and to feed the animals which took goods to market.[186] Cheese and butter making, and other crafts often practised in dairying regions,[187] must, too, remain supplementary to tending cattle and milk production in complex peasant household economies. So would textiles and mining where land, however infertile, remained abundant relative to population in remote areas. All over Europe there were bocage and mountain regions where as late as the twentieth century craft production remained firmly fixed within complex peasant households still engaged in agriculture, where

ways of behaviour do not resemble very much those of industrial workers. Work has still an eminently social character; it is done in work groups of neighbours, of young people ... or in the family. Likewise, it serves certain social goals ... rather than the maximisation of income.[188]

If the domestic industrial labour force comprised women, the continuation of the outwork system seems to have been usual,[189] not only in textiles regions like the Oberland and Pays de Caux, but also in English arable areas where women and children were progressively pushed out of the agricultural labour force and took up such crafts as lacemaking, straw plaiting and net braiding after the late eighteenth century.[190]

The full working-out of the proto-industrialization process also required markets and raw material supplies big and reliable enough for agriculture to be dispensed with, and sufficiently abundant capital to fund first this trade, and then the building of factories. Even where textiles remained closely tied to farming the eventuation of the proto-industrial process depended on urban merchants shipping raw materials in and finished products out.[191] Specialization in craft production and farming at the expense of other peasant household activities also meant that other goods and services had to be bought into farm households from towns and itinerant dealers.[192] The penetration of outside merchant capital into peasant domestic production usually went even deeper. When merchants both supplied raw materials and distributed finished goods to and from particular manufacturing households, the *Kaufsystem*, in which artisans paid for raw materials and sold finished goods, often gave way to the *Verlagssystem*, in which merchant capitalists retained ownership of the materials and paid artisans a wage for working them up.[193] This was the fiercest spur towards complete proletarianization of the workforce, although in some regions of putting-out where tenurial and other legal conditions proved suitable, domestic industry maintained contact with peasant farming and survived even this mode of destruction, as in the Canton of Zurich[194] and other parts of North-western and Central Europe, where 'worker peasants' remained common.[195] In fact, the peasant matrix of domestic production only seems to have been destroyed where the presence of water power and then coal were combined with privileged access to an enormous share of the world market, which in the eighteenth and early nineteenth centuries occurred primarily in Britain.

Many craft by-employments were dependent on large-scale plant for semi-finished goods or the completion of production. Woollen cloth pieces ended up in fulling mills, while the locks, hinges, bolts and nails made in farm smithies were at the end of a production process whose earlier stages had passed through heavily capitalized blast furnaces, forges and slitting mills, all needing water power and capital.[196] By the early eighteenth century, in England at least, the smithies' fuel was supplied by collieries too deep for natural drainage which needed expensive pumping engines. Where iron ore and coal were in the hands of small-scale peasant freeholders or the Crown such development was less likely to occur than where they were owned by landowners imbued with the spirit of estate development. This was unusual in Europe, but generally the case in Britain.[197] These large plants did not only supply raw materials for peasant crafts but, like saltpans, glassworks and potteries, they also provided work off the farm. Due to their need for spring water power, early iron smelting and forging were seasonal,[198] as was mining until steam pumping and means of all-weather transportation were perfected.[199] Employment there fitted snugly into the peasant household subdivision of labour. In the seventeenth and early eighteenth centuries, too, most collieries lasted less than a decade and 'most ore deposits were small and were quickly exhausted. Few could support a smelting works for more than a decade or two'.[200] Iron smelting and forging were heavy consumers of fuel which exhausted square miles of timber in a few years of operation. A few skilled workers migrated in the wake of the shifting mining, smelting and forging industries, but for most of the labour force such work could only provide temporary and supplementary employment. Technological change and concentration on larger bodies of ore eventually meant that these works provided full-time proletarian livelihood positions outside the peasant economy, with an effect parallel to that of factory production in erstwhile domestic industry. Unlike the latter, however, their growth owed little to the peasant household economy, but depended on a complex combination of resources, landowner initiative and government policy.

Peasantry could succumb to agrarian capitalism, to 'Z goods', to proto-industrialization or to the growth of large-scale industry outside the peasant economy. These were not mutually exclusive or independent processes. The last two simply represented different routes, appropriate according to the natural ecology and culture of particular regions, to the peasant goal of more secure household subsistence, however far the path eventually led, in the wake of merchant pied pipers, away from peasantry altogether. But whether any, some or all of these processes occurred to transform peasantry into a capitalist market economy depended on the patriciate.

The role of the state

Peasantry could not exist without political coercion:

it is the appearance of the state which marks the threshold of transition between food cultivators in general and peasants ... [I]t is only when a cultivator is integrated into a society with a state – that is the cultivator becomes subject to the demands and sanctions of power-holders outside his social stratum – that we can appropriately speak of peasantry.[201]

The feudal state was based upon 'the fusion of vassalage–benefice–immunity to produce the fief',[202] for which the feoffee owed homage and loyalty to the grantor, and within which he had autonomous jurisdiction over those below him. At the apex of the hierarchy of homage was the sovereign. People were fixed in the feudal hierarchy by oath, not contract, and the duty to perform obligations gave automatic and indisputable liberties over the resources needed to fulfil them and over persons lower down the hierarchy. To the extent that vestiges of the feudal state survived, so did vestiges of serfdom and the protection and regulation of other rights and duties of the Estates of the Realm, which did not end in many parts of Europe east of the Rhine until the nineteenth century.[203] The process began on the northern, western and south-eastern flanks of Europe, where in the late fifteenth century centralized royal power over territory, or absolutism emerged.[204] With it came endemic warfare[205] and the system of government by monarchs through parliaments of Estates: 'on the continent of Europe, west of Russia and the Balkans, the five centuries before 1789 might well be called "the Age of Estates", the great link between the age of feudalism and the modern world'.[206] It was within this system of mutually antagonistic polities with continually changing boundaries, organized into Estates under monarchs, that most European peasants lived until the second half of the nineteenth century.[207]

Three political features were common to all. First, the principle (if not the practice) of supreme royal power in legislation, taxation and military affairs, embodied in centralized royal courts and bureaucracies and in standing, usually mercenary, armies and navies. Secondly, a parliament or a number of provincial parliaments in which the Estates met under the sovereign. The landowning nobility, whose role was to serve the monarch in war and administration, was usually dominant in these assemblies. It was in them and the royal bureaucracies that the renaissance of Roman Law, which legitimated royal power and allowed the definition of private property rights, was implemented.[208] Thirdly, by the seventeenth century mercantilist economic policies were generally followed by European states as an alternative to war as a means of increasing their wealth and therefore military might. Warfare itself completely disrupted normal economic interaction and was enormously destructive of peasant property over the areas where it was fought, not least because it was everywhere the duty of peasants to quarter troops. However, the net direct economic effect of almost incessant warfare in early modern Europe was not necessarily negative:[209] bocage and forest areas largely produced military supplies, and they were major beneficiaries, often at great distances from the actual fighting. War in France during the second half of the sixteenth century made the country dependent on imports of manufactures from England, where there had been almost continuous peace,[210] and the demands of the Dutch Republic for ships and munitions in its war with Spain in the seventeenth century spun off into the development of the Swedish iron industry under the tutelage of Dutch armament suppliers.[211]

Financing these forces and the lavish courtly display appropriate to martial society was the main function of state administration.[212] In most parts of Europe the bourgeoisie and the peasantry paid. Amongst the former, bankers and merchants were also significant beneficiaries, but the peasantry had few equivalent compensations. Even where a relatively large urban Estate existed to help pay the cost of war, the vastly more numerous peasantry bore a heavy share; where it did not, the burden was crippling. But the obverse of this was the need clearly to define what was inalienably peasant land, which sovereigns west of the Elbe were able to keep in place as a secure source of the income they needed for war, administration and display. In Eastern Europe, landowners managed to maintain complete control over the taxation of peasant commodity surpluses and over peasant surplus labour through serfdom, and the Crowns were weak. Or, as in Brandenburg–Prussia and Denmark, landlords almost completely lost power over taxation and military affairs to the crown, but maintained a firm hold over peasant labour through serfdom. Further west, landowners lost some of their control over peasant surpluses to direct royal taxation, and serfdom was rare. Indeed, in most parts of Western Europe landowners even lost control over the money rents they could charge for their land. And increasingly powerful kings prevented them from reasserting power over what was also their own major source of income.[213] But landowners remained able, and jealously guarded their ability, to escape taxation of their land and to exploit the feudal powers they retained over their peasants in non-capitalist ways through the privileges accorded to landownership in the system of Estates. The right to hunt over peasant land, the arbitrary administration of justice, the extortion of death and marriage duties, and the imposition of tolls and monopoly distilling and milling rights militated against the emergence of land as an alienable marketable commodity almost as surely as serfdom did. The 'bauble thesis' could not operate if, as in Eastern Europe, landowners could commandeer or fix the prices of merchant goods, or as in both east and west they could get the money to pay for them via the feudal power left to them by sovereigns keen to preserve the peasantry as a fiscal resource.

The absolutist state fixed the peasantry firmly between the rock of royal power to tax, and the hard place of noble privileges of *dominium directum*. In the east, one or the other was more adamantine; in the west, they were more balanced; east and west, the strength of both varied from polity to polity. But in all cases the peasantry was *fixed*: held in existence so that the income needed by states and landowners could be squeezed from the only place where it was available in an agrarian society. Meanwhile, more quietly but equally effectively, the church Estate took its one-tenth of peasant produce, and although the burghal Estate was obliterated throughout Eastern Europe, in the west it waxed as a second basis of royal taxation on the royally sanctioned prohibition to everyone else of manufacturing for sale and dealing for profit. The economic characteristics of peasantry were in early modern Europe as much a defence against these non-market feudal means of exploitation as they were innate within peasantry itself.

It is usual to set Britain (and perhaps the Netherlands and Switzerland) against these variants of the absolutist state. Only there did absolutism and the Estates system fail to develop. In the absence of an ability to tax land, and therefore without a milch-cow peasantry, the English Crown was chronically short of money, however unified its control over administration and jurisdiction within the state, and however culturally unified that state became under the national established Protestant church and a single parliament with a wide electoral franchise.[214] Other fiscal means had to be devised. One was the ability of the Crown to request, and of parliament to grant, subsidies for extraordinary purposes such as war which were raised on the value of movable goods. For normal purposes, certain other sources of revenue were prescribed for the use of the Crown. Primary amongst them were customs and excise duties, so that the English crown always had a strong vested interest in increasing the volume of trade, which was the major motivation of the Navigation Acts restricting colonial trade to English ships.[215] Overseas trade was also stimulated by granting monopolies to companies and corporations within the bourgeois Estate, and granting bounties and premiums to encourage imports and exports.[216] However, for foreign trade companies to function to maximum scale and profit, many internal privileges of the bourgeois Estate were abandoned. In consequence, 'Z goods' and distant markets for manufactures were more readily available than they would otherwise have been, and regional specialization of farming and manufactures could occur.

Of course, the English state was not categorically different from all others in all of these respects. Virtually all European states tried to encourage foreign trade and the largest possible net inflow of bullion in order best to guarantee their maximum military strength: they all subscribed to mercantilist economic doctrines and followed policies appropriate to them. But large trade balances are best achieved as an outgrowth of dynamic internal trade. The survival of peasantry and the feudal privileges of landowners, clergy and bourgeoisie inherent within the system of absolutism and Estates operated strongly against this by limiting commodity exchange and the penetration of price-fixing market relations into the exchange of land, labour and capital. Encouragements to increase trade on the continent were constrained within the system of Estates, usually by the crown granting letters patent for special privileges over the particular aspects of production and trade that were being encouraged to some precisely defined group within the bourgeois or landowner Estates. In England, too, especially in the first part of the seventeenth century, newly encouraged economic projects were made the prerogatives of particular individuals, companies or corporations. However, not only did this restrictive policy fall out of currency under pressure from parliament,[217] but projects 'flourished best when they passed beyond the pioneer phase, and had entered the deep interstices of the economy, establishing themselves in places that were out of sight of authority and government'.[218]

Absolutism and Estates ensured that vestiges of feudalism kept peasantry in place as the basis of taxation and prevented price-fixing markets emerging in the exchange of commodities, land and labour over most of Europe. However, the extent to which these checks occurred varied from state to state because of differences in the apportionment of taxation and political privileges.

Conclusion and agenda

However autarkic and conservative their aspirations, however variegated from ecotype to ecotype, and however well hidden their interactions, peasants could not escape history because they were linked to other parts of society. They must be in process. Their attributes were as much responses to the feudal structures within which they were embedded in medieval times, and to survivals of those structures through the *ancien régime*, as they were innate within an independent socio-economic entity. Political geography was as responsible as environmental differences for the ways in which peasantry varied from one part of Europe to another.

Our examination of the nature of peasantry and of the ways in which it was transformed into a capitalist farming system has taught some salutary lessons. On the one hand, even whether or not peasantry existed must be difficult to diagnose because many attributes of peasantry are empirically indistinguishable from those of capitalist farming. Whether varied farm sizes, employment of non-family labour, production of craft goods, savings, buying and selling of land, or regional differences in farming practices signify the existence of peasantry or not depends upon where they were tending through time. Only a study of dynamics can reveal whether they were becoming progressively more marked, and therefore components of a market-driven farming system, or constant or cyclical, and therefore components of a system geared towards secure self-provisioning. To understand why these features existed, and the reasons why what they signify was occurring, we need to look at flows of products, labour and cash between peasant farmers in different ecotypes, and the nature of the non-peasant elements of society and the form of their links with the peasantry.

In the following case study we have attempted to satisfy these requirements in four ways. First, we have chosen a region with varied physical environments (see figure 3.1). Secondly, we have not confined our treatment of farming to the peasantry itself, but have included the patricians who were scattered through the countryside. Thirdly, we have included non-agrarian production, which is examined both for its own sake and as an agency which affected the nature of the peasant and patrician farming systems. Fourthly, each chapter begins with a general outline of the distinctively Swedish attributes of the subject with which it is concerned. In these ways, we hope to be able to show exactly which of the processes described above were at work in our region of study during the early nineteenth century.

The imperative to study dynamics needs a longitudinal approach. Given the range of contents just outlined, and the resulting plethora of possible connections between economic agencies and across space, this raises acute methodological difficulties. It is easy to propose that various aspects of the peasant and patrician, urban and rural, agrarian and industrial economies of neighbouring contrasting ecotypes be studied as they interacted through time; but how can a scheme of analysis be devised to do it?

2

Conceptualizing regional change

Stasis and process

Concepts, methods and empirical findings cannot be separated: what is discovered depends on what is looked for, and what can be looked for depends on how we choose to look. How can we know whether peasantries are inherently homeostatic or not? Obviously, we cannot claim to know if the way we ask our questions presupposes particular answers: the sight of past 'stages' punctuated by 'transitions', whether of peasantry transformed into commercial farming, feudalism into capitalism, or whatever, might owe as much to the methodological lenses through which the past is examined as it does to what actually happened. This is not to claim that we can look at the past (or anything else) without distorting lenses. Rather, it is to recognize that we must be aware of the shape of our methodological glasses, and deliberately choose a set which is appropriate to our purpose. A search for peasantry as process needs a different set from that used by people who have seen economic stasis and cultural recidivism in the peasantries of the past.

The extent of and reasons for stability and change in human geographical patterns can only be understood by direct study of the processes responsible for them. We must study how things behave to remain or become what they are. Even for the most ardent advocate of this quest 'it is extremely difficult to verbalise as theory history as process'.[1] Often there is no alternative to drawing inferences from cross-sections, interpolating suggestions about the processes that may have been operating between each statically portrayed level,[2] because this is all that surviving sources of information allow. But in consequence it is often difficult to relate hypotheses about causes of change, couched in concepts relating to continuous processes, to empirical studies, framed in terms of static cross-sections. This is a tricky research problem because 'the number of conceivable models [of a process] is literally infinite and a lifetime may be spent in exploring possibilities: by supplying the proper stage directions at the proper time, we can specify any sort of sequence development desired and may find that there is almost no empirical content in the theory being expounded'.[3] Models and concepts of process are necessary for historical understanding; but it is difficult to relate them to empirical situations, which is the only way properly to formulate and test them. How can an empirical study be deliberately specified in those terms?

Longitudinal studies, time-geography, and the depiction of process

The longitudinal analysis of micro-level units of economic and social organization offers one solution – 'longitudinal' meaning 'tracing an entity over time'. Such studies have to be based on data concerning *individuals* as study objects; they need to keep the same population of units continuously under observation through time because without an intact study population we cannot properly specify the changes which occur. The only way to be sure that the studied population remains intact is to keep each of its individual units under separate observation. Observing the changes occurring continuously through time to separate unitary individuals, defined at any scale, within a population (which might be no greater than one) is what longitudinal analysis is about.

In geography, this perspective has been most developed and used in time-geography, originally introduced by Hägerstrand in the early 1970s.[4] Most later discussion of this work is concerned with the cartographic and diagrammatic techniques used to depict changes,[5] and their value as a kit-bag of tools for use in sociology.[6] We are concerned with the conceptualization which underlies time-geography, not its techniques. Hägerstrand's own aim was set firmly within the human ecological tradition of geography which we hope to exemplify: it was to create a method of achieving 'a contextual synthesis in the man-environment area'.[7] He wanted to devise an alternative to the, at that time, overwhelming tendencies to dissect the subject-matter of human geography along lines defined arbitrarily according to 'systematic' topics (such as social, economic, political, demographic and so on), and to be preoccupied wholly with the geometry of the spatial patterning of what had been abstracted.[8] His objective was to think in more holistic, ecological terms about the aspects of people's lives which systematic geography pulls apart, together with the milieu in which they occurred;[9] 'to try to turn human geography into a study of the conditions of life in a regional setting'.[10] He argued that this must be done by working at the level of the smallest components it is reasonable to distinguish in geographical enquiry.

The basic elements of the 'socio-environmental web model'[11] he devised to do this are few and simple. Human activities occur at specific sites, termed *stations* (for example places of work, schools, dwellings or concert halls). As they go about their lives, people trace a web of paths through the framework created in space and through time by these stations.[12] The people whose paths intersect at stations come together there to participate in *projects*: that is, 'the entire sequence of simple or complex tasks necessary to the completion of any intention-inspired or goal-oriented behavior'.[13] Stations thus comprise not simply positions or buildings, but institutions for the organized interaction of people and the consumption or transformation of material resources, and projects function as allocative mechanisms in time-space for human effort and material resources.[14]

Although the ambitions of its proponents have risen progressively, difficulties arise when we try to apply this scheme to produce a coherent account of how individuals, society and milieu combine into processes operating over longer time spans and at larger geographical scales. Hägerstrand's own early suggestions were modest,[15] but by the mid to late 1970s it was being claimed that time-geography could uncover 'structural patterns and outcomes of processes which can seldom be derived from the laws of science as they are formulated today ... by focussing on the local connectedness of real world phenomena, as well as the connections between spatially separate configurations of locally connected phenomena'.[16] By the 1980s time-geography was being proposed as a key for human geographers to enter the realm of social theory: a 'way to move from critique to *active* participation in the theoretical debates and developments now occurring within the social sciences'.[17] However, the empirical work so far carried out under the auspices of time-geography provides no more than *illustrations* of how 'wider social processes'[18] were reflected in, for example, the daily lives of nineteenth-century Swedish farmers[19] or the day-to-day free-time activities of workers during the industrial revolution.[20] A wide gap has grown between the claims made in theoretical work and empirical achievement, which has not got much further than Hägerstrand's initial presentation.[21] As soon as empirical illumination is called for, the large-scale and long-term are forgotten and back come the paths evident in short-term daily activities. This is of little help in depicting longer-term processes of the kind with which we are concerned.[22]

In the usual short-term conceptualization of time-geography, sets of stations with fixed and immutable attributes at given locations are considered merely as points on a plane, or as pipes in time-space, which provide a grid-like structure within which sets of constraints operate to produce webs of paths. Somewhat perversely given Hägerstrand's original intentions, it is the geometrical patterns traced by paths which have become the focus of interest. People simply move between points, or through and between pipes, according to a 'choreography of existence' which is determined by the given unexamined needs of different projects and a set of specified constraints on human behaviour.[23] The whole scheme concentrates on the constraints which structure human behaviour in the short term, and the predication of fixed stations as the 'pipes' which give form to the patterns of paths and webs created in response to these sets of constraints is simply an expression of this innate characteristic.

This is not appropriate to long term problems, notwithstanding claims to the contrary.[24] Space is considered to be *compositional* rather than *contextual*.[25] Instead of places, we have abstracted spaces; instead of *a room* with an ensemble of actual furniture, we have abstract 'room': a platform of boards with indelible marks at fixed locations between which people dance in response to forces beyond their control. In some cases changes in the shapes of these patterns consequent on the process of socio-economic modernization have been considered,[26] but such 'before and after' comparisons do not mean that the changes have been explained or that time-geography's method has been used to test theories of modernization. The 'choreography of existence' cannot be understood unless the stations are considered as more than points on a plane, and unless the social rules which constrain and permit behaviour of different kinds by different people in relation to the resources allocated at these points are fully and explicitly introduced into the analytical scheme.

'The root problem, then, is obviously that of connecting the contextual to the compositional; of showing how space-time configurations of paths, projects and domains flow from and feed back into developing structures of social relations and systems of social practices.'[27] This cannot be done in a programmatic way, in terms only of hypothetical schematized conceptions. The analytical perspective of time-geography, if it is to be truly ecological, must be applicable to the full richness of particular contextual spaces in such a way that the richness of the possibilities that they contain is fully conveyed, not lost in a crude translation whose purpose is simply to show that a translation of any and every historical situation into the basic terms of time-geography is possible.[28]

Livelihood positions and stations

Although there are weaknesses in conceiving of resources in terms of economic categories such as land, labour and capital, we cannot ignore them completely in favour of an exclusive preoccupation with human time, space and energy.[29] Without some consideration of what is actually transformed and allocated in the projects carried out at stations, the reasons for people being there at certain times must remain obscure. It is what is done at the stations that gives them a place in an individual's web of paths, fits individuals into the structures of social relations at particular stations, and contains the seeds of a different future. The room can only be furnished by specifying the productive and allocative relations within the stations: through a definition of the projects. It is because of this neglect that time-geography has been unable to formulate questions about changes in societies, but only consider empirical formulations of geometrical shapes and shifts of paths and webs as consequences of external influences.[30]

The concept of *livelihood position* provides a way of elaborating the treatment of projects and stations so that their productive, allocative and social structuring functions can be explicitly considered;[31] of disaggregating the station concept into a structured set of components which are commensurate in scale with the time-space paths traced by individual people. A livelihood position is the set of resources, divisible and indivisible, owned or otherwise acquired, which sustains the livelihood of an individual or a household.[32] A person might possess rights over these resources as an individual or as a member of a collective group, such as a village community or a family. These requirements can never be satisfied at a single location in space. The basic premise of time-geography is that people must move about continually in order to acquire them.[33]

Many people's activity systems do not include places of work where they are paid to labour, or supervise labour, on resources owned by others. The very young, infirm and aged are supported by the work of others. On the other hand, some people possess sufficient resources, frequently derived from inheritance – such as land, a fishing boat, tools and access to common forests, equipment and a workshop – to produce most, if not all, of their material livelihood on their own account based in their own households. This is so for peasants and the petite bourgeoisie. The proportion of these livelihood positions declines as modernization proceeds, and as that of proletarian positions increases, eventually to become overwhelmingly predominant. Even positions of the peasant and petty bourgeois kind often require the labour of others besides the owner of the means of production, although it is usually supported as part of the owner's household rather than by wage payments. The rights of the different people involved in particular projects over the resources being processed, whether at specialized places of work or in domestic households, vary considerably. The social relations between them reflect and depend upon this, and they will obviously differ considerably between peasant and petty bourgeois stations, on the one hand, and at places of work owned by capitalists and employing wholly proletarianized labour, on the other.

A livelihood position can be conceived of as a set of components of projects, connecting a number of stations into the activity system of an individual, a unit in terms of purposive social behaviour. A project can be viewed as a unitary component of a livelihood position. Connecting projects to livelihood positions specifies a subset of projects as the operations performed on resources to make them capable of helping collectively to sustain people at a variety of stations. Similarly, stations are structured bundles of (usually partial) livelihood positions and it is with these component positions that individuals are actually linked. An itemization of the livelihood positions that the project at a station helps to maintain, wholly or partially, defines its *production structure*.

The complementary interactions between the structures defined by the webs which represent the complete livelihood positions of individuals on the one hand, and those defined by the projects at the stations where they intersect on the other, cannot be visualized except as a tangle of seething complexity. But it is in the tensions generated in these mutually sustaining but inevitably conflict-prone interactions that the dynamism of society lies. It follows from this that the possibilities for change that exist in any socio-spatial system must to some extent be expressed in the production structures that represent the projects at the stations through which the system is sustained and organized. Certainly, there can be no significant economic or social change without a change in these structures of production, and any such changes must be related directly and intelligibly to complementary changes in the pattern of livelihood positions.

'Entitation is far more more important than measurement . . . only when the right things have been found to measure are measurements worthwhile'.[34] Time-geography defines a number of primary entities and some of their properties which can be measured. The stations of a peasant society are obviously its farms. This banality carries some important consequences. It means that we must be concerned with functioning economic enterprises, not just with cadastral data about who owned what land, or with the disposition of fields, which often constitute the basic information of historical studies of peasant societies and the development of commercial farming. But also, so that all the paths in the webs people traced can be discovered, we must be equally concerned with all other stations at which livelihoods were derived or consumed: urban workshops and rural mills, landless cottages and estate owners' mansions must all be included if we are to follow the precepts of time-geography. Then we must discover the nature of the projects carried out at the stations: what was grown and made on farms and estates, for consumption and sale; how the rural landless got their livings; the significance of non-farm work in the countryside, and the occupations and investments of townsfolk. Finally, we must discover who derived what kind of livelihoods from these projects: what production structures were expressed in the ensembles of livelihood positions at the stations of the region? Again, everyone who participated in production or consumption must be included: not just the nuclear stem families in peasant homesteads and urban workshops, but the aged, infirm, and any other non-family members who lived with them; those who came to work for them from homes of their own, and those who went out from the homestead to work elsewhere.

It is neither fortuitous nor contrived that this set of preoccupations matches closely with the issues we identified in our discussion of the nature of peasantry and the transition to capitalist farming: farm size and other resources belonging to farms relate to stations; cropping régimes and craft work to projects; labour proletarianization to livelihood positions. The imperative to examine stations and their attributes longitudinally reinforces our conclusion that the significance of many of the customary indicators of the transition from peasantry to capitalist farming lies in their directional tendency through time, rather than their absolute values.

Our excursion into time-geography has confirmed the validity of these preoccupations and fitted them into a coherent scheme of enquiry. But that is not the end of our encounter with it. Within time-geography these entities are defined and their properties are measured to allow the reconstruction of patterns of flows of people, through their daily activity systems and their life-paths, between them. This gives a means of pursuing another of our main conclusions about the process of transformation from peasantry to capitalism: it depended upon the ways in which peasant farming was linked to the activity systems of other members of society.

Flows, paths and the processes of economic and social change

Wholesale differences in the patterns of many kinds of flows across the landscape were implied in our diagnosis of the differences between peasantry and capitalist farming. Indeed, the transition between them can be neatly characterized in these terms. Peasantry aims to minimize the need for and impact of flows of goods, money, ideas and labour across space; capitalist farmers aim to maximize the flow of goods out through the farm gate and the inward flows of money and consumer goods. To do this they generate flows of labour, ideas, equipment and capital. All of the processes of transition from peasantry to capitalism outlined in chapter 1 can be expressed in terms of changing patterns of flows in the same way. For example, the proto-industrialization model specifies increases in the volumes and changes in the patterns of flows: between the scattered producers of goods in rural manufacturing regions and international markets – first sucked into, then spread out from the warehouses of merchant capitalists; between the domestic manufacturers and their sources of raw materials; and between different craftsmen each concentrating on one small part of the total manufacturing process. Mendels' basic model of the transition from a proto-industrial to a factory system of production relies heavily on the fact that flows of semi-finished goods become too expensive and intricate to control in an expanding domestic industrial system.[35] When we come to flows of people during the process of economic change, the problem is more complicated. On the one hand, as projects at stations are begun, transformed and ended, individuals may only be able to maintain their current livelihood position by moving elsewhere. On the other hand, when people flow over space they frequently change the nature of their livelihood, and therefore their social and economic position. Whether spatial movement is associated with social stability or change depends on whether it involves a change in livelihood position; it is not inherent within the simple fact of movement itself.

This is why we need to specify the nature of the projects at stations and the production structures comprised by the livelihood positions they support. The development of capitalist economy meant the creation of many new livelihood positions: not just those which evolved out of pre-existing kinds, but completely novel types as well. Many were concerned with the production of commodities for sale on farms, in urban workshops, in workers' homes, or in factories. But many others were of a kind which facilitated the circulation of resources, and many more fed from this in positions made possible by the greater specialization that larger volumes of more rapidly circulating wealth allowed. Greater flows of labour, raw materials, capital and goods were effected through new positions, such as merchants, financiers, carters, brokers, retailers and lodging-house keepers. Increased wealth amongst the occupiers or controllers of the stations concerned with primary production and the positions which ministered directly to their needs created further possibilities for producers and retailers of consumer goods and for the providers of services such as entertainment, medicine and education. The profitable investment of surpluses of capital, the switching of capital from one activity to another and the acquisition and disposal of land consequent on the commercialization of farming, needed lawyers and clerks in greater abundance, and so did the greater volume of more rigorous organization and administration that greater economic productivity and complexity brought. In short, the growth of capitalist economy generated large multipliers into secondary and tertiary activities which a self-provisioning peasantry tried completely to eschew.

Changes from peasantry to capitalism must, therefore, be expressed in the production structures which represent the projects at the stations through which the economic system is sustained and organized. There can be no significant economic or social change without an alteration in the livelihood positions which constitute them. These specifications also make it possible to analyse flows of commodities between farmers, merchants and markets outside the region or the various livelihood positions in consuming households within it; flows of capital, ideas and expertise between those who controlled economic operations at stations; reciprocal flows of wages and commodities between resource-owning stations and landless households through proletarian positions at the former. The reconstruction of livelihood positions at stations also allows us to examine how the social and economic mobility or fixity of individuals was related to their spatial mobility or immobility. In particular, we can look at how the degree and permanence of the proletarianization of labour 'freed' from the tie of family land was related to their spatial mobility, and draw up the life-paths of people as they successively occupied different livelihood positions at different stations. Thus, our modified version of time-geography allows us to show how flows of people were linked to a changing system of resource exploitation and allocation.

This conceptualization differs significantly from that which is normally used in time-geography, where attention is focussed on the web of links between *all* the stations within the activity system of individuals at one particular time of their lives. But a shift of attention to the ways in which people move through a sequence of livelihood positions over their lifetimes engages more readily with changes in the economic basis of social organization, and thus with the dynamic processes of history. This shift of emphasis also makes it possible to analyse systematically the actual changes in the material basis of ordinary people's lives during the course of economic change, rather than simply to produce schematic exemplifications of all aspects of time-use for 'typical' individuals before and after economic change has taken place.[36]

Towards a scheme of regional analysis

The headiness of theoretical wine is always diluted by an admixture of empirical ingredients. Our work is no exception. It is undoubtedly true that conceptualization of change in a region in terms of the longitudinal progress of the livelihood positions, production structures and projects at all of its constituent stations, and of changes in the patterns of flows between the component livelihood positions at those stations, would provide us with a rich and coherent depiction of the historical processes operating within it. But a moment's reflection demonstrates the utter impracticability of such a task. In order to include different physical environments in our study, the region we defined is necessarily a large one. It contains arable plains and a town with medieval roots, infertile shieldland and a rapidly growing rural industrial area, comprising the whole of western Östergötland. The twenty-five parishes or parts of parishes of this region contained 21,300 inhabitants in 1860. In order to span a possible transition from peasantry to capitalism, our fifty-year period of study is long in terms of time-geography's usual concern with daily time-budgets and networks of population flows. There must have been over 400 million daily activity networks in the region over the period, and as many daily specifications of complete sets of livelihood positions. This is to say nothing of specifying the resources attached to each station and the projects performed there, or the networks of other flows than the daily or annual movement patterns of people. Clearly, some drastic modification of what is in principle desirable is needed.

Rather than reconstructing continuous series of annual data on the resources, projects and livelihood positions of all stations, we chose to assemble only six sets of data, one for each decadal year from 1810 to 1860. These cross-sections were connected through complex linkage procedures to acquire some semblance of properly longitudinal study, and a number of peasant farms and agricultural estates were traced through each of the cross-sections to provide more detailed longitudinal material. We believe this to be a better way of imposing practicability on our research than by restricting ourselves to a small sample of stations traced continuously through every year of the period. This information is presented in three chapters, dealing first with the resources belonging to farms and estates of the region; secondly, with the production systems and livelihood positions supported by farms and estates, and thirdly with the resources, production systems and livelihood positions of non-agrarian stations in the region, including rural industries and urban crafts, services and manufactories. Our direct study of flows is also a heavily abbreviated version of what is in principle desirable. Inferences (and some data) about commodity flows are included in the three chapters mentioned above. We decided to devote separate chapters to capital and population flows, the former because of the importance of the question of when flows of money between peasants became flows of capital between commercial enterprises, and the second because of the equal significance of the process of labour proletarianization in the emergence of a capitalist economy. Even thus limited, it was only possible to reconstruct flows of capital and labour for samples of stations if we were not to drown in masses of data.[37]

The outcome of these decisions about how to limit our task to a feasible scale may not resemble the time-geography we are used to seeing. But the replication of well-worn schematized proposals was not our aim, which was to use ideas from time-geography to devise a way of systematically studying the process of economic transformation in a region; to guide us in the abstraction of data from impossibly copious sources of information towards a coherent empirical statement which might help to answer unresolved questions about the nature of peasantry and the process of transformation from peasantry to capitalism. As time-geography has hardly ever been used for such analytical (as distinct from exemplary) purposes, the lack of correspondence between what follows and time-geography's stereotype is not surprising. We trust that the result will be welcome.

3

Agrarian land ownership and farm holdings

Setting the scene

Introduction

Swedish agrarian development during the early modern period has been interpreted in two ways. First, changes have been attributed to government decisions and alterations in law; secondly, the government actions themselves have been seen as responses to pressure stemming from changes in the agrarian system. According to the former view, such things as enclosures, sales of Crown land, and permission for commoners to buy noble land exempt from tax[1] were 'changes from above' which precipitated an agrarian revolution. Modern research has pushed these changes further and further back in time.[2] As the gradualness of the process has become apparent, the notion of agrarian revolution has become less tenable. We will consider this development as an interaction of changes coming simultaneously 'from the bottom up' and 'from the top down'.

This chapter is concerned with land holding patterns: that is, with the formal descriptions of the stations of the agrarian economy given in taxation records. Who owned and who farmed the land, and how did the ways in which land was brought together into units of ownership and use vary through space and change through time? We will present our empirical findings for western Östergötland after considering recent research into these matters in Sweden generally during the seventeenth and eighteenth centuries.[3] We hope to derive explanations of development in our region which are specifically regional or human ecological. Thus, we seek to discover the degree to which farms became differentiated in size across space, any variations in the objectives of production that this differentiation might imply, and the degree to which this was related to natural environmental conditions.

50

Swedish agrarian society in about 1700

During the seventeenth century, Swedish agrarian society experienced many changes in terms of land ownership and tenancy rights.[4] They laid a foundation for further changes in the eighteenth and nineteenth centuries, which (somewhat inappropriately) have been called the Swedish agrarian revolution. We will give only a brief outline of the seventeenth-century changes as background for a more detailed analysis of developments during the following 150 years.[5]

The first two-thirds of the seventeenth century saw an increasing dominance of the nobility in Sweden, particularly in the countryside, where nobles accumulated estates. This was achieved through transfers of Crown land to deserving noblemen and purchases of taxed land by noblemen with fortunes captured in war.[6] The increasing dominance of the nobility was facilitated by two long regencies led by representatives of the peerage. Manors[7] were established on much of this newly acquired noble land. The rise of the nobility was so marked that in some parts of Sweden virtually no land belonged to the Crown by the 1660s. Crown land which was not converted into noble estates was reserved to pay for government services.[8]

This trend was broken by the confiscation (*reduktion*) of former Crown land by Charles XI in the 1670s. It was also ameliorated by the emergence of a firm land taxation system dependent on peasant ownership. This, in turn, was related to the establishment of a system for keeping the army which involved the provision of farms for the officers and crofts for the ordinary soldiers (*indelningsverket*).[9] Crown land recovered during the 1670s, together with that which had been retained, was leased out to tenant farmers or used to remunerate military officers, priests or local public officials. For the tenants, land rent paid to the state was quite substantial. The duties of Crown tenants were in principle greater than those of freeholders of taxed land and tenants of noble estates. From the late 1600s, the last seem to have been in about the same position as ordinary freeholders in terms of duties, although their security of tenure was weaker. The difficulty of increasing the dues of tenant farmers discouraged the nobility from keeping vast tenanted demesnes, even if their own level of tax exemption was still high at the beginning of the eighteenth century.[10] The freeholding peasants' position was at least not weakened relative to other groups during the late 1600s, which may have been linked to the strength of royal power and its dependence on peasant taxation for revenue and the support of the Peasant Estate in parliament.[11]

It has been calculated that one-third of the land belonged to each of the freehold, noble and Crown categories in about 1700.[12] This does not mean that each of these categories of land was owned and tended, respectively, by peasants, estate owners and Crown tenants. Part of the freehold taxed land was owned by town traders and by noblemen, and smaller areas of the noble land – perhaps 8 per cent of ordinary noble land and 3 per cent of the manors – were in commoners' hands.[13] There was substantial variation over the country. Northern Sweden had virtually no noble land, while in Stockholm and Södermanland provinces it accounted for 55 per cent of all land by the mid-eighteenth century. The proportion and size of true manors also varied geographically: large estates dominated recently conquered Skåne, where environmental and social conditions were similar to those of Denmark, while Östergötland and Småland had relatively large proportions of ordinary noble land and fewer manorial estates. The distribution of Crown and freehold land also varied. Many Crown properties originated in the substantial monastic and church estates taken over after the Reformation in 1527, or they comprised tracts of uncultivated land, which automatically belonged to the Crown. In western Östergötland numerous Crown farms devolved from the holdings of Alvastra monastery, and nunneries and monasteries in the towns of Vadstena and Skänninge.[14]

The way that land was farmed varied according to its status, from the almost 'manorial' operation of vicarage farms on Crown land and smaller noble estates, to peasant tenants on large noble demesnes and freeholders on taxed land. On the former, the main unit was run with hands and maids employed by the year who lived and worked on the manor farm. On the demesne too, were a number of crofts (*torp*), each with its own cultivated area, varying between a few acres and more than fifteen. Their occupiers paid rent to the landowner in the form of labour and grown, picked or home-made produce like berries, fungi, brooms or whisks, so that crofters contributed to both the labour force of the manor farm and to its provisioning. Vicarage or officer's farms on Crown land and noble estates were similar in these respects. However, the latter also normally contained a vast property tended by tenant farmers,[15] who either worked on stipulated days at the manor farm or paid rent.

The distribution of inputs depended on whether the *Grundherrschaft* system was used, with mainly rent-paying tenant farmers and a small manor farm worked by full-time labour, or the *Gutsherrschaft* system, with a large manor farm using day labour from crofters and tenant farmers.[16] Estate production in seventeenth-century Sweden was supposed to have moved in the direction of *Gutsherrschaft*[17] due to the increased power of the nobility and to an increase in market demand for agricultural produce. Rent-paying tenants were still common on noble land at the beginning of the eighteenth century,[18] but the establishment of a large number of crofts on estates during the late seventeenth and the early eighteenth centuries allowed estate owners to increase their labour forces without reducing the rental income of the estate. Crofts were very cheap to establish as the land used was normally marginal and soon augmented by more reclaimed by the crofters themselves.[19] The position of crofter was not unattractive, even to peasants' sons, as some estate crofts were at least as big as small peasant farms and estate crofters were exempted from conscription to the almost continuous wars of the seventeenth century. Whatever the contributory causes, it is clear that estate labour forces increased, and that this indicates a development towards *Gutsherrschaft* and market production on estates.

Peasantry was both a component of estate agriculture and a system of production in itself. All Swedish peasants in 1700 paid land tax or rent to the state, or rent or day-labour to an estate owner. Production probably did not vary much between different farms in a village, irrespective of whether they were on freehold, Crown or noble land. The duties, however, did, and therefore so did the *intensity* of production. Tenant farmers of taxed land owned by nobles were in general less plagued by duties than Crown tenants during the late seventeenth century. Freeholders were also advantaged in this respect, while tenant farmers providing day-labour on estates had to relinquish most of the surplus they produced beyond their own subsistence.

Methods of production in Sweden were typical of Western Europe. The family supplemented by living-in hands and maids formed the basic labour force, assisted by temporary day-labourers at times like harvest. The extent to which temporarily hired labour was used is unknown, but it cannot have been substantial due to the small supply of able-bodied labour outside farm families during this period, and legislation which severely restricted the employment of free wage labour.[20] Until 1755 the establishment of untaxed smallholdings like crofts on other land than estates was prohibited because the state wanted to incorporate as many settlements as possible into the systems of taxation and military recruitment. However, the large number of legal disputes over such matters[21] suggests that some croft colonization and cottage-building occurred, and this might have provided part of the peasants' agricultural labour supply.

Cultivated land was subdivided into arable and meadow, the former tended as one, two or three common fields. Outlying land was predominantly forested, common and used for cattle grazing, though some had already been cleared for private meadows and arable patches.[22] The proportions of these different kinds of land varied between different regions and in different physical environments. Early on, the plains of central Östergötland, the Lake Mälaren basin and Västergötland had a large proportion of arable, while the forested shieldland surface of most of the country did not support much more than patches of meadow and grazing. Farming methods of the time did not allow a complete absence of livestock, but the number of cattle per unit differed substantially between plains and forest. Where the land surface did not permit the cultivation of large continuous areas, settlements of single farms or small hamlets were the norm. The plains, on the other hand, were dominated by small open-field villages. Because settlements were small, really large common fields never existed in Sweden. In principle, agriculture was legally regulated since the medieval County Laws about *solskifte*,[23] although these strict rules for the spatial organization of arable and meadow were not consistently obeyed. In settlements of more than one or two farms, the norm was to subdivide the arable into parcels so that every farm owned at least one in each field, usually many more. Subdivision of arable and meadow was not confined to villages, but also occurred on single farms and estates. Until the covered ditch technique was introduced, it was necessary to break up cultivated land with deep ditches in order to drain the fields.

The interlocking of cultivation on the different farms of a village through cooperation in various tasks restricted the amount of change open to an individual peasant. In villages and larger hamlets, it was impossible to utilize more complex crop rotations, amalgamate the strips of one farm into larger areas, or grow more market-orientated crops without the entire community taking part. Peasants with only enough land for subsistence would follow a strategy of minimizing risk, whilst peasants with larger holdings might be concerned to specialize and speculate in response to the market, but the intrinsic inertia of the traditional agricultural system tended to slow down changes. Thus, an individual peasant would have had most room to expand market production on isolated patches of arable on outlying land. Such independent production cannot have been more than marginal and supplementary in the larger villages. Common outlying land seems to have been amicably subdivided into private holdings by agreement when these arable patches were colonized, but government regulations (*skogsordningar*) acted against substantial arable production there.[24] During most of the seventeenth century most arable colonization was, however, inside the infield (*inägor*), where large areas were still not utilized for arable production. Often, reclamation occurred in areas of little importance for animal husbandry, and it is therefore not possible to argue that the extension of arable areas on the plains in the seventeenth century represented a transfer from pasture to grain production. Because reclamation largely took place in the infield, within the village community, we can assume that there were no differences between farms of different land status. Within a particular village everyone must have participated in the infield reclamation. It is more difficult to describe the development of estates, which do not appear in land tax assessments and were rarely mapped. Some arable reclamation occurred on a large estate in Blekinge county in southern Sweden, though this was not reflected in the quantity of seed sown or in the harvest.[25] If this estate was typical, manor farm arable would also have increased in many places during the seventeenth century.

In general, the seventeenth century witnessed substantial arable reclamation. It was more intensive in areas where the quantity of arable was small at the beginning of the century[26] than on some central plains, where more than 25 per cent of the land was already arable in the early 1600s, and it increased by only 5–10 per cent over the following century. At the same time, in certain parts of the mountainous Bergslagen area, where mining and pig iron production were important, the arable area decreased during the late seventeenth century,[27] which might indicate regional differences in population growth and regional economic specialization according to comparative advantage.

Eighteenth-century developments

Peasant farming
The eighteenth century can be called 'the Peasants' Century' in Swedish agrarian development, in the same way that the seventeenth was the nobility's. It was the heyday of 'the free peasant economy',[28] which formed the basis of the capitalist agriculture which eventually emerged. Research has tended to attribute agrarian change to external stimuli, rather than regarding it as being part of wider socio-economic processes operating over a longer span of time, including displacements in the power and economic influence of the peasants and the appearance of economic and social stratification within the peasant population. The former belongs to the eighteenth century, the latter to the nineteenth. The political power of the peasants, especially those with freehold tenure, increased progressively from the late seventeenth century onwards.[29] At the same time, there was an increase in the market for agricultural products as towns and specialized secondary production grew, and as overseas territories were gained and lost in long periods of war. Population growth from the early parts of the eighteenth century also had an influence, not only as extra demand but because it meant that labour shortage no longer constrained production.

There was no incentive for a farmer to increase production unless he had the right to consume or sell the extra produce. If landowners raised the rent or duties of tenants at the same rate as production increased, the only likely result was greater peasant political unrest.

The eighteenth century saw a marked change in the land policy of the state, related to the changing power of different social groups. To an increasing extent, it sold off its landed property, converting it to freehold taxed land in the process.[30] Between 1700 and 1815 the government sold off 70 per cent of its farmland,[31] which had accounted for one-third of that of all Sweden. The tenant of a Crown farm had a strong bargaining position when it came to its tax purchase, even if somebody else wanted to buy it. Although some Crown farms were tended by persons of standing and tax purchased by them,[32] three-quarters of the sales were to peasants. It is difficult to generalize about their geographical distribution. Naturally, they were most common in areas with a large proportion of Crown farms, such as erstwhile monastic demesnes like those of Skänninge, Vadstena and Alvastra in our study area. It seems that areas with a more market-orientated agriculture had the earliest and highest degree of tax purchase, whilst a change in the regulations in 1723 which gave priority to ironworks influenced the amount of tax purchases in the shield-lands.[33] In western Östergötland, many tax purchases had been made by the mid-eighteenth century, but some were also made as late as the early nineteenth century.[34] Because the province contained so much Crown land, the proportion was still almost 33 per cent in Östergötland in 1755, while the freehold proportion was 29 per cent.[35] Economically, tax purchases were favourable to buyers. Prices were set below the going market rate for similar freehold farms, more so as the century wore on.[36] If the tax purchaser was previously a Crown tenant, his change of status brought total farm dues lower than those for normal freehold land tax. Freeholders and Crown tenants who had kept a cavalry soldier (*rusthållare*) with the support of other Crown farms (*stödrotar*) had priority in the purchase of the latter, even if they were leased out to tenants. In this way, tax purchases allowed land to be accumulated by peasants.

The eighteenth century also witnessed a gradual break up of traditional ownership patterns on noble land. Formally, commoners had always been forbidden to purchase noble (and therefore tax exempt) land. Gradually, this ban was removed by parliamentary acts, and by the early nineteenth century commoners were free to buy even manorial noble land. In Östergötland, 10 per cent of manorial and 15 per cent of ordinary noble land was in commoners' hands by 1755; in 1811, 20 per cent of the first and 26 per cent of the second had been sold by the nobility. However, this cannot automatically be interpreted as increased peasant ownership of estate demesnes at the expense of the nobility. Of the 20 per cent of noble manorial land in commoner hands in Östergötland in 1809, 85 per cent was owned by persons of standing such as vicars, military officers, lawyers and so on, and the remainder by town citizens.[37] Only 1 per cent belonged to peasants. In the same year, peasants owned about 11 per cent of the ordinary noble land, persons of standing a little more, and town citizens 5 per cent.

Although tax purchases and the acquisition of noble land increased independence from the Crown and nobility, many regulations and restrictions on the freedom of peasants remained in the eighteenth century. Free utiliza-tion of common outlying land was not permitted, and the establishment of crofts on peasant land was prohibited until 1755.[38] In addition, the forest regulations through which the government directed what was permitted outside the village infields, limited forestry both in the outlying land of villages and in more distant forest commons allocated on the shield to villages on the plains.[39] New forestry regulations introduced in 1734, largely to secure the charcoal needs of iron production, brought even greater restrictions on such practices as slash-and-burn farming, leaf gathering and wood cutting. The legal limit on farm subdivision to ensure that peasants could pay land taxes was lowered during the course of the eighteenth century.[40] There was strong pressure from the peasants to subdivide their farms, due partly to population growth, and partly to the intensification of agriculture and lower taxes and rents in real terms, which made it possible for peasant families to subsist on smaller acreages. The legal restrictions on subdivision simply meant, in fact, that farms were subdivided in reality but recorded as whole units in taxation registers and terriers, the reality only coming to light when the first enclosure was carried out and actual ownership and tenancy relations had to be examined by the surveyor. Due to the more rapid population growth, the pressure towards subdivision was much stronger in forested and other less fertile and less long-settled areas.[41] There, official valuations of farms had been low, so that although the room for expansion seemed smaller than on the plains, in fact it was greater, and there were also more opportunities for supplementing livelihoods outside agriculture. The increased subdivision of farms, of course, implies that the surplus per farm would be reduced despite the growing proportion of land in peasant hands. According to Heckscher, the survival and increasing power of the freeholding peasantry in Sweden was 'an important reason for the slow material progress of the country ... an interesting contrast to general European development'.[42] This was not the case in all areas, however, and where subdivision was negligible we can assume that peasants' wealth increased. It is also reasonable to assume that the property bought by persons of standing and town citizens was subdivided to a smaller extent, so that differences between these groups and the peasantry increased, leading to conflicts over production objectives, agricultural methods and cropping systems between these different groups in more complex villages. The traditional view is that persons of standing and townsfolk were more inclined to produce for the market while peasants were bent only upon minimizing the risk to their annual subsistence.

It used to be thought that persons of standing and townspeople would be the first to try to dissolve traditional farming structures to get rid of impediments to innovation. An alternative strategy for this group would have been to buy farms only outside the larger villages, where innovations could be introduced without any need to consider the village community. Another way of avoiding the village community's supposed inertia was to have it dissolved through some kind of enclosure. Like other important components of agrarian development, enclosure was in Sweden thought to have been instigated by the government.[43] In principle this must be correct, because it was not previously formally permitted to break up the medieval open field structure of villages. However, it has become apparent that the kind of enclosures formally introduced in 1748 were not really innovatory.[44] Already during the 1730s and 1740s, reorganizations of arable occurred in Södermanland and Östergötland.

It used to be thought, too, that persons of standing and estate owners were the initiators of enclosure, which only gradually filtered down to the peasants, who accepted it unwillingly. The earliest form of Swedish enclosure, *storskifte*, cannot directly be compared with radical English enclosure, but simply meant that the number of strips of arable and meadow were reduced to a few per farm. In 1762 a maximum of four parcels of arable and of meadow were permitted at the behest of peasant members of parliament. At the beginning, enclosure required a total consensus in the village, which meant that very few were actually carried through. In Östergötland, only about ten are known.[45] From 1755, probably in consequence of the small number of *storskifte* enclosures accomplished, a request from a single landowner to the county authorities could start the formal enclosure process. This was usually accompanied by loud protest from neighbours. Frödin called it 'a coarse outrage against the individual's rights of ownership and disposition', implying that it was imposed by those with authority, such as persons of standing and estate owners, on peasants who wanted to continue old ways.[46]

Enclosure became much more common after 1757. In Östergötland alone, 255 enclosures were accomplished during the following decade.[47] Were they an outrage against the peasants, and did they favour higher social strata? The second question cannot yet be answered, but we have reliable knowledge about the first, especially for Östergötland, where the most recent detailed studies of *storskifte* have been done.[48] The first time it was discussed in parliament, it was actually the Peasant Estate which raised the difficulties of agriculture resulting from small parcels of land and common fields. The peasant members for six counties, including all those from Östergötland, and single members from another two demanded that *storskifte* should be carried out at all future subdivisions of landed property.[49] Peasant resistance from some counties generally stemmed from two concerns. First, a worry that reorganization would mean that some peasants got uniformly poor, instead of at least partially good, land. Secondly, peasants feared that in counties where the medieval *solskifte* system had never existed some farms would lose out because no certain measurements of the extent of farm properties existed in these areas. No powerful opposition was expressed against the actual principle of enclosure, and opposition to it was actually stronger from other Estates in parliament than from the peasants.

Even so, it has been argued that applications for *storskifte* usually came from persons of standing, estate owners, and town citizens with farms producing mainly for the market.[50] More recently, it has become obvious that this was not so. In Lysing hundred in western Östergötland, peasants alone applied in 37.5 per cent of the enclosures up to 1772; peasants together with persons of standing in 28 per cent, and persons of standing alone in 35 per cent.[51] In Väversunda village it was the peasants with the smallest farms who applied for *storskifte* in 1765,[52] and in Ekebyborna parish, in our research area, the peasants dominated, with about two-thirds of the initiatives, which roughly corresponded to their share of landholding.[53] Clearly, peasants were from the start well aware of what an enclosure was and that it was in their interest to accomplish it.

Östergötland enclosures were earliest on the central plains, in areas with large proportions of freehold and Crown land. In forested areas *storskifte* enclosures did not become common until well into the 1780s. One reason for this lag is that the proportion of noble land was greater on the shield. The large share of farms belonging to a small number of proprietors meant that the need for enclosure was less pressing, and the much larger number of isolated farms also diminished the need for enclosure on the shield compared with the plains.

Peasants did protest against particular *storskifte* enclosures. Between 1748 and 1772 there were protests in 14 per cent of those in Östergötland. But in only twelve out of eighty-nine cases were they protests against the enclosure as such. Predominantly they were about perceived injustice in the proposed redistribution of land (sixty-eight cases), or because the suggested number of strips per owner was too few (nine cases). Thus, peasant protest does not support the theory of resentment at an 'outrageous' reform. Furthermore, in 75 per cent of the cases of complaints about proposed enclosures, the complaining peasant was allowed at least a partial correction on the adjudication of the Hundred Court or County Governor. It seems reasonable to expect peasant protests to have been greatest during the first *storskifte* period; but this has been shown to be wrong, too.[54] Rather, the number tended to grow over time. The protests changed character, however, when more strips were allowed in the fields. From being concerned mainly that the reorganization was too radical, later they expressed opposition against the percieved unfairness of the surveyor's suggested field division in terms of land quality. It is obvious that while the peasants of the plains were in general not opposed to altering the medieval cultivation system, they were not ready for a wholesale redistribution of the the land, fearing that it might lead to unfair adjustments in the distribution of fertility across the newly assigned holdings. The *storskifte* reform was not so radical, anyway, as to lead to the complete dissolution of the traditional village. On the contrary, on enclosure in many villages the tofts were reorganized according to the medieval *solskifte* rules. *Storskifte* enclosures eventually occurred in the forested shieldland. Arable land there existed only in scattered patches, and in small hamlets with only a few farms it was normal for each farm to have separate fields. But even where the farms had fields in common, hamlets were small and there could hardly have been substantial restrictions on a peasant who wanted to reform his cultivation. The interest in *storskifte* enclosure increased markedly when outlying land was allowed to be included in the process,[55] in which case the common forest could be subdivided and assigned to individual farms. In the shieldlands arable farming could not reliably produce subsistence foodstuffs: income to cover food shortages had to be generated by other forms of production. Hence, the interest of individual peasants in privatizing timber resources for charcoal burning and for sale. The timing of increased demand for forest products, and not lack of information, backwardness or unawareness of market opportunities, was responsible for the marked differences in the timing of enclosures on the plains and in the shieldland.

Two opposing motivations have been suggested for these enclosures. First, in stark contrast to the usual argument about the effect of enclosure, it used to be suggested that they enabled peasants to subdivide their infield arable land to provide extra subsistence holdings for more numerous sons. 'Whereas the old narrow strips could not be further divided without great inconvenience, the consolidated fields now created were more easily divided.'[56] Recent calculations show that the carrying capacity of a single farm could have been increased enough by enclosure for this to be possible.[57] Thus, the apparent consolidation of holdings brought by enclosure could simply have been a prelude to further subdivision amongst a self-provisioning peasantry, not a basis for the emergence of integral market-orientated farms. Secondly, it has been suggested that the enclosure occurred because the infield was fully reclaimed and it was necessary to reorganize village holdings to enable large-scale reclamation of outlying land.[58] Many villages which had already accomplished a *storskifte* enclosure on their infield underwent another in the 1780s or 1790s in which the entire area of infield and outlying land was included. Thus, enclosure was not necessarily brought in to dissolve a static system, but became important when it was necessary to reclaim arable from the outlying land. Of course, there is no necessary contradiction between the argument that enclosure was inspired by population pressure amongst peasants and findings that enclosure mainly occurred in order to consolidate outfield land. But such consolidations could equally well have been inspired by a desire to increase market surpluses. This would certainly be consistent with the changes brought by *storskifte* towards more individualized patterns of ownership, although it must be remembered that the farming system still involved the intermixture of the (now fewer) parcels of different farmers across common fields.

Estate agriculture
Did the advancement of peasant freehold agriculture mean a retrogression of estate farming? The answer depends on definitions. A majority of the estates sold by the nobility were bought by ironworks owners and other persons of standing, and we must decide whether to define an estate legalistically, as something the nobility practised on their tax exempt land protected by strong privileges, or practically, simply as very large-scale agriculture with a demesne and tenant farms. It seems more reasonable to do the latter, in which case large areas did not necessarily slip out of estate control as tax exempt land moved into non-noble hands. Although estate farming is now often denigrated as 'feudal', and therefore backward,[59] the traditional view that estate agriculture

played an important innovatory role, to be followed by the (dumb) peasants, cannot be entirely groundless.

Heckscher maintained that *Gutsherrschaft* estates, with the concentration of production on a manor farm, were developed to a larger extent during the 1680s after the confiscation of noble land,[60] those nobles who lost large parts of their demesnes tried to compensate by farming what was left more intensively. The profitability of *Gutsherrschaft* was still an oft-discussed issue in the eighteenth century, not least in connection with the introduction of married hands (*statdrängar*) on manor farms. Up to the eighteenth century, hands seem, as is customary in peasant societies, to have been unmarried and in the early adult stage of their life cycle, after which they got married and took over a farm or a croft of their own. However, it was asserted by several authorities in the eighteenth century that a large manor farm with married hands was more profitable than any other form of estate management.[61] In Central Sweden, estates with large tenanted demesnes tended to have small manor farms; corvée duties from tenant farmers were used only to a small extent on manor farms, and the products paid as rent were mainly different sorts of grain for sale, although money rents were also paid.[62] Such estates had the largest surpluses, suggesting that the large-scale agrarian organization of *Gutsherrschaft* was less economical than traditional small scale peasant farms of a *Grundherrschaft* estate.

At the beginning of the eighteenth century, before labour supplies became more plentiful, it must have been difficult to expand and turn tenants' dues into day labour. Moreover, it seems that entire farms' rents were rarely paid as labour duties. This could have been because of the effectiveness of opposition to increased duties from tenants, or it could have resulted from estate owners' search for higher and more reliable cash income, in terms of money rents or grain to be sold on the market, pushing many of the risks of cash cropping onto the tenants' shoulders. In order to expand manor farm production and area without greatly reducing land rents, more labour was needed. But hands or day labourers were costly in the eighteenth century. The Swedish solution was to allow croft settlements on manor and tenant farmland in return for heavy crofters' labour duties.[63] This arrangement had existed at least from the sixteenth century. But with an increased tendency towards *Gutsherrschaft*, estate crofts grew markedly in number. The crofters produced their own subsistence, and their days' work at the manor farm were thus almost costless for the estate owner. Parallel with the increased number of crofts, the area tended by them expanded because crofters were normally permitted to reclaim arable, which made it possible to increase their labour duties. It became normal for a crofter to employ a hand who worked full time at the manor farm to fulfil them. Often, when crofters retired, their crofts had more arable than they had originally been allocated,[64] and hence the crofts could be divided and given to two crofters simply by building another dwelling. This Swedish model of *Gutsherrschaft* seems to have increased steadily in extent during the eighteenth century, along with the normal tenant farm version, only to be gradually replaced by large scale direct farming, with more and more married hands and new farming techniques. The profitability of such hands in estate production seems to have been exaggerated by their promoters, or their calculations were based on areas where it was possible to convert the entire demesne to one enormous manor farm. As this could be done in few parts of Sweden, it is not not surprising that traditional ways of organizing estate production were retained.

Even if a *Gutsherrschaft* estate proper was not a fully capitalistic enterprise, this does not imply that it lacked contacts with the market. Market production was substantial in the 1750s on a number of estates in Central Sweden,[65] which sold grain to large towns and ironworks in relatively distant areas. Many new ironworks established in Sweden during the latter half of the seventeenth and the first half of the eighteenth century were owned and run by estate owners who channelled part of their grain output into the subsistence of their industrial workers. Ljung estate in central Östergötland, for example, sold 82 per cent of its produce in 1750 to Stockholm and to other ironworks under the same ownership. In 1768 another ironworks was established in the immediate vicinity of Ljung manor farm and supplied with grain from it. Large parts of the estate's agricultural output were used in pursuit of secondary production. In order to keep the costs of this down, it was obviously important to supply inputs as cheaply as possible, not only agricultural produce, but also charcoal from its forests, burnt by crofters and tenant farmers as part of their labour duties. Thus, Swedish 'agrarian' estates contained large volumes of secondary production which were not linked to agriculture by flows of capital and goods through the market, but by internal transfers within the multi-faceted economies of large unitary estates.

Land-holding in western Östergötland at the beginning of the nineteenth century

The area selected for our study provides, as nearly as any small region can, a microcosm of Sweden as a whole, with areas of both fertile plainland and rugged forested shieldland. It contains twenty-three parishes grouped into the two hundreds of Aska and Dahl (see figure 3.1). Dahl in the south covers part of the fertile Vadstena plain. Its calcareous soils based on Cambro-Silurian bedrock remain chemically neutral in this generally acidic environment, and have always been considered to be amongst the most productive in Sweden. The southern part of Aska hundred has similar excellent prerequisites for agriculture, but further to the north, towards Motala Ström, is an intermediary zone where outwash gravels make the surface less favourable. Further north again, beyond Motala Ström, is rocky shieldland with meagre acidic

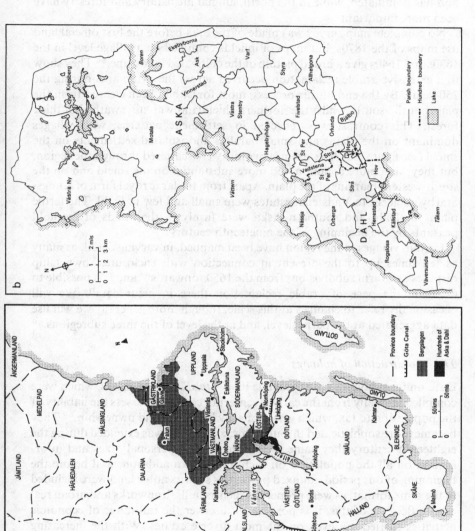

Figure 3.1. (a) The location of the hundreds of Aska and Dahl in Southern Sweden.
(b) The constituent parishes and part-parishes of Aska and Dahl hundreds.

soils. These diverse conditions have always been reflected in the agricultural activities of the region. In the south on the calcareous plains, arable production has dominated, while in the north, animal husbandry and forestry have been more important.

No complete map survey was made of the area before the first official land use maps of the 1870s, but maps of infields compiled on a village level in the 1630s and 1640s give a broad picture of the cultivated landscape.[66] They show that extensive arable reclamation occurred across the entire area during the 250 years. By the end of this process, most forest had been cleared from the plains of the south, whilst the northern shieldland was still swathed in thick forest. This contrast was reflected in settlement patterns, with villages dominant on the plains and small hamlets and isolated settlements on the shield (see figure 3.2). Tax-exempt manors were scattered through the region, but they were more common and more substantial on the shield and on the south-western margins of the plain. Apart from the large royal farm of Kungs-Starby adjacent to Vadstena, estates were small and few in Dahl. The fertile plains of Dahl and southern Aska were firmly in the hands of freehold peasants by the beginning of the nineteenth century.

Certain villages in the region have been mapped, in varying detail, as many as ten times up to the present in connection with enclosures, ownership disputes and farm subdivisions from the 1650s onwards[67] and it is possible to calculate the pace of arable reclamation there in great detail. We will occasionally refer to changes at this scale, though more generally we will use data aggregated at the parish level, and at the level of the three subregions.[68]

The reconstruction of holdings

Data on farm holdings are available in *mantalslängder* registers, which were compiled annually from the early seventeenth century to assess the liability of the population to tax, which was primarily based on land ownership.[69] They became more sophisticated through time: as taxation was extended during the eighteenth century they came to include various personal taxes and information on all the population on each holding. In addition, well before the beginning of our period, untaxed properties such as noble land were included in the same format, as were properties such as mills, ironworks and urban real estate. Because of this, it is possible to discover the full range of economic activities carried out even on the most diverse estates. With the increasing complexity of the system of taxation, it became necessary to have a second system of listings, known as *taxeringslängder*, which contained the monetary liabilities of individuals as calculated from the detailed descriptions of taxable and untaxable property and its value set out in the *mantalslängder*. Our choice of period was to some extent based on the fact that these two systems of registration had been fully developed according to a completely systematic

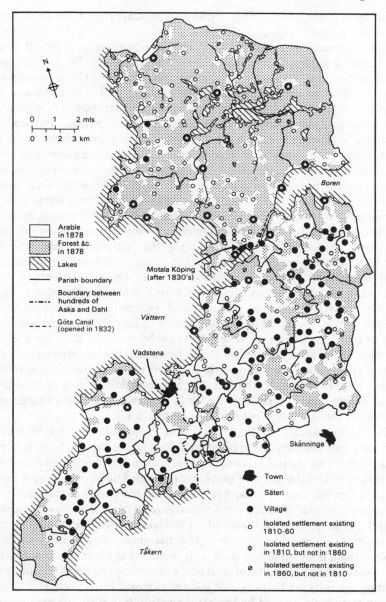

Figure 3.2. Forest and plainland in Aska and Dahl in the 1870s and settlements 1810–60.

scheme by 1810 (although values of non-agricultural properties were unfortunately not recorded until after this year).

Thus, information is available for each year of our research period on the extent of farm holdings, their values,[70] and the people who lived on and were supported by them. The latter includes not just members of farm households, but everyone who lived in crofts and cottages as well. These registers are complex and the reconstitution of farm holdings from them is complicated. They are organized as follows. Each parish is listed separately by hundred, parishes split between hundreds being partially entered under each. Under the parish name, each *hemman* or *helgård* was given a separate entry (a *hemman* was an ancient homestead, originally adequate for the support of one taxpaying peasant household). *Hemman* were listed in sequence, their names normally comprising that of the village followed by a farm name such as *Holstensgården* (the farm of Holsten), *Mellangården* (the middle farm) or *Källegården* (the farm by the well). Outlying *hemman* in our region normally had single names which often contained the suffix *torp* (that is croft), indicating that they were originally developed out of medieval croft colonizations. By the opening of our period most of these *hemman* had been subdivided, so that entries usually took the form of numbered parts (see figure 3.3). These parts form the basic units of land registration in the sources, which we have termed 'plots'. There could be three or four times as many of these plots as there were *hemman*, reflecting the subdivision of holdings over the centuries. This fragmentation, which has frequently been used as an indicator of farm subdivision,[71] continued through our period. Considerable further complexity was created by the fact that before enclosure each of these plots would, in fact, have comprised pieces of land in each field of the village, as in the plots of the farms of Apelby village on figure 3.3.

The extent of each of these plots was expressed in terms of *mantal*. Originally, one *hemman* was in principle valued at one *mantal*. Thus, each plot was a fraction of a *mantal*. It is important to realize that these *mantal* figures do not simply reflect the areas of plots. More fertile land would have a higher *mantal* value per unit of area, but because arable land was stressed in the evaluations, *mantal* values do not accurately reflect even the carrying capacity of land. Although the assessment of the monetary value of each plot given in the registers bears some relationship to the *mantal* figure, the correlation between them is by no means perfect. It is clear that the taxation authorities took the full economic value of each plot's arable capacity into account when deciding upon its tax liability, not just the *mantal* figure. These had become almost completely ossified by the early nineteenth century for each particular plot of land, whereas taxation values for plots which remained constant in size were frequently changed, even though it has been argued that they too became more and more fixed as the nineteenth century wore on and that they did not accurately reflect the market prices.[72] It is because of their better

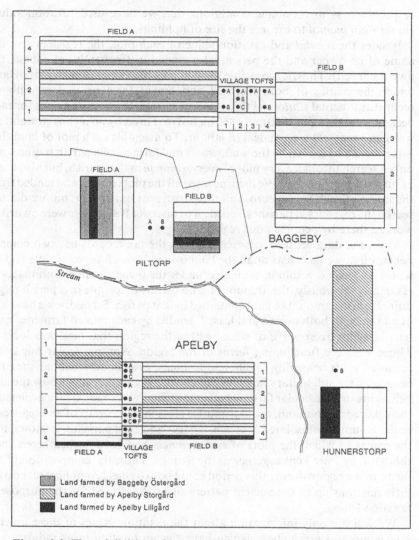

Figure 3.3. The subdivision of *hemman* into plots, and the amalgamation of plots into farms: a hypothetical example.

responsiveness to economic conditions that we have used taxation values rather than *mantal* to express the size of holdings.

Besides the *mantal* and taxation value of each plot, the registers give the name of its owner and the person who tended it (frequently, of course, the owner himself). Thus, it is possible to aggregate the values of each plot of land under the names of both freehold and tenant farmers; that is, fully to reconstitute actual *stations* of this agrarian society. We have done this for each decadal year between 1810 and 1860 inclusive. This is by no means straightforward and requires a great deal of labour. To assemble each plot of land that was owned and tended by the landowners and farmers of a parish requires not only a search through every plot of every *hemman* in that parish, but also those of surrounding parishes. We incorporated all the plots owned and tended from the farms of each parish across all the parishes of our region, but we did not search the registers of parishes outside it to discover if any plots were owned or worked there by people in our region.

Thus, the data set we have created from the tax records has two components. First, we have lists of all the land owned by each person in the region, wherever located within it, tabulated under the owners' names and places of residence. Obviously, the amount of land owned by people in a parish might differ from the amount of land contained in that parish. Secondly, we have lists of all the land, both owned and leased, tended by each named farmer of each parish, again irrespective of where within the region that land was located. Those were the functioning *farms* of the region. Any particular one could comprise plots belonging to different *hemman*, even in different parishes; few *hemman* were still unitary farms in our period. Figure 3.3 shows how the land belonging to a particular farm, whether that farm was situated on a *hemman* in the village or on outland, could comprise plots that were part of and registered under a number of isolated *hemman*, villages and even parishes. It shows, too, the extent to which the plots of each *hemman* had been dismembered, even though they were kept together in the listings. Thus, the composition of the farms in the region during this period could be extremely complicated, bearing little relationship to the ancient pattern of holdings used for the purpose of taxation listing.

We will use only information about the taxation values of these units of ownership and production, which have certain limitations as indicators of 'farm size'. They are not measures of area, but of economic value, which means that they changed with economic conditions, so that a given piece of land could rise or fall in value as prices rose and fell. A particular problem in this regard is the 50 per cent devaluation in 1859. We found it best to remove the dramatic effects of this from our data by redefining the limits of size categories used in farm classifications. Despite the rather abstract nature of taxation value as an indicator of farm sizes and the way they changed, it has the great advantage of measuring the economic value of the resources owned and worked by the farmers of the region, and reflects the resources available at stations better than area measures of farm size.

Land ownership in 1810

The total taxation value of agricultural land in the region in 1810 was almost 2,000,000 *Riksdaler* (henceforth Rdr). Less than 2 per cent of it was owned by peasants from outside the region who lived in its immediate neighbourhood. Some 5 per cent was owned by outside persons of standing, in part living nearby, but also in distant towns like Stockholm, Norrköping and Jönköping. Within this group Ulfåsa estate, which at 95,000 Rdr accounted for 5 per cent of the total taxation value of the region, was owned by a nobleman resident in Stockholm, and some plots in Vinnerstad were owned by a factory proprietor from Norrköping. In total, 11 per cent of the regional land taxation value was owned from outside. Of the remaining 89 per cent owned by people resident in Aska and Dahl, some belonged to people outside agrarian society, such as citizens of Vadstena, the only urban centre in the region at the time, who owned about 1.5 per cent of the total agricultural taxation value of our region, mainly in parishes close to the town.

In all, 87.5 per cent of the agricultural land in the region was owned by its rural occupants. The southern and central parts were dominated by freehold land; after the eighteenth century tax purchases only one or two Crown farms were left in each parish. In 1704 the Crown had owned eight of the then ten farms in Väversunda, most of them having descended from Alvastra monastery. In 1810 it retained only two, the vicar's and district clerk's farms. In the large parish of Motala, two similar Crown farms remained plus the Crown estate of Bispmotala, originally one of the medieval properties of the Bishop of Linköping. Obviously, only Crown farms of the kind which could not be tax purchased remained unsold by 1810. The only large holdings still in Crown hands in 1810 were Kungs-Starby in St Per parish; Bispmotala estate, and the farms which belonged to and supported the old soldiers' home in Vadstena, originally part of the Vadstena nunnery demesne. There was more tax-exempt noble land than Crown land in 1810, comprising estates and individual farms. Even though noble land was no longer exempt from land tax, it still carried some privileges: for example, noble landowners need not support soldiers' crofts. It was most common in the northern part of our region, and although considerable areas of taxed land had been acquired by estate owners by 1810, it was mainly in the same northern parishes. Peasant freehold tenure dominated in all parishes, overwhelmingly so on the plains of Dahl and southern Aska.

In 1810 there were 1,080 plots of land. In principle every plot could function as an independent farm, and the ownership of each one was registered separately in the tax rolls. Table 3.1 shows the numbers of owners and plots in each parish of the region. In general, the ratio of plots to owners was much

Table 3.1. *Numbers of plots and local landowners in the parishes of Aska and Dahl in 1810*

Parish	Number of plots	Number of resident owners	Ratio of plots to owners
Västra Ny	79	47	1.68
Kristberg	14	11	1.27
Motala	94	53	1.77
N. Aska total	187	111	1.68
Vinnerstad	49	24	2.04
Ekebyborna (part)	23	12	1.92
Ask	28	17	1.65
Västra Stenby	93	65	1.43
Hagebyhöga	72	54	1.33
Varv	44	27	1.63
Fivelstad	51	39	1.31
Styra	37	31	1.19
St Per (Aska)	31	22	1.41
Orlunda	44	28	1.57
Allhelgona (part)	44	31	1.42
Strå (Aska)	24	12	2.00
Hov (Aska)	13	12	1.08
Bjälbo (part)	6	4	1.50
S. Aska total	559	378	1.48
Nässja	33	21	1.57
Örberga	81	49	1.65
St Per (Dahl)	1	1	1.00
Strå (Dahl)	27	16	1.69
Rogslösa	79	55	1.44
Herrestad	55	29	1.90
Hov (Dahl)	8	4	2.00
Källstad	26	26	1.00
Väversunda	25	22	1.14
Dahl total	335	223	1.50
Grand total	1,081	712	1.52

Sources: Data on all tables are taken from the *Mantalslängder* and *Taxeringslängder*, Aska and Dahl hundreds, Landsarkivet, Vadstena, unless otherwise stated.

higher in the north. This suggests that land ownership was more concentrated there than in the south, but this was not necessarily so because it was quite common for people to own land in more than one parish. The residents of some parishes possessed more land than the parish contained, whilst those of others owned less than the value of the land of the parish in which they lived. The estate demesnes were the most striking example of this tendency, and this is what caused the stronger concentration of ownership in the north.

The differences from parish to parish in the ratios of plots to owners were not as wide as we might expect, given the much greater prevalence of estates in the north. The reason for this is that in the north, apart from the estate owners, most people owned only one plot. In the plains to the south, on the other hand, numerous people owned several plots, although rarely more than two or three, and peasant land holdings were clearly quite heavily fragmented. For example, a third of all peasant freeholders owned more than one plot in Herrestad, where already in 1810, as in the rest of Dahl hundred, an overwhelming proportion of land was under peasant freehold ownership. In Herrestad, seven of the nine (out of a total of twenty-nine) with more than one plot had only two; the two owners with more than this each had a third plot in a neighbouring parish. In the northern parish of Västra Ny, the situation was markedly different, even though its plot/owner ratio was very similar. Only three peasant freeholders owned more than one plot in the parish, although two of them also owned a third plot in a neighbouring parish. The total number of peasant owners in Västra Ny was similar to Herrestad, so the difference in their plot/owner ratios was due to persons of standing and estates. An intermediary group of parishes in central Aska had low owner/plot ratios. The dominance of estates and persons of standing was missing there compared with to the north, but unlike the plainland parishes further south they contained few peasants owning more than one plot. For example, in Styra only four out of twenty-nine owners did so, two of them owning their second plots outside the parish.

This picture of regionally differentiated patterns of ownership is very simplified. Some land was owned by estates in the intermediary and southern parts of the region, even if the proportion was far smaller than in the north. However, our major conclusion still stands: in the southern part of the region the concentration of ownership of agricultural plots into relatively few hands was mainly because many peasants owned more than one plot, whilst in the north it was based on ownership by estates.

The southern subregion, where peasant ownership was most concentrated into multiple plot holdings, also contained the really early *storskifte* enclosures. We know that they were initiated by ordinary peasants,[73] implying either that the peasants there were early to move into production for the market, or that they were soonest to reorganize in order to split up their holdings between more numerous heirs.

Of course, tenancy could completely change this picture of regional contrasts in land holding. How were these units of ownership distributed amongst functioning farms?

Farm structure in 1810

Often, cadastral information of the kind discussed above is all that is available. For this reason alone, much of the work on peasant farming and the development of agriculture in Europe as a whole, including Sweden, carries the implicit assumption that land which was unitary in ownership formed a unit of use: that cadastral and functional definitions of land holdings were the same, and indeed, that the latter simply reflected the former. This elision between land ownership and its use is not only a consequence of the type of information which has survived. It is also encouraged by some aspects of the way in which farming was organized, before and after the changes usually termed 'the agricultural revolution'. In common field farming systems, each family farmstead had 'its own' land, scattered in strips across the open fields, whether held in freehold or leasehold tenure. Diagrams showing the disposition of the strips belonging to each farm are commonplace. So are comments about inefficiencies due to this scattering, or about how it compensated for local topographical and pedological inequalities, times of ploughing, and so on. But the scattering of the strips owned by a farmer only has these consequences if units of ownership were unitary in use. Similary, the ring-fenced farms introduced after enclosures, again whether in leasehold or freehold tenure, are thought of as entities in which unitary ownership was necessarily reflected in unitary use. The usual intimation is that in both cases these units of ownership were kept integral over long periods of time, passing through generations as the continuing bases of family subsistence or landowners' rental incomes. Although they might be broken up at enclosure or the succession of sons, what was broken up was a system of integral units in which use reflected ownership, and what was created after the break-up was another such system, with the lines between units simply drawn in different places across the cadastral map.

There are good reasons for these assumptions in what we know about attitudes to land in both peasant and commercial farming systems. In the basic model of peasant societies, each family must possess sufficient land to ensure its subsistence and any dues it must pay. The peasants themselves were obsessively attached to their ancestral family land, and spent most of their non-working lives scheming to 'keep their name on it'. Whether the farm was in strips or ring-fenced is immaterial to the strength of this bonding. Thus, we have the commonplace model implicit in most work on peasant societies: a system of hardly differentiated holdings, each family with its own plot of land (which could, of course, comprise a number of scattered strips). The ancient homesteads (*hemman*) of Sweden were just such units, the integrity of which would be guarded as assiduously by the Crown to guarantee its tax base as by the peasants to guarantee their subsistence. Estate owners, too, had a strong interest in leasing their land in units of a size and disposition which would best guarantee the size and continuity of their income. A farm was both a unit of ownership and of use in commercial as well as peasant farming.

Many generalizations about historical farming systems are only valid if this correspondence between units of ownership and use actually occurred. Only then would the subdivision of family land through partible inheritance necessarily cause the fragmentation of farms, increased scattering of shrinking plots of land through the generations, and smaller and less efficient farms, or would the consolidation of ownership create larger and more efficient ones. But it is important to stress that most of what we believe about these processes in the past is based solely on cadastral information. Most of the work in Sweden, too, until recently, charted cadastral changes and simply drew inferences from them about how farming must have been changing in consequence. It is in order to compare our findings with such work, as well as to discover the changing distribution of ownership for its own sake, that we have dealt with cadastral matters here. However, because information exists in Sweden on who tended, as well as who owned, each plot of land, it is possible to go further and discover how much of this land was leased out, and therefore the exact nature of the units in which it was actually farmed.

As with ownership structures, we reconstituted the total areas of land cultivated by each peasant, including owned and leased plots, for each decadal year. We have called these groupings of plots under the cultivation of one person 'farms', although this should not be taken automatically to imply that we are assuming that all the plots worked by a single person necessarily comprised a single, integral farming enterprise. Indeed, the degree to and ways in which distant plots were incorporated into the operations directed from a peasant homestead is a question which will be pursued in some detail in a later chapter, when we come to examine the sorts of livelihood positions that existed in agriculture and the way they changed.

As on figure 3.3, the constituent plots of most *hemman* were worked by different peasants. Often, a single *hemman* contained a number of farmsteads on its plots, each the dwelling of a different peasant family which worked the plot on which its dwelling was situated. In some cases, a single plot of this kind was all that a particular peasant cultivated. In other cases, a peasant also worked plots on other *hemman*. Whatever the particular case, it was these groups of plots or single plots tended by one peasant that comprised the 'farms'. They, rather than the *hemman*, were the *stations* within which unitary production structures were directed by farmers as single enterprises.

The 1,080 registered plots in 1810 were owned by 702 owners resident in the region. However, because much of the ownership concentration was due to estates in the northern parishes, if estate demesnes were tended by tenants rather than from manors, the number of farms might in fact exceed the number of owners. Regarding the peasants who owned more than one plot, we are on unknown ground. If a particular peasant simply leased out a superfluous inherited plot as a separate farm to another peasant, besides owning and tending a plot himself, it would be of less significance in terms of indicating the emergence of commercial attitudes within the system of family farming than if the owners of more than one plot tended all of them, or if tenants assembled large groups of plots from many different owners.

The farms have been classified into three groups according to their total taxation value, irrespective of the number of plots they contained. The first group consists of farms with a tax value of 3,500 Rdr or more, the second is an an intermediary group with a tax value of between 1,500 and 3,500 Rdr and, finally, the third is farms with a tax value of less than 1,500 Rdr. Estates and larger farms could have taxation values which considerably exceeded 3,500 Rdr in value. However, we would be unlikely to find an archetypal peasant family farm above this limit, geared mainly towards the subsistence of a labour force comprising only the family and a living-in hand or maid. Given that a large majority of farms were of 1,500 Rdr or less in value, these must have represented such 'subsistence' peasant farms, and, thus, farms in the largest category must have been geared mainly to producing for the market. Holdings vastly in excess of 3,500 Rdr in value must have been entirely different kinds of enterprise from peasant farms. We have therefore distinguished holdings worth above 25,000 Rdr. It seems reasonable to consider these very large holdings, which belonged to nobles and persons of standing, as the functioning 'estates' of the region, irrespective of whether or not they were based on formally designated manors on noble land.

The area of arable land represented by particular taxation values varied over the region. In principle the taxation value was related to the arable land and potential forest produce was excluded from the assessment. Poor soils were assessed at a lower value than fertile ones, and thus a farm of 1,500 Rdr value in the north might well contain a larger area of arable, as well as much more forest, than one in the fertile south. In Väversunda, on the southernmost part of the Dahl plain, each 1,000 Rdr's tax value corresponded to about 10 acres of arable land.[74] The total taxation value of Björka and Björketorp hamlets on the shield in Motala parish was 5,000 Rdr, only one-eighth that of Herrestad on the plain. Björka/Björketorp was three times as big as Herrestad in area, but had only one-third as much arable land. Farms there had less arable land per 1,000 Rdr value, but much more forest and grazing. Clearly, the taxation values reflected only the value of a farm's arable land, rather than arable acreage or total resources.[75]

There were 810 farms in the entire region in 1810 – about a hundred more than there were land owners. Table 3.2 shows that the proportion of large

Table 3.2. *Plots, numbers of farms in different size categories and average sizes of holdings in the three subregions in 1810 (values in 000s of Rdr)*

	Dahl	S. Aska	N. Aska	Total
Plots	335	559	187	1,081
Category I farms				
Number of farms	47	70	9	126
Total value	309	457	107	873
Average value	6.6	6.5	11.9	6.9
% of total farm value	51	46	44	47
% of total number of farms	21	16	6	16
Category II farms				
Number of farms	91	134	14	239
Total value	209	319	36	564
Average value	2.3	2.4	2.6	2.4
% of total farm value	34	32	15	31
% of total number of farms	41	31	9	30
Category III				
Number of farms	83	222	140	445
Total value	89	220	98	407
Average value	1.1	1.0	0.7	0.9
% of total farm value	15	22	40	22
% of total number of farms	38	52	86	55
Total				
Number of farms	221	426	163	810
Total value	607	996	242	1,845
Average value	2.7	2.3	1.5	2.3
% of total value	33	54	13	100

farms was greater in Dahl hundred than in Aska, but the proportion of really big farms was much higher in northern Aska, given the large value of the holdings in that category. At the same time, small farms accounted for 41 per cent of the total taxation value in the north but only 15 per cent in the south, in spite of the fact that their mean value was higher in the south. Clearly, there were more complex structures of tenancy in the south, and in the north the lower limit of category I was generally only exceeded by estates, which were much bigger than the large peasant farms of the south. Thus, it is possible to characterize the northern area as comprising an amalgam of traditional peasant agriculture and estate production, while gradually, the number of peasants tending multiple-plot farms increased progressively southwards.

As we have already shown, in Herrestad parish, on the southern plain, there

Table 3.3. *Numbers and values (in 000s of Rdr) of farms in different size categories in a sample of parishes in 1810*

	Herrestad	Strå (all)	Hov (all)	Nässja	Styra	Motala
Category I						
Number of farms	7	7	5	2	0	5
Total value	32	60	56	12	0	71
% of total value	36	62	77	31	0	52
Category II						
Number of farms	19	12	5	12	13	6
Total value	47	27	11	23	31	15
% of total value	53	28	15	58	62	11
Category III						
Number of farms	9	10	5	4	21	69
Total value	9	10	5	4	19	52
% of total value	10	10	7	10	38	37
Total						
Number of farms	34	29	15	18	34	80
Total value	88	97	73	40	50	138
Average size	2.6	3.4	4.8	2.2	1.5	1.7

were twenty-nine ordinary peasant landowners, of whom nine owned more than one plot. There were thirty-four farms (see table 3.3). The proportion of large farms was, at 36 per cent, smaller than in Dahl hundred as a whole, and the centre of gravity lay on the intermediate units. Figure 3.4 shows that many of these category II farms worked more than one plot, and that multiple-plot farms were more numerous than multiple-plot units of ownership: no less than thirteen of the thirty-four farms worked more than one plot in 1810. Thus, there was a considerable difference between the distribution of land ownership across the parish population and the distribution of land cultivation. Some farmers did not own any land, but just leased one or more plots; others owned one plot and tended another as well, and some plot owners did not farm any land at all. Some Herrestad village farms included plots in other hamlets of the parish, and even in other parishes.

How representative was this of Dahl hundred and the southern plains in general? Table 3.3 shows that the proportion of farms in the largest category varied from 74 per cent in Strå and 89 per cent in Hov to 29 per cent in Nässja and 36 per cent in Herrestad, despite its numerous multiple-plot farms. Strå lay at the opposite extreme in terms of farm size. Did the complexity of its land holding structure also differ? In the two parts of Strå, which straddled the boundary between Aska and Dahl hundreds, there were twenty-eight land-

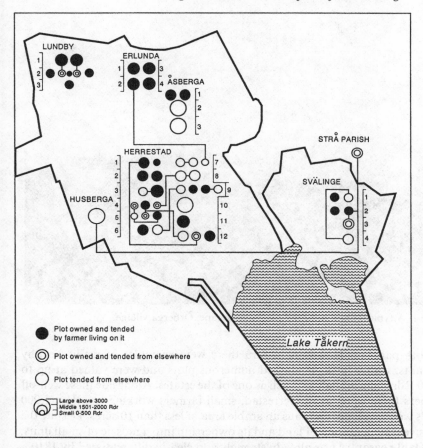

Figure 3.4. Ownership and tenancy of the plots of the *hemman* of Herrestad parish in 1810. The numbers on the vertical scales for each settlement in the parish represent *hemman*.

owners and twenty-nine farms: the difference between the two was smaller in Herrestad. There were seven large farms, twelve middle-sized and ten small ones. Both the number and the value of the intermediate group of farms was smaller in Strå than in Herrestad, and those of the small farm category almost identical. Unlike Herrestad, Strå contained two estates in 1810. These estates plus the Crown farm of an army quartermaster represented three of Strå's category I holdings. Apart from these holdings, the situation was similar to that in Herrestad: eight peasants in Strå had more than one plot on their farms, although they did not always own all of them and sometimes owned none. Between one-third and one-quarter of the farmers tended more than one plot

Plate 1. A typical Dahl hundred plains landscape: Örberga village.

in both parishes. However, in Strå there were examples of really wealthy peasants, whose farms contained numerous plots and were valued at up to 6,500 Rdr – almost half as much as one of the estates. Alongside these well-off farmers there were, as in Herrestad, small farmers working less than 1,000 Rdr's worth of land, and thus an arable area of less than 10 acres. Despite the massive fragmentation of land and its ownership into a mosaic of small units, the Dahl peasantry was already strongly and clearly differentiated by 1810.

Was southern Aska's more widely dispersed land ownership reflected in a wider spread of use-rights across more numerous small farms than in Dahl? Table 3.3 shows that in Styra in southern Aska there was not even one large farm in 1810, as many category II farms as there were category I farms in Dahl parishes, and many more category III farms. In Styra, too, there were few multiple-plot farms (only two tenants tended more than one plot) and the mean taxation value per farm was lower. Even when estates are removed from the Strå figures, Styra farms have a considerably lower mean value. Differences between peasants were smaller: no farmer worked more than 3,500 Rdr worth of land in Styra and although there were also some very small farmers in the parish, they were fewer in proportion than in Strå and Herrestad, where the mean value of category III farms was lower. Thus, the peasants of Styra were much less differentiated than those of the parishes further south: few farmers had holdings which would produce much more or less than family subsistence.

Land ownership was more concentrated in Motala parish, as elsewhere on the shield, than on the plain. This mainly reflected the existence of a few big estate landowners; otherwise ownership was not very concentrated at all. Although the proportion of large farms was small in Motala parish, they were very large and controlled 52 per cent of its land area. Category II contained only of 11 per cent of the farms (see table 3.3). Units consisting of more than one plot were few and far between: there were only six, of which three were large farms. Of the three others, only one was tended by an ordinary peasant. Among the large farms, estates of the nobility and persons of standing were utterly dominant. The largest, Carlshult estate owned by the nobleman Chamberlain Fleetwood, consisted of six plots and was valued at 31,000 Rdr. The whole estate included another manor farm which was managed separately from Carlshult by a tenant. The second largest Motala estate was Lindenäs, owned by District Judge Duberg, a commoner person of standing, who also tended 16,500 Rdr worth of other land comprising sixteen plots of widely varying value scattered across Motala and the neighbouring parishes of Vinnerstad and Ask. In aggregate, 41,000 Rdr worth of land was farmed from Lindenäs, which was one of the largest holdings in the region. Apart from five very large estates, land in Motala parish was tended by sixty-nine small peasant farms, most of which were assessed at under 1,500 Rdr (see table 3.3). Only one peasant tended more than one plot.

The difference between this shieldland parish and those on the southern plains is striking: estates dominated land holding, but there was a heavy numerical preponderance of category III farms. However, it is not possible simply to infer from this that capitalist farming had penetrated the peasantry in the south, whereas in Aska hundred, particularly the northern part of it on the shield, a traditional peasant society with purely subsistence production on freehold land and noble estates still survived. Before drawing such a conclusion, we need to know much more about production on individual farms, and whether the peasants of the north had small single plot farms because of their agricultural practices, or because of by-employments and incomes from other sources, such as charcoal burning or metalworking on or off their farms.

Changes in ownership and tenancy, 1810–60

Changes in peasant ownership, 1810–60

More radical enclosure laws passed in 1827 (*laga skifte*), the removal of restrictions on subdividing holdings, and improved methods in agriculture are customarily considered to have been major instruments of agricultural development in the nineteenth century. External stimuli came from more rapid population growth and an increased export demand for grain, mainly oats. Population growth was not equally rapid over the whole country, but normally much more marked in forested areas than on the plains. However, it meant that the labour force available for agriculture increased at the same time as the demand for agricultural produce. Population growth was not evenly distributed socially, either. It was mainly within the land-holding group that the number of children increased.[76] Hence, it has traditionally been assumed that a strong demand for agricultural land led to an increasing degree of farm subdivision, which reached its peak during the nineteenth century when it became permissible to subdivide holdings almost without limit. The effects of this were reinforced by the *laga skifte* enclosure regulations. In principle these should have ensured that each farm in a village came to have one continuous area of arable, meadow and outlying land, which made subdivision much easier. Even so, the number of available peasant positions was outstripped by the number of peasant sons, and it was the proletarian stratum of the population that grew most, together with the interstitial stratum of crofters, as peasant children were proletarianized.

We have already shown that substantial reclamation from the forest had taken place in several areas during the seventeenth century, so that traditional open-field cultivation was not a barrier to agricultural changes of this kind. Land reclamation became more marked as the *laga skifte* enclosures progressed. However, because this coincided with the introduction of new crop rotations, the beginning of urbanization, population growth, and strong foreign demand for grain, it is difficult to attribute land reclamation simply to *laga skifte* enclosure. It is, of course, also easy to conclude that land reclamation was associated with farm subdivision: that what would otherwise have been smaller and smaller farms could, in fact, succeed in producing enough to provision their households because of it. Does the concentration of both ownership and worked land among the peasants of the southern part of our region imply that farm subdivision was not as intense there, and in consequence that land reclamation, enclosure and population growth were not as prevalent?

Figure 3.5 depicts how the number of plots in our area changed between 1810 and 1860. Generally, it grew up to 1830, although Dahl hundred in the south witnessed a small reduction. The *laga skifte* enclosures only began in 1827, and only a few villages in our region had been enclosed according to this scheme before 1830. The number of plots was subsequently *reduced* in the entire study area up to 1840, least markedly in the south and most markedly in the north where, apart from on the estates, the principle of one freehold plot per peasant farm had predominated in 1810. Later changes were not dramatic, as the plot total increased to 1,265 in 1860, decreasing slightly in the south, and increasing slightly elsewhere. The moderateness of these later changes suggests that the *laga skifte* in Dahl and the southern part of Aska, at least, did not lead to plot subdivision. Whether it led to land reclamation can only be discovered through analysis on a village level, to which we will return later. A 20 per cent

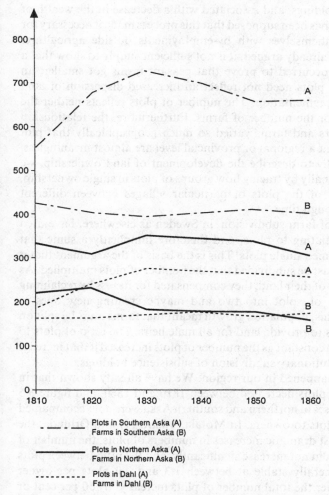

Figure 3.5. Numbers of plots and farms in each subregion, 1810–60.

increase in the number of plots over half a century is not remarkable. However, the aggregated figures hide much more dramatic changes at a local level. Orlunda parish almost doubled its number of plots from 43 to 81, and the number in Motala parish grew from 94 to 144. The majority of the other parishes developed in a similar way to the subregions in which they lay: plot numbers were gradually reduced in all the parishes of Dahl, and most of the parishes in southern Aska saw a slight increase after that year.

If development over time is analysed simply in terms of plot numbers, it is tempting to conclude that the increase in the number of plots was caused by the subdivision of landholdings, and associated with a decrease in the wealth of peasants. Recently, it has been supposed that this process made it necessary for farmers to support themselves with by-employments outside agriculture. However, as we have already argued, it is not sufficient simply to show that a subdivision of plots occurred to prove that peasant farms got smaller: an increased number of plots need not reflect an increased dispersion of land ownership or farm fragmentation. The number of plots reflects neither the number of owners nor the number of farms. Furthermore, the relationship between plots, owners and farms varied so much geographically that plot numbers aggregated at a regional or provincial level are almost meaningless. In order more usefully to describe the development of land ownership, we examined it longitudinally by tracing how groups of plots in single ownership, and the distribution of the plots of particular villages between different owners, changed through time.

Most discussions of farm subdivision, in Sweden as elsewhere, have used cadastral evidence relating to plots, and therefore inmplicitly assume that freehold peasants farmed single plots. This is the basis of the argument that it was because farms must be subdivided that the number of plots multiplied. As peasants sold off part of their land, they compensated for its loss by reclaiming more, splitting their old plot into two and maybe creating new ones in addition. On top of this, population growth meant that farms must be split on the death of peasants to provide land for all male heirs. The ratio of plots to owners would remain constant as the number of plots increased if that increase were due to the involutionary subdivision of subsistence holdings.

This is not what happened in our region. We have already shown that in Dahl the number of plots decreased between 1810 and 1860, and figure 3.5 shows that the increases in northern and southern Aska were not accompanied by stable ratios of plots to owners. In Motala, Västra Ny and Orlunda, the parishes with the most dramatic increases in numbers of plots, the number of resident landowners did not increase significantly, and the ratio between plots and owners was generally stable at between 1.3 and 2.0 plots per owner irrespective of whether the total number of plots increased by 50 per cent or 100 per cent (see figure 3.6). The levels and changes in the ratios of plots to owners were remarkably similar over the entire region, in spite of its heterogeneity in terms of ownership patterns and conditions for cultivation. The purpose of increasing plot numbers, where it occurred, cannot have been the subdivision of farms for the reasons usually suggested.

In Herrestad the number of plots fell after a maximum in 1830, which implies that the number of one-plot owners or the total number of owners must also have fallen after 1830. This is what happened, but at the same time a concentration of ownership was under way, so that the number of landowners with more than one plot grew from nine to fourteen between 1810 and 1830 and was reduced by only two up to 1860, in spite of a 20 per cent fall in the

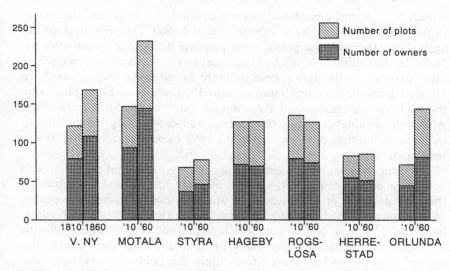

Figure 3.6. Numbers of plots and owners in a selection of parishes in 1810 and 1860.

number of plots. Whereas in 1810 the peasant landowners of Herrestad held fifty plots, they had fifty-five in 1830, but thirty-three were controlled by multiple-plot owners in the latter year compared with twenty-four in 1810. Despite the number of plots being greater in 1830 than in 1860, the total controlled by multiple-plot owners was the same in both years. Another measure of this process of ownership concentration is the number of plots owned by Herrestad peasants in other parishes. In 1810, there were only two, while seven plots in the parish were owned by farmers from outside. In 1830 Herrestad peasants farmed five plots in other parishes and in 1860 nine, whilst the number of Herrestad plots farmed from outside the parish changed from seven in 1810 to eleven in 1830 and five in 1860. This indicates both a concentration of ownership in the southern part of the region and an increase in the spatial spread of the plots belonging to particular farms.

Styra parish in southern Aska, the intermediate part of our region, had thirty-seven plots in 1810. Of these, four were owned from outside and Styra peasants owned two plots outside the parish. In total, four farmers owned more than one plot, but nobody owned more than two and so twenty-seven owned just one. In 1860 the number of plots had increased to forty-five, and a similar development to that in Herrestad occurred, though smaller in scale and later in time. In 1860 eight Styra peasants owned more than one plot and one owned three. In sum these eight controlled seventeen plots, 38 per cent of the total, compared with the 60 per cent of Herrestad plots in the hands of multiple-plot owners. The majority of this increasing concentration of owner-ship occurred between 1810 and 1830. The number of plots owned from outside Styra varied considerably over the period, and in general the owner-ship structure fluctuated a lot between 1810 and 1860. This, of course, speaks against any survival of a peasant one plot/one household system after the 1820s. In 1830, thirteen of Styra's forty-nine plots were owned by farmers from other parishes, while Styra people still only owned two outside plots. As in Herrestad, the balance was more even in 1860, when six Styra plots were owned from the outside and three outside plots were owned from Styra. Although the number of one-plot owners was slowly being reduced by the increase of multiple-plot ownership, in 1860 twenty-five one-plot owners remained.

The mean value of plots declined as they increased in number in Herrestad and Styra up to 1830. Later, subdivision must have been limited because the average value of plots was quite stable in both parishes, irrespective of the numerical trend. In both parishes differences in the size of land-holdings increased. However, several complications prevent us from safely inferring anything about what was happening to farm size from these trends. For example, the plots which went into multiple-plot holdings might have been those which underwent subdivision, whilst the plots of one-plot owners did not change at all. Tenancy complicates the picture further: one-plot owners might have tended plots other than their own, while multiple-plot owners might have tended one only and let the others out. Thus, one-plot owners could have bigger farms than multiple-plot owners, even if the latter controlled more fixed capital. Only when we include both ownership and tenancy can we say anything definite about whether *farm* size grew or shrank.

The increased number of multiple-plot owners is unlikely to have resulted from peasant inheritances and marriage settlements, but must mean that the sale and purchase of plots was frequent, and that a market existed for agricultural property. Motives for the purchases could, of course, have varied. It could have been to convert surplus wealth into real property, or to compensate for the inheritance of a plot that was too small to support a household. Alternatively, given the legal requirement for partible inheritance in Sweden, peasants with several heirs could secure their inheritances only by acquiring extra plots, leasing them out until the death of the household head. Or, the reason might have been that suggested by Chayanov: that the changing size of the family labour force through the life cycle made it necessary to acquire more plots as children grew older, and then dispose of them as they left the parental home to form their own families. Whatever was activating the land market, whether land was being treated as marketable property capable of storing wealth and yielding income, or as something with only use-value that was acquired and disposed of as the demands of family subsistence and heirship required, can only be discovered by examining the nature of the livelihood positions it supported. The answer will depend on whether only family labour was used to work the land.

In any event, it is apparent that landowners in the intermediate subregion of our study area, as well as those in Dahl to the south, were deviating progressively further from the simple model of traditional peasant society: more and more plots were owned by particular owners. However, this was not the case on the shieldland in the northern part of the area, as exemplified by Västra Ny parish. Taken at face value, the degree of concentration of ownership there was substantial: more than half of the plots were in the hands of owners with multiple plots already in 1810. But, as we have shown, this was mainly due to estate demesnes in the north generally: in Västra Ny, those of Bona and Medevi owned fifteen and seventeen plots respectively. Apart from these estates, only six landowners out of thirty-nine owned more than one plot in 1810.

The number of plots in Västra Ny increased rapidly, but unlike in Herrestad and Styra so did the number of one-plot owners, which climbed from thirty-three to fifty-five. The number of estate plots fell and the number of other multiple-plot owners increased from six to nine. They were largely persons of standing outside the nobility, including a sergeant major (two plots), the vicar (four plots, of which the church itself owned two) and two non-noble part-owners of the third estate of the parish, Blommedahl, who had two plots each. Unlike in the south, only two peasants became multiple plot owners. Indeed, subdivision of peasant farms must have occurred as the number of plots in the parish grew from 79 to 103 and the average plot value did not grow in line with the general increase of the taxation values of agricultural property, but remained almost constant. The number of plots continued to increase right up to 1860 in Västra Ny after 1830, but the number of multiple-plot owners did not. Paradoxically, the number of one-plot owners did not grow after 1830, either, because the number of Västra Ny plots owned by residents of the parish was gradually reduced in spite of the growing total. In 1810, three plots were owned by landowners from outside, two of them ordinary peasants. One peasant in Västra Ny at the same time owned a plot in Motala parish. In 1830, seven plots were owned by outsiders, and Västra Ny peasants owned four in other parishes. Four of the former belonged to non-noble persons of standing, while three peasants in Motala completed their multiple-plot holdings with a plot in Västra Ny. Only one of the outside plots owned by Västra Ny inhabitants belonged to an ordinary peasant.

Ownership over longer distances was apparently developed further in Västra Ny between 1810 and 1830, too, although unlike in Herrestad the net effect was a negative balance of plots. This tendency increased further by 1860, when no less than 17 of the 109 plots of the parish were owned from the outside, compared with the four plots in other parishes owned from Västra Ny. To a large extent, it was persons of standing and other people with a footing in non-agrarian activities who extended their ownership: a tanner from Motala Town owned one plot in Västra Ny, a big mill owner from Motala another, a white-collar employee of Motala Works owned two, and so on. The number of ordinary peasants living outside who owned plots in Västra Ny had increased to six, half of them from adjacent Motala and half from more distant parishes in the study area. Thus, although for one reason or another the opportunity to buy plots in Västra Ny increased, the peasants of the parish hardly took part in this process of building up multiple-plot holdings. However, in the region as a whole, a larger and larger proportion of plots were owned by people who lived in the parish where they were located, even though they were often organized into complex multiple-plot holdings.

On a smaller scale, how did the ownership of land in particular villages and hamlets in the sample parishes develop over time? From Herrestad parish, we selected the largest village for study, the church village with the same name as the parish. It had twenty-eight plots in 1810, making it one of the biggest farming settlements in the region. The number grew to thirty in 1830, then fell to twenty-one in 1860. The trend was thus similar to that of both Herrestad parish as a whole and Dahl hundred in general, and we can be confident that Herrestad village was not atypical of the villages of the southern plains. A greater degree of concentration of ownership than in the parish as a whole had occurred by 1810, when sixteen of the twenty-eight plots were already parts of multiple-plot holdings. No fewer than seven of these were owned from outside the village, four from outside the parish. The remaining nine therefore belonged to peasants of the village who had extended their holdings. They usually tended all of them, and many also leased further plots, often those owned in Herrestad by peasants from other parishes.

All owners of land in Herrestad were peasants, and the development of holdings through the generations suggests that at least some vestiges of a traditional inheritance system survived, even if the existence of multiple-plot holdings is not consistent with what is usually thought to be conventional peasant behaviour. An example of this is the holding of Samuel Andersson (see figure 3.7), who in 1810 owned three-eighths of Källegården and a quarter of Klockaregården, as well as tending another eighth of Klockaregården. In total, his property was assessed at 3,125 Rdr. In 1820, Samuel Andersson was dead and his son Anders Samuelsson, who already owned and tended a farm in Vilseberga, Rogslösa parish, had taken over one of his father's two owned plots, whilst the other was still controlled by the trustees of his dead father. There were family links to several other plots in Herrestad village, and especially those of Källegården and Klockaregården: in 1810, three-eighths of Klockaregården was owned by the heirs of Anders Andersson in Vilseberga, Rogslösa and tended by Gustav Andersson, who in 1820 had taken over this plot plus the one which the heirs of widow Cajsa (obviously the relict of Anders Andersson) had owned and which Samuel Andersson had tended in Vilseberga in 1810. Gustav Andersson also tended the plot owned by the heirs of Samuel Andersson.

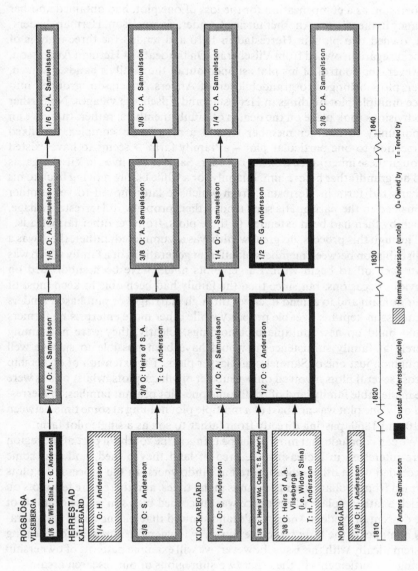

Figure 3.7. The holdings of Anders Samuelsson and his close relatives in Herrestad.

ROGSLÖSA
VILSEBERGA

| 1/4 O: Wid. Stina, T: G. Andersson | 1/6 O: A. Samuelsson | 1/6 O: A. Samuelsson | 1/6 O: A. Samuelsson |

HERRESTAD
KÄLLEGÅRD

| 1/4 O: H. Andersson |

| 3/8 O: S. Andersson | 3/8 O: Heirs of S. A. T: G. Andersson | 3/8 O: A. Samuelsson | 3/8 O: A. Samuelsson |

KLOCKAREGÅRD

| 1/4 O: S. Andersson | 1/4 O: A. Samuelsson | 1/4 O: A. Samuelsson | 1/4 O: A. Samuelsson |

| 1/8 O: Heirs of Wid. Cajsa, T: S. Anders'n | 1/2 O: G. Andersson | 1/2 O: G. Andersson |

| 3/8 O: Heirs of A.A. Vilseberga (i.e. Widow Stina) T: H. Andersson |

NORRGÅRD

| 1/8 O: H. Andersson | 3/8 O: A. Samuelsson |

1810 — 1820 — 1830 — 1840

Anders Samuelsson   Gustaf Andersson (uncle)   Herman Andersson (uncle)   O= Owned by   T= Tended by

By 1830, Anders Samuelsson had moved to Herrestad village and taken over the plot which had previously been owned by his deceased father's trustees, though he kept his property in Vilseberga. His uncle Gustav Andersson, as a compensation for the loss of one plot, had obtained another tenancy in the village. Another uncle of Anders Samuelsson, Herman Andersson, owned two plots in Herrestad in 1810 and tended the three-eighths of Klockaregården owned from Vilseberga. On the death of Herman Andersson, however, the control of his plots slipped out of the family's hands. Even so, seven plots belonged to grandchildren of Anders Andersson, grouped into three multiple-plot holdings in Herrestad and Vilseberga villages. No further subdivisions took place on the deaths of family members: rather, this gave an opportunity to another member to add a plot to his complex. No fixed connection to one particular plot – a 'family farm' – seems to have existed through these inheritance transfers. Anders Samuelsson lived in Vilseberga, as had his grandfather before him, but only for a while before moving back to his father's old farm in Herrestad, from which he later moved to yet another farmstead in the parish. His sons tended their property in Herrestad village, which by then had been extended to three plots, from yet other farmsteads.

Through this process, no great wealth was accumulated; rather, there was a redistribution between members of different generations of a family which was quite well off to begin with. Larger plots might have been subdivided on previous occasions, but since then the family had been able to keep most of their domain and to extend it, either through marriages or purchases. Anders Samuelsson kept his sizeable property while other more enterprising farmers could build up new multiple plot holdings. Clearly, they were not aiming merely at family subsistence: it would have been possible to survive well enough on just one of Samuelsson's larger plots. The extension of ownership across several plots involved only a sub-set of village plots, whilst others were the inalienable family land of traditional one-plot peasant families; in Herrestad only one plot was not part of a multiple plot holding at some time between 1810 and 1860, passing directly from father to son as a single plot farm.

We can conclude that most of the peasants in the southern part of the region were interested in increasing the area of land they owned, and that some succeeded while others disappeared as landowners. In the process, the plots farmed from a village changed hands many times, and there were few plots on which a father's place of residence was succeeded by a son's. The example of the descent of Anders Andersson's land showed that leasing could considerably complicate the farm structure shown by ownership alone. Before dealing systematically with this issue, however, we will examine patterns of ownership change in settlements in the other two subregions of our research area.

Our example from Styra parish in southern Aska is the hamlet of Övre Götala. In 1810 it comprised six plots, only one of which belonged to a multiple-plot holding. The appearance of more complex ownership structures was, as for the entire parish, slower than in the southern subregion. Even so, five of the eleven plots in Övre Götala in 1830 were owned in multiple-plot combinations, three of which were based outside the hamlet. We have previously argued that development towards capitalistic agriculture began later in southern Aska than in Dahl. This would lead us to expect fewer Övre Götala plots to be part of multiple-plot holdings, and more plots than in Herrestad to be owned by successive generations of single-plot peasants. However, this was true of only a couple of plots; all the others were at some time in the period part of multiple-plot holdings. Even so, a build-up of multiple-plot holdings did not occur. Such holdings were not only few in Styra, but they contained a maximum of two or three plots.

We showed earlier that, contrary to received wisdom, enclosure (particularly through the *laga skifte* procedure) did not lead to the greater subdivision of holdings. Rather, the *laga skifte* was used by already wealthy farmers to consolidate their property and push out some of the peasant freeholders. This was the case in Herrestad village and in Övre Götala. Indeed, where changes can be attributed to enclosure, they again took the form of consolidation. Instead of having a number of plots in different places, the peasants tried to gather them together in space. In Övre Götala for example, one plot was owned by a peasant from a neighbouring village, who was the applicant for the enclosure, through which he forced the other landowners to accept that his plot should be located on the hamlet boundary, next to his land in the adjacent hamlet.

Examples from Västra Ny parish on the shield put developments in the two more southerly subregions into a new perspective. Whereas plots assessed at between 1,000 and 2,000 Rdr were common elsewhere in 1810, and peasants often owned more than one plot, in the north plots were normally worth less than 1,000 Rdr and in single ownership. Of the thirteen plots existing in the hamlets of Järsätter, Kavlebäck, Övre and Nedre Lid, only one was assessed at more than 1,000 Rdr, whilst ten were valued at less than 500 Rdr – less than any in Herrestad or Övre Götala. In spite of this, relatively few landowners in Västra Ny tried to compensate by building up a multiple-plot holding. In total, only two of the owners in the four hamlets had a second plot somewhere else in the parish, and one plot in Övre Lid in 1810 was also owned from the next parish to the north, Hammar. The proportion of plots included in complexes was then further reduced, and only one of the twelve plots was owned from outside in 1820 and 1830. The situation changed radically later, however, in a somewhat unexpected direction. The number of plots decreased, to only ten in 1860, as Övre Lid came into the hands of a single owner, and in Nedre Lid plots were amalgamated by an owner living elsewhere. In total, four of ten plots in the four hamlets were part of ownership complexes by 1860, and if the plots are recalculated to the structure of 1810, seven of the thirteen were. It is apparent, then, that a large part of the increase in Västra Ny's total number of plots by thirty over fifty years occurred outside the traditional hamlet settlement structure, where the opposite situation was being created. A great part of the build-up of multiple-plot holdings in these parts was due to purchases from peasant owners who lived in other parishes. In total, this accounted for almost 20 per cent of the total number of plots in 1860. Peasants in the northern part of the region were not completely uninterested in building up multiple-plot holdings. Indeed, the intensity of such activity in 1860 was similar to that in Styra. At the same time, however, four of the original thirteen plots of our sample hamlets were kept within the same families over the period without ever being owned together with others, which suggests a stronger survival of peasant family land than on the plains.

Our analysis of land ownership shows that ordinary peasants all over the region built up holdings comprising numerous owned plots. The process started earlier and was most marked on the fertile southern plains, where grain cultivation was most advantageous. In the northernmost subregion, more peasants stuck to a traditional pattern of land ownership, with one peasant per plot and plots passed from father to son, even though the assessed values of the plots were much lower there. This indicates either sheer traditionalism, or the possibility of non-arable production. Generally, it is striking how mobile plots of land were between owners, the plots containing their dwellings as well as those which were wholly farmland: all could be changed completely over a peasant's lifetime.

Agricultural holdings of townsfolk

How did the build-up of multiple plot holdings by some peasants affect other categories of landowners? Certainly, the number of peasant landowners decreased; but was the activity of their expansionary neighbours also at the expense of estate demesnes or the agricultural holdings of townspeople?

It certainly does not appear to have been at the expense of townsfolk's agricultural holdings. Over the entire region the agricultural land-holdings of townspeople increased in value from 29,000 Rdr in 1810 to 64,000 Rdr in 1860. However, as table 3.4 shows, the progression between these totals was not smooth. The holdings of people from Vadstena decreased sharply in the last decade of the period, whilst those of people from Motala moved in the opposite direction. The net effect was to release 6,000 Rdr's worth of land into the hands of rural farmers in the 1850s; on the southern plain, where the majority of the land owned from Vadstena was situated, 15,000 Rdr's worth was transferred to countryfolk.

The citizens of Motala Town generally worked their rural land, whereas apart from in 1810 more rural land was owned from Vadstena and rented out to local peasants than was worked by its residents. This might have been because Vadstena had its own townland arable and over 100 grass plots within

Table 3.4. *Agricultural land controlled from towns in western Östergötland,
1810–60 (in 000s of Rdr)*

	1810	1820	1830	1840	1850	1860
Vadstena						
Number of farms	—[a]	31	37	31	27	21
Value tended	19[b]	49	63	44	56	41
Value owned	29[b]	53	77	65	80	45
Motala Town						
Number of farms	—	—	—	7	5	8
Value tended	—	—	—	19	12	21
Value owned	—	—	—	26	21	19
Total						
Value tended	19	49	63	63	68	62
Value owned	29	53	77	91	101	64

Notes:
[a] Indicates no information.
[b] Farmland in Vadstena itself not valued.

its boundaries, whereas the rural holdings of Motala's citizens were the only
farmland they possessed. From the 1840s onwards a large proportion of
Vadstena's townland was owned from outside the town; by 1860 half of it. In
the 1840s and 1850s most of the externally owned townland was in the hands of
the Lallerstedt estate based at Kungs-Starby, which bounded the town to the
south, although some was also owned by peasants. In 1860 the biggest holding
of Vadstena townland was that of Royal Forester Abolin, who occupied large
farms at Kvissberg, a detached part of St Per parish completely surrounded by
Vadstena townland, and half of Huvudstad, which adjoined Vadstena in St
Per. By 1860 the people of Vadstena had divested themselves of a large part of
their rural holdings and their townland. Nonetheless, a considerable amount
of land was still farmed from Vadstena, on its own townland and on the
surrounding plain. Over half as much was farmed by citizens of Motala Town
in the adjacent parishes of Motala and Vinnerstad. Altogether, there were
twenty-nine farming enterprises in the two towns in 1860. In 1830, at the height
of farming from Vadstena, there had been thirty-six in that town alone,
working as much land as was contained in one of the moderately sized parishes
on the plain.

All the 'farmers' of the two towns were involved in other occupations. The
farms operated from Motala Town in rural villages of the surrounding
parishes contained farmsteads which housed the workers who did the farming
under occasional supervision from the town-dwellers. Much of the rural

farmland of Vadstena was run in the same way, comprising (sometimes large) village farms with farmsteads housing the workers. However, even though few people from Vadstena who owned rural land did not also own townland, the latter seems to have been worked in a different way. Until the very end of the period there was only one farmstead in the parish of Vadstena outside the built-up area of the town, the rest of the townland being worked from the town itself. However, this was changing rapidly by 1860. Not only was much more townland worked by farmers and estate owners from surrounding parishes, but there were by then four areas of townland with their own farmsteads, presumably being worked as integral ring-fenced farms. To all intents and purposes, they were rural farms, contributing further to the passage of farmland out of the hands of Vadstena citizens into those of the increasingly complex holdings of the rural peasantry.

In sum, the total influence of urban holdings of rural land was marginal, and despite the creation of a second town in the region during our period, it decreased towards 1860. Either townsfolk could no longer compete for rural land against the rapacious appetites of peasants and estate owners, or they preferred to invest their surplus capital in other ways.

Estates and the farms of persons of standing

Did the peasants succeed in taking over parts of the estate demesnes, or did they remain exclusive to persons of standing? As we have mentioned, an estate can be defined in two ways: either as a *säteri* of the nobility on *frälse* land, or as any very large holding belonging to a person of standing, whether based on untaxed noble land or not. We have chosen 25,000 Rdr as the minimum value for the latter kind of estate, which we will discuss together with noble *säterier*.

Some of the larger estates were broken up during our period, but many of their constituent plots were taken over by other, newly developing ones. In 1830 a maximum of a dozen 25,000 Rdr estates between them controlled land with a taxation value of 766,000 Rdr – about a third of the total value of agricultural land in the region. In 1810, ten 25,000 Rdr estates had comprised only a sixth of the total value, and fifteen of them accounted for between a fifth and a sixth of it in 1860. Using the alternative *säteri* definition of estates gives a rather different picture. In 1810, there were twenty-eight of them, including the royal estates of Kungs-Starby and Bispmotala. Five belonged to owners of other estates, so that there were effectively twenty-three *säteri* estates, which contained all the estates defined as holdings of over 25,000 Rdr. Their taxation value in 1810 was nearly 600,000 Rdr, more than 25 per cent of the total taxation value of the land in the region. In 1860, there were twenty-seven *säteri* estates, accounting for the same proportion of the total land value of the region as in 1810. However, the 25,000 Rdr estates were no longer contained entirely within them: the latter had expanded off tax-exempt noble land onto peasant freehold land. Table 3.5 shows that different estates dominated in

Table 3.5. *Estates (*säteri *and holdings of over 25,000 Rdr in value) in 1810 and 1860 (in 000s of Rdr)*

Estate	Total value		Total number of plots		Tended value[a]		Tended number of plots		In noble ownership	
	1810	1860	1810	1860	1810	1860	1810	1860	1810	1860
Bergsäter	—[b]	13	—	1	—	13	—	1	*	
Bispmotala	—	23	—	3	—	23	—	3	*	
Blommedahl	3	14	1	1	3	14	1	1		
Bona	29	21	16	17	2	14	2	6		
Carlshult	37	35	7	5	31	35	6	5	*	
Karlsby	9	16	2	3	9	16	2	3	*	
Lindenäs	41	32	16	5	17	30	3	4		
Medevi	25	52	17	17	14	31	4	5	*	*
Säter	6	50	1	14	6	31	1	7	*	
Storeberg	19	25	2	1	13	25	1	1	*	
N. Aska total	169	281	62	67	95	232	20	36		
Biskopsberga	9	36	2	4	5	36	1	4		
Börstad	—	52	—	7	—	52	—	7		
Charlottenborg	37	29	8	1	21	29	2	1	*	*
Egeby	29	19	12	1	10	0	5	0		
Huvudstad	5	—	1	—	5	—	1	—		
Klosterorlunda	18	21	1	1	18	21	1	1	*	
Medhamra	10	29	1	1	0	29	0	1		
Kvissberg	18	30	5	10	0	22	0	7		
Rävsjö	14	12	3	1	12	12	2	1	*	
Rocklunda	11	—	4	—	11	—	4	—		
S. Freberga	—	57	—	4	—	57	—	4		
Sörby	42	21	8	1	23	0	3	0	*	*
Stavlösa	—	45	—	3	—	45	—	3		
Ulfåsa	95	146	18	10	45	96	2	4	*	*
S. Aska total	288	497	63	44	150	399	21	33		
Åbylund	24	25	4	3	17	25	1	3	*	*
Arneberga	30	11	9	1	23	11	4	1		
Kalvestad	15	22	3	1	15	22	3	1	*	
Kungs-Starby	51	138	1	14	51	86	1	6	*	
Dahl total	120	196	17	19	106	144	9	11		
Grand total	577	974	142	130	351	775	50	80		

Notes:
[a] Tended as part of a manor farm. If a manor farm was tended from elsewhere, the tended value is counted as 0.
[b] Indicates that the estate concerned belonged to another estate at the time.

1810 and in 1860, although some, like the noble estate of Ulfåsa in Ekeby-borna parish, remained virtually undisturbed.

In 1810, the property which District Judge Duberg assembled around his base on the little *säteri* of Lindenäs in Motala parish comprised nineteen plots in Motala, Vinnerstad and Ask parishes, worth about 40,000 Rdr taxation value. In 1820, the number of plots had grown by three, spanned five parishes and were worth 66,000 Rdr. Lindenäs stayed at this size, and incorporated urban real estate, mills and quarries, until it dissolved in the 1850s after Duberg died in the 1840s. The net increase of just three plots 1810–20 belies considerable flux. Five of the plots of 1810 were sold and eight new ones bought: plots were sold in Vinnerstad and Ask, and the Södra Freberga estate in Västra Stenby parish plus a number of other, smaller plots in Västra Stenby and Ask were bought. Two of the plots were sold to peasants and three to persons of standing with substantial properties. Södra Freberga and part of Kvissberg estate were bought from persons of standing, but six other plots were bought from peasants. Between 1810 and 1830, the number of plots owned from Lindenäs remained constant at twenty-two, but four were bought and four sold. There was a pattern to these purchases and sales in that the plots seem to have become more concentrated over space in Motala and Västra Stenby parishes, and in larger holdings like Bromma, Norra and Södra Freberga, which were small estates on their own account.

If any general conclusions can be drawn about the very big landowners from this example, the first is that their holdings were extremely fluid in composition: an estate of twenty plots exchanged more than half of its stock of plots in only twenty years. Secondly, estate demesnes were not entirely made up of erstwhile land of nobles and persons of standing. Small plots of the kind from which ordinary peasants built their multiple-plot holdings were also bought and sold by estate owners, apparently as pure merchandise. Peasants and estate owners were obviously operating in the same land market.

We will examine two more examples of estate development to test these provisional conclusions, one of the seemingly stable 'traditional' estates in the north and another relatively dynamic one, this time on the plain.

Our second example was a traditional estate based around a *säteri* on *frälse* land. Unlike Lindenäs, it contained a gradually diminishing number of plots during our period. Ulfåsa estate in Ekebyborna parish dated from medieval times. During our period it had several owners, most of whom lived in Stockholm and had a manager to run the estate. Ulfåsa's manor farm was located on the edge of our study area, and parts of the demesne were situated outside it. Our account is limited to the holdings within our region, where Ulfåsa consisted of eighteen plots in 1810. The two containing the *säteri* formed the nucleus of the estate and were tended as an integral farm by the estate manager. In 1860, the estate contained only ten plots, but its total value was almost unchanged from 1810. The Ulfåsa demesne in 1810 was quite different from that of District Judge Duberg. Apart from the two *säteri* plots, it consisted of tenant farms extending around the manor in Ekebyborna, Ask, Varv and Fivelstad parishes, all on noble land which had belonged to the estate for centuries. As the number of constituent plots fell, the proportion tended directly from Ulfåsa increased from 47 per cent to 65 per cent of the total value. The estate thus seems to have experienced what are supposed to have been classic trends in Swedish estate development in the eighteenth and nineteenth centuries: remoter plots were sold off and larger proportions of the demesne were cultivated directly from the manor farm. However, closer inspection shows that most of the reduction in plot numbers is chimerical. In fact, plots were gradually amalgamated into more substantial tenancies or into larger plots tended from the manor. Six separate plots ascribed to the estate in the taxation registers in 1810 were identical in aggregate to three listed in 1860. Thus, only two of the 1810 plots had actually been sold by 1860, those farthest away from the manor to peasant farmers. Five large old plots had been included in the manor farm but counted as only two separate units in the registers, while three other plots had been added to outlying farms in order to create larger tenancies. Even if changes in property value were small at Ulfåsa, there was considerable reorganization.

Our third example of an estate, that of District Clerk (later War Commissioner) Lallerstedt based at Kungs-Starby in the heart of the southern plains, is more difficult to get to grips with. Not only did it contain owned and tended land, but also the substantial royal *säteri* where Lallerstedt came to base his operations, but which he did not lease until the early 1840s during the process of expanding his holdings in our region from 27,000 to 87,000 Rdr in taxation value.[77] He began to accumulate farms in the parishes of St Per, Orlunda, Fivelstad, Allhelgona, Väversunda and Vadstena in the 1820s and 1830s. After getting hold of Kungs-Starby, which filled the southern part of St Per parish, Lallerstedt began the process of concentrating his holdings in the other part of St Per parish and in Vadstena, which separated it from Kungs-Starby. In 1840, he owned and tended another five plots in St Per and had sold off his most remote Väversunda holding. By 1850, when the whole estate was worth 97,000 Rdr, he had sold off another two remote plots and fourteen of the remaining eighteen were in St Per. Like Duberg of Lindenäs, Lallerstedt also owned urban property in Vadstena, although in his case it was at the heart of his ever-concentrating rural holdings, comprising real estate, a warehouse and the large Crown distillery. Also like Duberg, he bought most of his plots from peasants, although his initial purchases in the region at Orlunda were distant plots belonging to the disintegrating Bromma estate in Motala parish, and by 1840 he controlled three *säterier* on the plains. After Lallerstedt's death in 1851, three plots were sold back to peasants and the remainder was split into three substantial persons of standing's estates. Two were held by his heirs and the other by Royal Forester Abolin of Kvissberg in St Per. Each was based on a *säteri*, with other plots scattered around it, and valued in total at over 25,000 Rdr.

Our three examples of estates developed along two distinct lines in terms of the plots they owned and tended. Ulfåsa can be termed 'traditional'. It had existed for centuries before 1810, comprised only noble land, and during our period was fairly stable in extent. However, such estates were a wasting phenomenon. Of the twenty-three *säteri* estates in the region in 1810, thirteen were owned by noblemen, and all the 25,000 Rdr estates were based on noble land; only seven of the twenty-four *säteri* estates of 1860 were owned by noblemen, and only two of the seven 25,000 Rdr estates. The rest had been incorporated into the enterprises of persons of standing like Lindenäs and Kungs-Starby. These represented a new kind of estate which was far more unstable in plot composition, incorporated peasant as well as noble land, and was obviously part of a much more dynamic economic enterprise including urban and industrial as well as agricultural property, even at Kungs-Starby on the plain. The instability of these 'modern' estates was also reflected in the fact that whereas Bona and Ulfåsa existed throughout our study period, even though both changed owners quite a few times, the Lallerstedt and Duberg holdings were accumulated during their creators' lifetimes, and both broke up after their deaths. In this, the modern estates were similar to the peasant multiple plot holdings, which were created and dissolved during the span of a peasant lifetime. The structure of land ownership between 1810 and 1860 was much more turbulent than one embodying traditional estates and peasant farms.

The Crown owned one or more plots in each parish of the region, amounting to about 7 per cent of the *mantal* in Dahl. In 1810 this land constituted parts or the entire farms of persons of standing who were state employees such as vicars, district clerks and judges in local courts. Duberg and Lallerstedt used their Crown farms as stepping-stones to the accumulation of very large estates. What of those persons of standing who did not progress to estate ownership?

Crown *hemman* were not normally as heavily subdivided as those in peasant hands, so that the single plots of state employees could be substantial farms. The vicarage farms of Vinnerstad and Hagebyhöga, for example, were valued at 6,000 and 5,000 Rdr in 1810, a lieutenant's farm in Varv at 5,000 Rdr, and a county forester's farm in Rogslösa at 9,000 Rdr. Only a small minority of the Crown land farms were not in our category I. If they continued to farm on their own account in our period, persons of standing who were not estate owners would have formed a substantial proportion of the category I farmers; if they leased their land out, large areas would be made newly available to peasant farmers. The first policy characterized two of our subregions, the second the other.

In northern Aska, holders of Crown land continued to farm it and supplemented it with the tenancies and purchases of taxed land. In 1810 persons of standing other than estate owners farmed 16,000 Rdr's worth of land, but none was a category I farmer. By 1860 they had more than doubled their holdings and three of the nine had category I farms. In Dahl, the opposite occurred. In 1810, eleven persons of standing farmed land worth 85,000 Rdr, ten of them category I farms. By 1860, there were nine Crown land farms, seven in category I, worth 54,000 Rdr. Whereas in northern Aska persons of standing took in progressively more land, in Dahl, like the urban landowners they divested themselves of it to the peasantry.

However, not all parishes followed the trend of the subregion in which they lay. In Herrestad in Dahl in 1810 the district attorney lived at the village Crown farm, assessed at 3,750, and the vicar at Husberga, valued at 5,000 Rdr. The vicarage farm was leased out to a local peasant in 1810, and so was that of the District Attorney in 1860. Meanwhile, Sergeant Major Enander had extended his holding of Crown land in Svälinge hamlet in Herrestad by two plots, and by two in Väversunda and Strå parishes, amounting in total value to 8,000 Rdr. In consequence, although two out of three of them ceased farming, the proportion of land in Herrestad farmed by persons of standing stayed at about 5 per cent. In Styra in southern Aska, there were three Crown farms of persons of standing in 1810. One was leased out, but the other two accounted for 8 per cent of the total taxation value of the parish. In 1860, persons of standing farmed only 2 per cent, and the rest of their Crown land was leased to peasants. In Motala parish on the shield, 6 per cent of the land was farmed by persons of standing who were not estate owners in 1810. After 1840, they acquired some of the land of the estates which were breaking up, accounting for about 14 per cent of the land in the parish in 1860. Almost as much erstwhile estate land had passed into peasant hands.

In aggregate, noble estates, the holdings of townsfolk and the Crown farms of public servants shrank at the expense of peasant and non-noble persons of standing's land-holdings. Despite the continued existence of Crown farms and some noble estates, and the efflorescence of vast holdings like those of Lallerstedt and Duberg, the peasantry did not release their grip on the region's farmland. They increased it significantly in Dahl (see table 3.6). A significant proportion of the plots which they assembled into multiple plot holdings came from Crown land and estates, which divested themselves of most land in Dahl.

In this discussion of estates of nobles and persons of standing, we have dealt with ownership and tenancy together, whereas we dealt only with peasant land ownership earlier. We will now return to what was happening on peasant holdings to discover if changes in patterns of ownership were associated with changes in the mosaic of farms into which the plots of the region were assembled for the purpose of cultivation.

Changing farm structure in Aska and Dahl 1810–60

A complex pattern of tenancy rendered the 1,080 plots of the region, distributed between 702 owners, into 810 farms in 1810. By 1860, the number of plots had increased by about 30 per cent and the number of landowners to

Table 3.6. *Estates and other farms in 1810 and 1860*

		Noble estates	Person-of-standing estates	Person-of-standing farms	Peasant farms
1810					
Dahl	Total value	68,000	37,000	85,000	417,000
	% of value	11	6	14	69
S. Aska	Total value	101,000	78,000	62,000	756,000
	% of value	10	8	6	76
N. Aska	Total value	75,000	10,000	16,000	130,000
	% of value	31	8	7	54
1860					
Dahl	Total value	22,000	96,000	81,000	668,000
	% of value	3	11	9	77
S. Aska	Total value	125,000	218,000	170,000	1,017,000
	% of value	8	14	11	67
N. Aska	Total value	44,000	243,000	48,000	241,000
	% of value	8	42	8	42

848, but the number of farms fell to 774. Generally, the greatest changes in farm numbers were in the 1810s and 1820s. Figure 3.8 shows that the trend was quite different in northern Aska and in Dahl. In the former, the total number of farms increased slightly over the period, while in Dahl it fell by more than 25 per cent after 1820. Southern Aska, in line with developments in plots and their ownership there, lay in between: the number of farms was reduced, but only by 5 per cent. Thus, although farm development was similar in direction to that of land ownership, it was much more extreme. We must remember, too, that at the same time the number of plots grew quite substantially and, in consequence, the development of farms was more drastic than is implied by numerical trends alone. Farms must have become more complex in plot structure as they grew, and despite the transfer of land into peasant hands from other social groups, some (few) peasants must have gained at the expense of (many) others during the period.

Although generally there was a break of trend in the 1830s (see figure 3.9), numerical changes varied between farm size categories and subregions. The number of category II farms fell by almost half in Dahl between 1810 and 1860, but doubled in northern Aska, and whereas the number of category I farms was stable in Dahl, it more than doubled in northern Aska. The far more numerous small farms of category III at first diminished in southern Aska, then climbed back beyond their 1810 number by 1860; they diminished then increased then diminished again in northern Aska, and subsided continuously

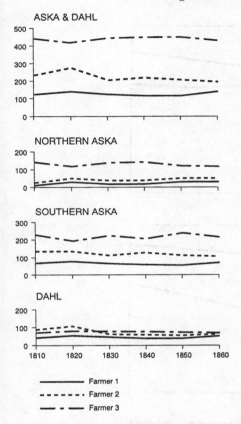

Figure 3.8. Farm numbers in each subregion, by farm size category, 1810–60.

though the period in Dahl. Thus, it was only in Dahl where large farms were almost as numerous as either medium-sized or small ones, and where they increased in proportion over the period. In northern Aska large and middle-sized farms were few in number and increased their proportions only slightly between 1810 and 1860. In southern Aska small farms were both heavily predominant in numerical terms and increased their proportion as large farms became slightly less numerous and middle-sized farms much less so whilst small farms multiplied. In aggregate, these changes represented a considerable shift of land into the hands of large farmers, who cultivated more than half the land in all subregions by 1860. This proportion was only marginally exceeded in southern Aska, but it was over 70 per cent in Dahl, where over the fifty years large farms, already more dominant than elsewhere in 1810, increased their grip over the land faster than in the other two subregions. On this level of resolution, however, it is not possible to see how plots and farms were

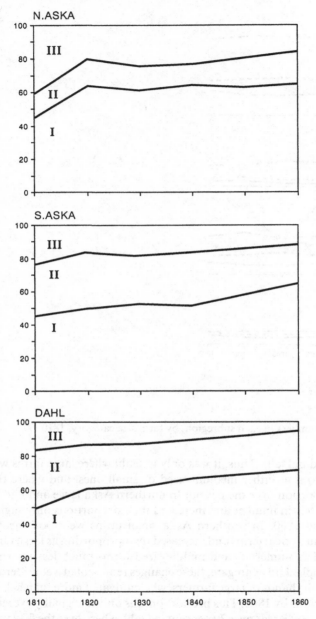

Figure 3.9. Proportions of land value in each subregion worked by farms of different size categories, 1810–60.

Table 3.7. *Plots per farm in Herrestad parish in 1810 and 1860*

| | Plots per farm | | | | | | |
	1	2	3	4	5	6	Total
Number of farms in 1810	21	9	3	1	0	0	34
Number of farms in 1860	13	5	8	0	1	0	27

redistributed, and especially not why. Our next step will thus again be to look in more detail at some parish examples.

Herrestad parish, which saw an increasing concentration of ownership and only moderate growth in the number of plots, contained thirty-four farms in 1810 and twenty-seven in 1860. The rate of reduction was typical of Dahl hundred. So was the growing proportion of multiple-plot farms and the shrinkage of category II farms, which must have been amalgamated into category I holdings or tended from outside the parish because the number of Herrestad's small farms grew by only one and their total value fell (see table 3.7). Naturally, it was the large farms which consisted of two or more tended plots, none of the farms with more than one plot was in category III. The total number of plots tended from Herrestad farms in 1810 was fifty-two, the same as fifty years later. The number was maintained despite some plots having been amalgamated because, in parallel with ownership, cross-parish holdings became more common. In 1810 only one Herrestad farmer tended land elsewhere, whilst in 1860 four farmers tended five plots in neighbouring parishes. Table 3.7 shows that at the same time more plots were concentrated in the hands of the large farmers: whereas there were twenty-one one-plot farms in 1810, there were only thirteen in 1860, and the three three-plot owners of 1810 had become eight in 1860.

As the land of Herrestad village comprised more than half of that in the parish as a whole, it is not surprising that considerable change also occurred on this smaller scale. In 1810 it consisted of twenty-eight plots, sixteen of which had farmsteads on them. The other twelve were worked by farms in the village or, in one case, tended from neighbouring Erlunda hamlet. The peasants of Herrestad village at this time did not farm land elsewhere, even though they owned seven plots outside. Seven of the farms based in Herrestad village worked more than one plot. However, many of them worked only two plots, which were located on the same *hemman* and would be adjacent in the common fields before the *laga skifte* enclosure. The process of change in Herrestad village was apparently less dramatic than elsewhere in the parish or in Dahl hundred as a whole during the next decades. Change was greater from the 1840s and 1850s after the *laga skifte* enclosure. The number of plots and farms in the village were both reduced by a third, though the number of multiple plot farms did not grow. The concentration of plots into fewer farms involved what had been a number of plots worked by a single farm, some owned and some leased, coming into single ownership and being reclassified as a single plot. For example, in 1840 the farmstead of Lars Larsson was located on one of the plots of Storgården *hemman*, from which plots in three other *hemman* in Herrestad and one in another parish were worked. In 1860, Larsson's farm was no smaller, but strategic purchases of plots on Storgården *hemman* after the enclosure had made it possible to group his plots spatially, and thenceforth they were listed as a single unit in the taxation registers. Whereas in 1840 the Storgården *hemman* had consisted of four plots, all worked by different peasants, in 1860 it comprised only Larsson's single plot. His total land-holding grew in value from 4,286 to 7,100 Rdr, although a count of plots would indicate that it had not changed in size.

Thus, peasants were able to create large farms, some as big as small estates, by assembling several plots. But this was a complex long-term processs, and in Herrestad it came late compared with the farms of some of the persons of standing in the parish. Figure 3.10 shows the development through time of the plots tended and owned by Larsson and two other large peasant farmers of Dahl. All three followed the same strategy. From one 'base' plot containing the farmstead, tenancies were acquired in the vicinity, usually in the same or neighbouring parishes. Gradually, the outlying tenancies were either bought or swapped for tenancies closer to the farmstead, which were subsequently bought. The final situation was one where most of the plots were close to the farmstead and owned rather than leased. The objective was obviously to avoid paying for numerous leases or travelling long distances between plots. The development of these multiple plot holdings seems to demonstrate consider-able capital accumulation among the peasantry. However, if they were bought on mortgage loans this would be fictional, and the instability of these large farms might be an indication of this. All the examples on the diagram, and very many others, were at least partly dissolved at the succession of the first heir and then dismantled completely on his death (usually not until after 1860), even if he had only one heir. Large complex peasant farms, like the 'modern' estates of persons of standing, seem to have rested heavily on the initiative of the individual farmer, rather than on accumulation within a stem family, as on traditional estates or in peasant society.

The process of farm expansion partly explains why there were more farms than landowners in 1810, but more landowners than farms in 1860. It became more common in Dahl for a peasant to supplement the owned plot on which his farmstead stood at first with leased plots, which were only gradually bought. The higher turnover of both ownership and tenancy to which this gave rise is illustrated by figure 3.11, which shows the changing ownership and tenancy of the plots comprising Herrestad Haggård *hemman*. It comprised three plots in 1810; the first plot had four different owners over the fifty-year

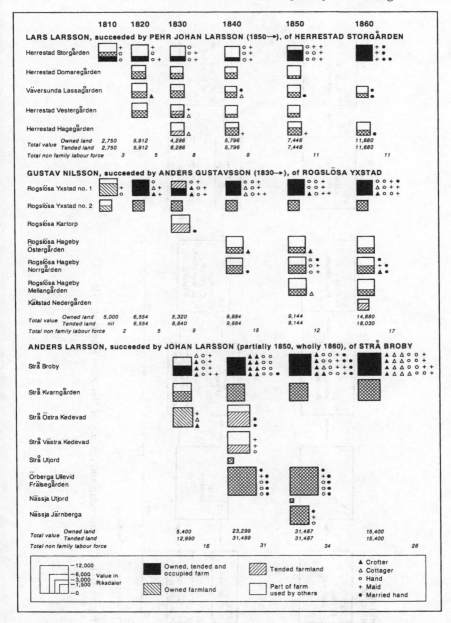

Figure 3.10. The changing plot composition of three expanding peasant farms.

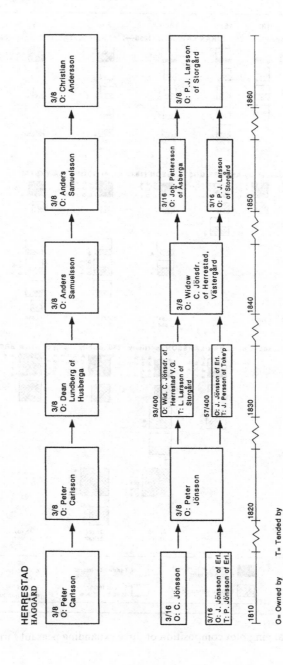

Figure 3.11. The ownership and tenancy of Herrestad Haggård, 1810–60.

Table 3.8. *Plots per farm in Strå parish in 1810 and 1860*

	\multicolumn{7}{c}{Plots per farm}
	1	2	3	4	5	6	Total
Number of farms in 1810	22	6	1	0	0	0	29
Number of farms in 1860	23	6	3	1	0	0	33

period, and the second and third changed hands even more often, eventually to be merged into the farm based on Herrestad Storgård. In 1820, each of the three Haggården plots contained a farmstead, but in 1830 they were all owned and tended from farmsteads on other *hemman*. Like the example of the Anderssons' farm in Herrestad we discussed earlier, this was by no means an extreme example: some plots changed ownership and/or tenancy in every decade. This massive flux implies that agricultural land was more and more being considered as a commodity, at first amongst some of the persons of standing and then amongst some of the peasants.

The concentration of tenancies was also marked over the period in Strå parish on the boundary between Aska and Dahl hundreds (see table 3.8). The number of farms with more than one plot grew from seven to ten, those with three or more from one to four. At the same time, the number of farms in Strå increased by four, and although the number of plots grew only from fifty-one to fifty-four, forty-eight were worked from farmsteads in the parish in 1860 compared with thirty-seven in 1810. As in Herrestad, peasants in Strå were gaining control over more of the plots in their home parish, and seven of them also farmed plots outside it in 1860 whereas none had done so in 1810. The noble estate of Åbylund was leased out to a peasant by 1860, though that of Kalvestad was still farmed by its noble owner. In addition, by 1860 Strå contained the category I farm of a non-noble person of standing which comprised a single large plot. Two contradictory processes were occurring in the parish. As in Herrestad plots and farms were being amalgamated into fewer large holdings, but at the same time some plots were being subdivided into smaller ones. In consequence, the number of category I farms remained unchanged at seven; the number of category II farms was halved to six, and the number of category III farms doubled to twenty. The mean taxation value of farms in categories I and II remained stable, whilst that of category III farms was almost halved. The three *hemman* of Bondorlunda 1 and 2 and Tistorp 1 accounted for almost the entire growth in the number of plots in Strå. In 1810 Tistorp 1 was already heavily subdivided into six plots, most of which were worked from farmsteads on other *hemman*; in 1860 it comprised thirteen plots, of which half supported farmsteads working land worth less than 300

Rdr. The plots of certain peasant *hemman* were virtually exploding into fragments while surrounding holdings were undergoing the opposite process. This seems to be a classic case of peasant differentiation, and shows how enclosure might actually have been linked with simultaneous consolidation and fragmentation.

To understand why this happened we must discover when the explosion occurred, how the *hemman* had previously been organized, and when the *laga skifte* enclosure occurred. Tistorp 1 comprised six plots, two crofts and two cottages in 1810, while in 1860 there were two crofts and no fewer than seven cottages as well as thirteen plots. Thus, it was not the case that pre-existing crofts and cottages had plots of land allocated to them, as frequently occurred on enclosure elsewhere. The taxation values of Tistorp 1 in 1810 reveal the existence of two kinds of plot: four had fairly normal values, but nine were worth less than 300 Rdr, less than one-third of the average value of category III farms in the parish. A superficial comparison indicates that the two normal units of 1810 which disappeared by 1860 had been transformed into small-holdings, probably during the *laga skifte* enclosure. However, this is not the entire story. As in many other parts of the region, the number of crofts and cottages grew in the 1810s; by the 1830s there were seven, each of which had a plot of land assigned to it in the enclosure. We can only speculate why the owners of two largish plots gave over their land to this allocation: it might, for instance, have been the best way to capitalize the value of a plot which was difficult to subdivide into functioning farms between a number of heirs. Certainly, all the crofters and cottagers of Tistorp 1 were not permitted to buy a holding from their landowner because a number of them still existed in 1860, mostly located on one of the smallholdings rather than on the larger plots. This suggests that small plots were more prone to subdivision into virtual hamlets of cottages, frequently (as without exception in this case) occupied by destitute and aged people.

Our southern Aska parish example, Styra, followed the same pattern as Herrestad in Dahl. However, the process seems to have begun later. In 1810 Styra contained just two multiple-plot holdings, each with only two plots. By 1860 nine multiple-plot holdings accounting for almost a third of the total number, two of them with more than two plots. Peasants in Styra also controlled a couple of plots in other parishes by 1860, and the total number of plots in their hands had increased from from thirty-six to forty-five (see table 3.9).

We saw earlier that there was quite a different ownership system in the north, with expanding and contracting estates intermingled with one-plot peasant owners. In Motala parish in 1810 only nine holdings comprising 10 per cent of the total had more than one plot. Amongst them only three estates and the innkeeper in Motala village tended more than two plots, and only two peasants had more than one plot. The remaining two plot holdings belonged

Table 3.9. *Plots per farm in Styra parish in 1810 and 1860*

	Plots per farm						
	1	2	3	4	5	6	Total
Number of farms in 1810	32	2	0	0	0	0	34
Number of farms in 1860	22	7	1	0	0	1	31

Table 3.10. *Plots per farm in Motala parish in 1810 and 1860*

	Plots per farm							
	1	2	3	4	5	6	7	Total
Number of farms in 1810	71	5	1	1	1	1	0	80
Number of farms in 1860	60	15	7	1	1	1	1	86

to a miller, the owner of a small ironworks and the district attorney, all of whom lived in the village alongside Motala Ström. Table 3.10 shows that the situation changed radically: in 1860 multiple-plot farms comprised a third of the total and fourteen peasants had more than one plot. By that time, Motala was in line with parishes in southern Aska: a third of the farms had more than one plot, and Motala peasants held plots in other parishes. However, Motala's land-holding pattern was still in some respects different from that of the parishes on the plain to the south. There was no expansion of the holdings of non-estate-owning persons of standing; indeed, only two of their seven farms were in category I, and both of these were the rumps of estates rather than accumulations made during the period. Moreover, the number of category II farms actually expanded through the period in Motala, and because there had been many more category III farms at the beginning of the period, there were still proportionately many more of them in 1860.

Much of the change in Motala parish was concentrated in the villages of Motala and Holm, on the narrow strip of land along Motala Ström and the Göta Canal. Motala village already contained several industries in 1810, and most of its nine plots were tended by millers and smiths who were obviously not interested in their tenancies for purely agricultural reasons. Plot values in the village were quite low at between 250 and 750 Rdr, but rights to the massive water power of the river, which largely belonged to the village plots of Motala and Holm, were unassessed in the taxation registers. On the opening of the Göta Canal, industrial pressure on the farms along the waterfalls increased, and although the number of mills, sawmills and smithies in Motala village did not grow much numerically, the quality and capacity of their plant were improved substantially.

By 1860 the Göta Canal Company and a merchant from Motala *köping* ('market settlement') just to the west each owned a plot in the village and the other plots, as before, were also held by people who obviously were more interested in industrial rather than agricultural potential; all were running secondary production as well as farms. Little amalgamation of plots occurred: one person owned two plots, whilst three out of eight also tended other small plots outside the village. However, two plot tenants had more than one mill while others combined milling with brewing and even an eel-fishery! The survival of smallholdings in Motala and Holm villages had little to do with farming and was mainly a response to the development of industry, which will be dealt with in chapter 5.

What of the large part of Motala parish on the shieldland north of the fault along Motala Ström, where abundant forest and water power provided alternatives to arable farming? In Björka/Björketorp hamlet, plots measured in taxation value were already small in 1810: four were assessed at 1,000 Rdr each, four at half that, and the two in Björketorp at 750 Rdr. None of the farmers in the hamlet had any land elsewhere. The pattern was thus radically different from that in Herrestad in 1810. However, the *total* area of the plots in Björka/Björketorp was much larger than those in Herrestad because they included large areas of forest, which became increasingly valuable during our period as a source of charcoal, tar, firewood and timber. The production potential of these farms was higher than their tax value would suggest, although not in agriculture. Apart from this, the chance of off-season non-agricultural employment was high in mills, ironworks, transport and handi-craft, which were all activities with the peak season in winter and spring. This would reduce the dependence of farmers on agriculture, and might explain why farms were small and why no amalgamation of plots occurred. The number of plots doubled in Björka/Björketorp over the course of our period. But this did not mean that farms became more complex. Whereas the hamlet's eight plots supported eight farms in 1810, it consisted of sixteen plots and sixteen farms in 1860. The only differences between 1810 and 1860 apart from the heavy subdivision of plots was that three farms were worked by tenants, although the owners still lodged with their tenants on the premises, and the tenants only tended these single plots. It seems that the old farms had made over portions of their land to crofters, who thus (as in Strå and Orlunda) had become smallholders with full ownership rights over their land. Consequently, of the sixteen farms in Björka/Björketorp, one had a value above 1,000 Rdr, nine lay between 500 and 1,000 Rdr and the remaining six had values between 90 and 450 Rdr. Two of the larger farms seem to have been informally subdivided as well: they contained crofts which were formally assessed in the taxation registers at values of about half the farm. If these crofts are subtracted from the larger farms of the hamlet, there were eight farms of over 750 Rdr in value and ten very small holdings in 1860, the former most probably being identical with the eight farms of 1810. In this respect, the hamlet was not similar to Strå and Orlunda, where just one or two plots exploded into smallholdings. Instead, it was possible for the Björka landowners to skim some cream off their holdings, selling off minor parts and still retaining enough for subsistence. Whether the newly created holdings could provide enough food for a family is doubtful, but the potential for secondary income was much greater on the shield. Places like Motala and Björka became increasingly difficult to treat in the same terms as purely agricultural villages in other subregions.

Conclusion

There was a significant difference between the landholdings represented by plots, tax register units (*hemman*), properties in single ownership, and functioning farms. These are generally considered to be synonymous (at least, it is rare that they are explicitly distinguished) in studies of change in peasant society. The picture we have presented is very far from the simple model of traditional peasantry in which a farm might be expected to comprise one owned plot used to sustain the reproduction of the same stem family over many generations. Similarly, estates were no longer legacies of noble power.

Between 1810 and 1860 the number of plots grew by about 25 per cent in our region. In a one-plot-per-peasant-farm society with partible inheritance, this would indicate that farms were being subdivided due to population growth, and that the productive capacity of each farm was reduced in consequence; that involution was occurring. Heckscher concluded that this was the case in Sweden generally at the time.[78] Indeed, his thesis that the survival and increasing power of the peasantry was 'an important reason for the slow material progress of the country' depended heavily upon inferences drawn from data such as this. However, this kind of argument ignores three aspects of rural land-holding which we have shown to be of considerable importance. First, it ignores the fact that peasants were not the only people involved in farming; secondly, it takes no account of the difference between units of land ownership and the actual structure of farms, and thirdly, it ignores environmental differences and therefore variations in agricultural potential from place to place.

Even in the heyday of the free peasant economy in Sweden, nobles, persons of standing and townsmen owned and worked rural land: on the ground, as in principle, peasantry cannot be separated from other elements of society. As we have seen, townsmen were never of great significance in our region. They increased their holdings until 1840, when they controlled about 3.5 per cent of the region's farmland. Later, the citizens of Vadstena got rid of most of their holdings, and the increase of the land held from Motala Town did not nearly compensate. The nobility were far more important. In 1810, 12 per cent of the region's farmland was worked from their estates. In 1860 the nobility still farmed about 6 per cent of the region's land. Estates which survived throughout the period were stable in plot composition and remained confined to noble (*frälse*) land. The third non-peasant category was more complex, varied and difficult to generalize about. Persons of standing ranged from curates holding small Crown land farms, through military officers, to rectors and district clerks, some of whom leased their land out and others of whom expanded their holdings across plots of taxed land, to people such as War Commissioner Lallerstedt and District Judge Duberg, whose large estates based on manors eventually incorporated considerable areas of both exempt and taxed land. In total, persons of standing worked 13 per cent of the region's farmland in 1810, and doubled their proportion by 1860.

The net effect of these countervailing changes was that the peasants' share of the land in the region remained almost constant at 72 per cent in 1810 and 70 per cent in 1860. The number of *plots* owned by peasants resident in the region grew, but our reconstitution of *farms* shows that their number fell at the same time. The units in which peasants owned land shrank, but the units in which they worked it expanded. Many peasants owned more than one plot, but many more owned fewer than they tended. The concentration of the ownership of plots into fewer hands was marked, but it proceeded less quickly than the concentration of owned and leased plots into larger farms.

The reorganization and concentration of plots occurred in all kinds of farming enterprise. Even on traditional noble estates, plots were amalgamated and their spatial distribution was increasingly clustered around the manor farm base. The land-holdings of non-noble persons of standing followed the same patterns: scattered plots of land were owned and leased in succession so that eventually a larger proportion was located in close proximity to their places of residence. This was true not only of very large estates of over 40,000 Rdr in taxation value, but also of holdings of under 10,000 Rdr. The process occurred soonest and at the largest scale on farms on the plains situated outside the close-knit structures of the peasant villages. However, it was also in full spate in even the largest villages of the plains and in the isolated hamlets of the shield after 1840. On the plains, the process at first created more numerous plots out of the traditional village *hemman* as farms based on one acquired land on others, which was often initially rented before eventual purchase. On the shield the multiplication of plots was even greater, but there its seems to have represented the expansion of arable areas on the mainly forested plots of small hamlets and isolated farms, which could then be split without reducing their arable capacity.

The great expansion in the number of plots occurred all over the area in the 1810s and 1820s, while ownership concentration became more marked in the

1840s and 1850s. It was traditionally believed that the *laga skifte* enclosure in Sweden brought about further plot (and therefore farm) subdivision, in tandem with relaxations in the laws governing the degree to which taxed plots could be split, and greater negligence by the local courts which were meant to supervise the implementation of these laws.[79] However, it is quite obvious that these enclosures, carried through during the 1830s and 1840s in the southern part of our area and later in the north, were *not* the cause of subdivision, which had already occurred. Instead, enclosures had the opposite effect, bringing amalgamation of both owned and farmed plots. Only in a few cases, after what can appropriately be described as the explosion of *hemman* in one or two villages, can fragmentation be connected with enclosure. In those cases it is obvious that the newly created plots functioned as smallholdings rather than subsistence family farms. At least half of the farms in both subregions on the plains worked more than one plot by 1860, in many cases four or more. The process was most pronounced in Dahl hundred, where it seems to have begun, and least so in the northern shieldland. There, many single-plot category III farms still existed at the end of the period.

The emergence of larger farms was even more marked when expressed in terms of taxation values rather than numbers of plots. Large category I farms already controlled more than 50 per cent of the taxation value of land in Dahl by 1810, which rose to more than 70 per cent in 1860. Although small category III farms increased in number in southern Aska, their share of taxation value fell from 23 per cent in 1810 to 14 per cent in 1860. Out of the fairly normal distribution of farm sizes in 1810, two groups had crystallized by 1860, one consisting of farms with many plots and high taxation values, the other comprising smallholdings, many of which functioned as allotments rather than fully fledged farms. The middle-sized farms of category II diminished most in numbers everywhere.

However, the balance of these groups, and their composition in terms of nobles, other persons of standing and peasants, varied quite widely between the subregions in 1860. Large farms owned and run by peasants were most predominant in the plainland villages of Dahl, where both middling and small farms were greatly reduced. In southern Aska, small farms grew in number and middling farms maintained their proportional share, whilst in northern Aska middling farms became less numerous but small farms remained very numerous. On the plains, and particularly their southern extremity of Dahl, peasant society was rapidly differentiating as the plots into which the continuous arable land was split up were incessantly reassembled into larger farms worked from the ancient *hemman* of the villages, which then dissolved as their land was leased and bought by still-growing farms. In the north, where new settlements were being opened up in ever remoter parts of the forest, hamlets and estates contained expanding farms, but many more still comprised single owned plots, and these small farms were smaller in taxation value than the category III farms of the plain. If anything like a traditional peasant society survived in our region in 1860, it seems to have been here on the shieldland. However, we cannot reliably draw this conclusion until we have examined the non-agrarian economy of the shield. The survival of small farms and the failure of large farms to increase their share of the land in the north was in part because of the break-up of some of the estates of Motala parish, which were involved in industry as well as agriculture. The fact that farms in the villages of Motala and Holm and the hamlets along Motala Ström were more valued for their water power rights than their arable land must also have encouraged the survival of small farms, as must the existence of off-farm work in sawmills, quarries, ironworks and forest activities generally on the shield.

Our findings have some clear implications in terms of the general models of the role of peasantry in agrarian change presented in chapter 1. Most basically, it is clear that a number of processes which are usually considered in isolation in general discussions were occurring simulataneouly in western Östergötland between 1810 and 1860. Many non-peasant estates continued to exist. But they did not follow the path to agrarian capitalism suggested by Brenner. Instead of introducing commercial leaseholds on larger farms, most continued to be run by their owners, either directly or through managers. Those accumulated by non-noble persons of standing changed their plots, broke up, changed hands and partially re-formed like patterns in a kaleidoscope. Through all this flux, though, a pattern does seem to be discernible in which plots were gradually concentrated in ever more rational patterns around the estate bases. The large peasant farms which developed in Dahl in particular had the same characteristics. Small peasant farms remained interspersed amongst the growing fissiparous estates and peasant farms of the plain, and became relatively more important northwards until they were numerically dominant on the shieldland of northern Aska. As the period progressed, large farms grew and small ones shrank. The progressive collapse in the number of middle-sized holdings in Dahl suggests that the peasantry was differentiating in the way which is usually attributed to the effects of the transition to capitalist farming.

As far as patterns of land-holding are concerned, therefore, two versions of agrarian capitalism seem to have been emerging side-by-side with involution. A land market must have been operating to allocate plots amongst estate owners and more and more of the enterprising freehold farmers had by 1860 passed well beyond the threshold which would make the term 'peasant' appropriate to their farms. How was the penetration of the market into land-holding related to the market-orientation of production and the proletarianization of labour? And how did the ever-shrinking small farms of the plains and the shieldland manage to continue to provide a livelihood to their occupants?

4

Agrarian livelihood positions, projects and production structures

Setting the scene

Introduction

Increased acquisition of farm labour through the payment of wages draws peasant and worker into the labour market. Like the purchase of land upon which to deploy these workers, this cannot occur unless more of the farm's output is sold on commodity markets. The balance of cash receipts and expenditure becomes the measure of the value of a farm to its owner. With the aim of maximizing the difference between costs and receipts, cost-saving innovations in farming techniques and the régime of labour deployment become more likely. So do changes in the crops and animals raised in response to changes in demand and technical possibilities. The much greater involvement of the farmer himself in organizing production and sale removes him from direct involvement in all farm work. At the same time, dependence on markets for agricultural inputs and the disposal of outputs, and the embourgoisement of the whole farm family, reduces the time, inclination and necessity for craft work in the farmhouse. In short, the whole nature of the farms's production project changes, and therefore the work content of the livelihood positions it supports. This cannot occur in a once-and-for-all transformation: continual changes in both projects and livelihood positions are to be expected in a developing capitalist farming system.

Similarly, peasant farms cannot shrink in size and continue to provide a living without changing the nature of their production projects and the work involved in the livelihood positions they support. Involution requires more output per acre, with more inputs of labour into farming or more craft work or paid employment off the farm. Each strategy alters the number of hours spent by different members of the family in different kinds of work: each alters the nature of the livelihood positions of at least some members of the peasant household.

113

These are the subjects we will be primarily concerned with in this chapter. In addition, however, a further topic must be included. As changes in livelihood positions occur, so do the family formation strategies of those who occupy them. To understand changes in the farming system we must examine demography.

Livelihood positions in Swedish peasant society

The traditional system

Many of the farms described in chapter 3 must obviously have employed labour in addition to the family occupying the homestead. Large estates and other holdings of persons of standing were important and dynamic elements within the farming system. At the same time, many peasant farms had quite clearly grown beyond the size at which they could be run with family labour, whilst others shrank to a size below what seems previously to have been necessary for subsistence. How did these changes affect the sorts of livelihood positions available, and how did this, in turn, affect the rate of population change?

On peasant farms   Traditional Swedish peasant society was not a homogeneous mass of equal farmers under the rule of feudal landlords. As we have seen, the land was to an increasing extent owned by the farmers themselves. In addition, Sweden's very varied physical environment resulted in a number of quite distinct peasant ecotypes.[1] Only on the fertile plains of the extreme south and the central lake-filled depressions was classic grain-based peasant farming possible. Elsewhere, subsistence based on fishing, hunting or dairy production made farming quite different.

It has long been believed that peasant farms in Sweden were traditionally rather like Balkan *Zadruga*, occupied by extended family groups consisting not only of three or four generations of a stem family but also of grown-up siblings of the farmer, maybe with their families, making a total group of a dozen or more.[2] Later research has shown that such extended families existed during the last few centuries only in 'sparsely populated regions where there was a high degree of economic differentiation within the household',[3] as in remote parts of Finland where large extended families were more common than among the Scandinavian-speaking populations.[4] No clear-cut development from extended to nuclear family patterns can be distinguished from the sources. Servants were not uncommon, as Scandinavian peasants balanced variations in consumption and labour power over the family life cycle.[5] They were mainly recruited from the peasant farmer group; as part of a life-cycle pattern, farmers' boys and girls were recruited as hands and maids outside their home farms for a couple of years before getting married and taking over the family farm.[6] The peasant household could include parents and other relatives as well as farmhands and maids, and people said to be 'living-in' with the family.[7] This last category sometimes included parents who had retired and transferred the farm to their heir, retaining specified rights to shelter and subsistence from the holding (*undantag*), as well as other usually older people who seem to have had no blood relationship to the farm family or a specific productive role within it. In addition to the peasant household itself, a farm might contain one or more crofts, the occupants of which were entitled to farm a patch of arable land in return for providing specified amounts of labour to the farm. According to Swedish law, everyone in the countryside must belong, as member of or servant to or crofter on, a farm from which they were entitled to subsistence, unless they were aged or disabled and supported by the village community as a whole, in which case they might occupy independent cottages.

The size of household was normally quite small. Thus, in By parish in Dalarne, farm households comprised between 5 and 6 people in 1750, perhaps slightly more in the late seventeenth century, with an average of between 5.5 and 6 people in the households of the larger farms and around 5 people in the smaller ones.[8] On large farms in By, the workforce was between 3 and 4 while on the smaller, it was about 3. The workforce consisted mainly of members of the nuclear family. On average, farms had only between 0.3 and 0.4 employees in 1750. By parish consists mainly of shieldland and is in Bergslagen, where iron smelting has been important since medieval times. Farm households there in part based their livelihood on iron smelting and were on average larger than in purely agricultural areas, with more hands and maids who were to some extent occupied outside agriculture.

Among peasants in Dala parish in Västergötland, a more fertile part of Sweden, the household structure was similar to that of By.[9] Thus, in 1780, about 80 per cent of all farm households in Dala consisted of a nuclear family while another 7 per cent had a parent living-in and only 8 per cent of households contained just one adult. Only 3 per cent of the households were extended families. The labour force on farms in Västergötland was apparently also small in the eighteenth century, although shieldland areas like Sjuhärads-bygden, where people registered as farmers did all sorts of things for a living, contained larger households than the plains and were in that respect similar to Bergslagen.[10]

The landless group living in crofts and cottages seems to have grown only slightly after 1700.[11] According to the 1751 census, there were 186,000 male heads of farm households and about 48,000 heads of rural landless households. However, hands, maids and people living-in were 'hidden' inside the farm households, so that of the total agrarian workforce of 700,000–800,000, between 20 and 25 per cent were landless.[12] On peasant land other settlements than farms were rare.[13] Cottages were subject to less legal proscription than crofts and seem to have existed here and there on taxed land, for example on village common land. Their number was small and their inhabitants usually non-productive old or disabled people. In 1781, the male cottagers accounted for 3.4 per cent of the rural male population, but not all of them lived on peasant land.

On estates and farms of persons of standing   By the eighteenth century, the 'normal' estate consisted of a central production unit run by an estate manager, with corvée labour from crofters and peasant tenants on the estate domain. For example Thureholm, a large estate in Södermanland, had sixty-four tenant farms and thirty-seven crofts in 1793, which together were obliged to work more than 7,000 days on the manor farm. In general, the tenant farmers had to deliver more goods or pay more rent,[14] but even within one estate the duties of peasant tenants could vary substantially,[15] although crofts always owed corvée duties to the manor farm. Many estates had combinations of peasant farms on more fertile land and crofts on marginal areas, frequently with the objective of further arable reclamation: potential crofters were granted patches of land on favourable terms provided that they reclaimed arable and built their own houses. Frequently, too, estates had a few hands and maids employed at the manor farm.[16]

Estate owners generally had firmer control of their domains than the peasant village communities had of their common land, over which many different wills wanted to rule. This appears to have affected the colonization process on peripheral parts; cottages rarely existed on estates before the 1700s. It was much easier for the estate owner to evict squatters than it was for peasants to get rid of elderly relatives of neighbours or disabled uncles who needed somewhere to live. Hence, estates had fewer 'unproductive' positions than freehold peasant villages, which were more similar to England's 'open' counterparts.[17]

The farms of persons of standing were frequently organized as small estates with numerous crofters, though smaller ones were more like peasant farms. The workforce varied with the size of the farm, but rarely comprised more than half a dozen crofters and more than two hands and maids at the same time.

Even though we can distinguish three main types of agricultural holding in the mid-eighteenth century, these contained other production projects as well; peasant farms had cottages, the inhabitants of which cultivated whatever small patches of land they were permitted, apart from paying land rent in kind or labour to the individual farmer or the village community. In the same way, an estate contained a number of different production projects, peasant farms and crofts, which apart from either paying land rent in kind or money or performing corvée duties, also produced their own output. When studying the changes in this agricultural society, it is necessary to include the changing structure of livelihood positions on both the main and the subordinate units.

Even without any change taking place, the structure of livelihood positions could vary a lot between purely agricultural areas of Sweden, depending upon the composition of basic units described here. Thus, in the lake Mälaren basin, where estates were more common than elsewhere in Sweden, the total number of positions included many more crofters in the eighteenth century than, for example, Gotland where peasant freehold farms were much more prevalent, or western Östergötland, where estates, although present, were not completely dominant.

Changes after *c*. 1750

In general About 185,000 farmers owned or tended land in Sweden in 1751. In addition, there were 75,000 hands and 48,000 landless heads of households. According to the censuses, landless positions grew most rapidly over the next hundred years. This group, containing hands, crofters and cottagers, comprised about 300,000 heads of households in 1850,[18] by which time the number of farmer heads of households had increased by only 22,000. Similarly, maids became more numerous over the same period, while farmers' wives did so only marginally. The biggest growth of all occurred among the crofters (by about 400 per cent), cottagers and living-ins (by about 450 per cent) and a new category, married hands, who lived in farm households but were obviously not occupying the life-cycle stage between leaving a peasant home on reaching adulthood and returning to inherit. Married hands, who received a small wage as well as their keep, were obviously much closer to fully fledged proletarianization than traditional single hands and crofters. There were no more than 17,000 of them in 1800.[19] Most of the growth in the number of productive positions occurred on estates and person-of-standing farms, which not only employed crofters but were also the initiators of married hand positions, first registered on estates in the Stockholm region in the 1760s.[20]

The distribution of categories of labour was quite different over the country, and became more so during the first half of the nineteenth century. The proportion of farmers fell much more in east-central Sweden (including Östergötland) than in Småland, while the proportion of hands and crofters was substantially larger in the former area. The concentration of land-holding was less intense in Western Sweden and, in consequence, the number of crofters and hands smaller in proportion. Thus, there was a clear relationship between a concentration of land ownership or tenancy and a high proportion of landless labour such as hands, crofters and cottagers.

On peasant farms: from family labour to crofters and married hands Legal changes in the 1750s permitted the establishment of (untaxed) crofts on the outlying areas of farms on taxed as well as noble land. The legal changes do seem to have resulted in the quickening of a process which had already started. In Sunnerbo hundred in south-western Småland, the number of crofts grew by 1,200 per cent between 1755 and 1855! No fewer than 128 of the 170 crofts in

Sunnerbo in 1750 were on estates, but of the more than 800 in 1815, 585 were on taxed land.[21] Crofters on estates were normally tied to the manor farm for six days a week, performing corvée duties. Peasant crofters rarely worked more than a dozen or two days annually on the landowner's farm and had often paid money to acquire long term contracts for their croft (*förpant-ning*).[22] Their payments in kind were also much smaller than those of estate crofters. However, peasant crofters were not necessarily much better off than their estate counterparts: estate crofts could contain a dozen acres of arable and frequently needed to employ hands, but peasant crofts rarely had more than one or two acres of arable and a similar area of meadow.[23] Because peasants' crofters were less closely tied to their farms they could spend more time on other things such as day-labour on other farms,[24] so that peasants' crofters provided not only an addition to the labour force on the farms where their crofts were located but also an increase in the day-labour workforce of an area. In the days of the open fields, most crofts were located on village common land and the crofter's relationship was to the village community rather than the individual peasant. When village common land was subdivided in the enclosures, the relationship became individualized. One result was that the number of crofts on individual farms varied a lot: some peasants were in favour of crofts, others were not.[25]

In Dala parish in Västergötland, the number of farms grew by 27 per cent between 1780 and 1850, although their proportion of the total number of livelihood positions was reduced. After 1810, the number of crofters grew by 67 per cent while completely landless cottagers more than doubled in number.[26] The landless people who lived within peasant households did not increase their numbers by much. Generally, as the number of farms and farmers also remained relatively stable, increased labour inputs came from outside peasant farm households. Crofters, cottagers and married hands became common as supplements to the stable household labour forces of the farms. However, there were marked differences between the plain and shield in Västergötland. On the plains peasants were reduced in number and the crofters increased by more than 100 per cent, while on the shieldland the peasants grew by a small proportion and the crofters more than doubled their number.[27]

With the more active agricultural attitude of the nineteenth century and the *laga skifte* enclosures from 1827 onwards, the freeholding peasantry became more interested in extending their cultivated area. This was especially so if their farm, or one of its constituent plots, had fared badly in the enclosure – for example from a compulsory move to previous meadow or grazing land which had to be reclaimed into arable if the plot was to survive.[28] Often, especially on the shieldland where there were few sites to which farms could be moved out from the common field and village, former croft sites were used. This was facilitated by the nullification of croft contracts by a *laga skifte* enclosure.[29] Thus, enclosure did not simply allow the amalgamation of more plots to each farm, but also got rid of 'parasitical' crofts and cottages, whose land was added to the farms. Frequently, crofters owned their houses, which they were permitted to move to some worthless spot in the village when their contracts were nullified. From a peasant's point of view, crofts were beneficial in that they guaranteed the availability of non-family labour and brought some annual money income as well as the initial *förpantning* payment, while the patch of farmland that the croft used was often reclaimed by the crofter anyway. Even as arable it would be of minimal value because it lay away from the open fields of the village on the plain, or maybe miles from the farmstead in the forest. With larger-scale farming, it became preferable to use all potential arable and meadow from the peasant farm and to employ either full-time or seasonal wage labour to work it. Even on estates, which had a much greater income from their more numerous crofters, this would be a reason for the number of crofts on the plains to decline rapidly.

The landless did not just grow in number, but also changed character. In 1850, cottagers were no longer all aged or infirm, but often in their thirties with families, some described as day-labourers and others as active workmen of other kinds. Although married hands were more common on estates where the demand for full-time labour was larger, they also started to appear on peasant farms during the late eighteenth century. Like crofters, they had annual contracts stipulating how many days per annum and hours per day should be worked and how much was expected from their wives. The contract also described the *stat*, the payment in kind which comprised the major part of the wages of married hands.[30] The people described as living-in also seem to have become more productive and younger over time. Did these changes simply reflect an intensification of farming, or were other influences at work?

Cottagers on peasant farms were, even more than crofters, loosely connected with the production of the farm.[31] Thus, even if a peasant farm had many crofts and cottages, this need not necessarily mean that there was a large labour force attached to the production projects of the farm itself. There was certainly labour available, but it had to be purchased like any other commodity, apart from the small corvée duties of the crofters. In the same way, the day labourers who started to appear during the nineteenth century with a more lax legislation concerning the necessity to be permanently employed, were not obliged to work only for the farms on which their cottages were located. Thus, intensified agricultural production on farms would be reflected in increases in the amount of labour used, but not necessarily in the labour force living on a particular farm. Increases could just as well take the form of casual employment on a daily basis of crofters, cottagers and day-labourers.

In peasant-dominated areas of Sweden, gradual proletarianization can be recognized from the mid-eighteenth century, but its scale differed from region to region, even from parish to parish. In general, plains areas seem to have become more proletarianized: completely landless cottagers and married hands were more common there and landowners and tenants were relatively fewer. On shieldland, crofts were more common. Naturally, it was in the interest of the budding capitalist farmer to try to keep as much of his produce out of the hands of the proletarians, to try to keep the labour force as small and efficient a group as possible and to try to avoid large settlements of non-productive or only partly productive cottagers on his land. This, in principle would apply to all peasants who did not stick to 'old-fashioned' ideals. In consequence, but in a way paradoxically, we can expect the more capitalist oriented farms to have the fewest crofts and cottages, the land of which could instead be used for the farm's own agricultural production. Thus, if it was not for other indicators, the most aggressively capitalistic farms could in principle look the least so and more peasant-like than a farm on which the establishment of crofts and cottages had been permitted, and where the day labour necessary to the operation of the capitalist farms lived.

This only applies to areas in which virtually all land could be cultivated, such as the Östergötland plains. On shieldland, where little patches of potential arable were spread out over large areas full of rocks and lakes, it was better for the farmer-cum-capitalist to let more remote parts to crofters and cottagers and in that way obtain an income from them, rather than to try and cultivate everything himself.

Rapid and complete proletarianization did not occur in peasant agriculture in Sweden. Development was much slower and more gradual, as incompletely proletarian groups like crofters and married hands superseded juvenile hands and maids. The proletarian labour force evolved smoothly, if as yet incompletely, from the traditional life-cycle position of juvenile hand and maid in the household of others in traditional peasant society.

On estates: from Grundherrschaft *to agrarian capitalism?*   The change from a *Grundherrschaft* system of management to *Gutsherrschaft* was not complete in Sweden, where numerous estates in the nineteenth century were still dependent on peasant land rent and corvée duties on the manor farm. Some estates close to Stockholm turned readily towards *Gutsherrschaft*, replacing small manor farms and numerous subordinate peasants' tenant farms with one large farm worked by full-time employed married hands and, to a lesser extent crofters who did substantial corvée duties.[32] If such changes were common, they could certainly have been one of the causes of the large-scale changes in population composition in Sweden during the eighteenth and nineteenth centuries.

The motive for increased profitability on estates could have been that they were no longer producing a surplus solely for the upkeep of a feudal lord but for an increasing market. Many estates also contained industrial plant which they supplied with raw materials and labour. Furthermore, the tax-exemption privileges of noble land were gradually reduced during the eighteenth century, so that an estate had to produce more simply to maintain its owner's wealth.[33]

To what extent and how quickly did this transformation occur? Normally it was easier for an estate manager to establish crofts on the manor farm domain, but on the outlying plots the tenants fought fiercely in order to avoid such an intrusion onto 'their' land.[34] A crofter labour force was thus first established on the manor farm and spread only gradually to the outlying plots tended by peasants. This process seems to have begun by the early eighteenth century. The further intensification of estate agriculture in the late eighteenth century involved abolishing crofts and employing an army of married hands. It is now known that this general model of the evolution of estate agriculture, proposed by Utterström,[35] is too sweeping: substantial traces of *Grundherrschaft* persisted throughout the eighteenth century. Tenants remained on estates, paying monetary land rent, although this was frequently transformed into day labour. In three parishes, dominated by estates, in the province of Uppland, peasants, mostly in the form of tenant farmers, still comprised about 20 per cent of the adult male population in 1780, whilst 14 to 20 per cent of the males were crofters and married estate hands were few.[36]

Another way of attempting to make estates more profitable was to merge tenant farms into larger units and raise money rents. This was, of course, the process to which great dynamism is attributed in seventeenth-century England. In Sweden, this method of reorganizing estate agriculture, without an intermediary croft stage, started in the late eighteenth century.[37] All the kinds of changes identified above have been recognized on the estates of Södermanland in the nineteenth century. Of 480 tenant farms on 21 estates, 138 were added to the manor farms, 60 amalgamated into larger tenancies and only 5 were sold off. At the same time, 56 new tenant farms were created by subdividing manor farms, 39 by subdividing old tenancies, and 1 through purchase. The number of tenant farmers was reduced in total by 107, to 373, while the crofters diminished only from 550 to 532 and were supplemented by an increase of married hands from about 80 to slightly fewer than 300.[38]

A commercially motivated estate owner would have responded more efficiently to an expanding market than the estate owners described above, by moving more fully to a system in which purely proletarian labour worked one substantial manor farm with more and more advanced equipment. These were, however, mainly noblemen, many of whom spent time at court, and had numerous estates which were managed on a 'satisficer' level. They might have seen their agricultural property as secure fixed capital with low running costs rather than the means to make a fortune (which they already had). Moreover, so many perturbations came from topography, micro-climate, soils and the social attractions of working in a family unit that it was often more productive and less risky to farm an estate through small peasant-run units.

Agrarian labour supply and demography
So far, we have looked from the perspective of the agrarian production units and how their labour organization changed with time. But what of the

opposite point of view: where did the landless labour come from? This must have been related to demographic changes and wages.

Peasants, proletarians and population growth   In the late eighteenth century Joseph Marshall claimed that Swedish population was regulated by the homeostasis which is supposed to exist in freehold peasantries generally: because the peasants owned their own land they regulated family formation to prevent holdings having to be split between more than one married son.[39] However, censuses show that population was already growing in Sweden at that time. Continual wars and famines due to poor harvests kept the population stable during most of the seventeenth century and up to the early 1720s, but it grew from less than 2 millions in 1750 to more than 5 millions in 1900, despite massive emigration to America in the latter half of the nineteenth century.[40] Like the population as a whole, this growth was overwhelmingly rural, but it occurred at different rates in different parts of the country. Sundbärg defined three broad regions: the west, with high fertility and moderate mortality and, hence, rapid population growth; the east, with low fertility and fairly high mortality and low population growth, and the north, with low mortality and extremely high fertility, giving it Sweden's most rapidly growing population.[41] This generalized pattern can be interpreted in different ways. It might be argued that Eastern Sweden's slow population growth implies the strongest survival there of the homeostatic demographic controls which Marshall attributed to the Swedish peasantry as a whole. However, it has been argued that this region of Sweden was, in fact, dominated by estates and had a high degree of social stratification.[42] If this were so, the regional pattern could imply that areas of freehold peasantry had the highest fertility, and areas where the landless were most common the lowest,[43] supporting Heckscher's argument that it was due to involutionary processes within the peasantry. However, all blanket explanations based on the categories of Sundbärg's crude map are implausible. Local studies have shown that fertility and mortality varied from ecotype to ecotype and between different social groups.

Although too little work has been done at these levels for reliable generalization, the differences which seem to have existed from 1780 to 1850 between social groups in a number of important demographic variables in Dala parish in Västergötland support an interpretation of Sundbärg's regionalization in terms of peasant involution rather than proletarianization. The landless did have fewer children who reached five years of age than the peasants. The average number of children per 100 married women was 274 between 1776 and 1805, and 324 between 1805 and 1850. The difference between the landed and the landless was more marked in the earlier period, when the former had 295 children per 100 and the latter 231, compared with 335 and 300 per 100, respectively, between 1805 and 1850. Peasant freeholders had only 310 children per 100 women, significantly fewer than tenants on estate land, which might imply deliberate limitation amongst the former and the breeding of more hands to do corvée duties among the latter. The rate of natural growth among the landless in the early part of the period was probably too low for the group to reproduce itself. Over time, too, the landless group, became more economically active as the percentage of old and infirm amongst them shrank. It also grew absolutely and in proportion to the total population. Additions must have come from migration or the local proletarianization of peasants, due to the number of farms available being insufficient to accommodate all peasant children. Tenants were evicted as their farms were incorporated into estate manor farms or sold off to people outside the peasantry, and whilst there was some subdivision of freeholds plots, others were amalgamated into larger farms in much the same way as in western Östergötland. The percentage of freehold and tenant farmers in Dala shrank from 56 to 48 per cent of the total number of households between 1780 and 1850. Population growth in Dala parish occurred mainly in the peasant stratum to begin with. In 1780 the percentage of farmers in Dala was *lower* than in the country as a whole, while in 1850 it was substantially higher. Was the lower fertility of Eastern Sweden, then, due to greater amalgamation of farms, a response to the declining availability of holdings amongst the peasantry, whereas in Dala the process of farm amalgamation was less marked, and the peasant demographic response had not yet occurred because the stimulus was absent? The numbers of children per landless family in Dala were smaller than amongst the peasants, but there were significant differences between crofters and soldiers on the one hand and cottagers, the living-in and hands on the other. The former had an average of about two children below fifteen years of age per household in 1780, the latter little more than 0.5. By 1850, the completely landless had 1.2, reflecting the addition of productive workers to the cottage population and the emergence of married hands, while crofters and soldiers had 2.25. The large and increasing families of crofters might indicate either that production on crofts was expanding or that more extensive corvée duties were being demanded by landowners.[44]

Åsunda hundred in Uppland, west of Stockholm, exemplifies what was happening in Eastern Sweden.[45] Here, tenant farms were incorporated into the manor farms, and married hands were employed to tend the enlarged units. Farmers of all kinds, who in 1790 comprised 50 per cent of the households, made up less than 28 per cent of the total in 1890. Variations in the household composition of different social groups were marked. In 1790 farmers had on average 1.45 children under fifteen years of age compared to Dala's 1.5 in 1780. In 1890, the number of children in Åsunda had fallen to 1.2, while in Dala in 1850 it had risen to 1.6. Crofters and soldiers had 1.2 children below fifteen in 1790 and about 1.5 in 1890. Cottagers and other landless had 1.0 children under fifteen in 1790 when their Dala counterparts had 0.4. Already in the eighteenth century, the landless seem to have been in a more active age group in Åsunda. When Åsunda household sizes and rates of fertility are compared with those in other parts of Sweden, they generally appear to be on the low side. The mean household size there was 4.2 in 1790 and 3.8 in 1890, while in Dala it was 4.5 in 1780 and 4.4 in 1850. The rate of fertility was low, especially among the peasants who clearly limited their number of children by the latter half of the nineteenth century, and it seems as if this was also the case of the crofters. The married hands were reproducing most rapidly.

From this comparison it is clear that demographic differences did not simply reflect the proportions in various social groups. On the one hand, farmers seemed to reproduce rather well in Dala, whilst the figures for Åsunda, low already in 1790, shrank further in the nineteenth century. The landless in Dala had very few children in 1780, some more in 1850 but never anywhere near the figures for either cottagers or married hands in Åsunda. Could this indicate that the demographic behaviour of peasant farmers was simply related to the availability of holdings, with high fertility where holdings could be increased in number and low fertility where they were being amalgamated, and that the demographic behaviour of the landless was similarly a direct response to changes in the the availability of livelihood positions?

Neither Dala nor Åsunda were known for much else than agriculture. Elsewhere there was significantly higher age-specific fertility in households engaged in home weaving compared with ordinary crofters or farmers.[46] This is obviously something that we must try to incorporate in our analysis of demographic changes in western Östergötland.

Wages and labour supply   The nominal wages of agricultural workers, deflated by the price of rye, were low during the first two decades of the nineteenth century then rose rapidly by 25 per cent around 1820, only to stagnate for almost the next three decades.[47] During the 1850s, they rose by as much as 40 per cent and then, with a series of minor setbacks, continuously over the remainder of the nineteenth century. Thus, an agricultural labourer in 1800 had to work for three times as long for a barrel of rye as he had in 1900. No other prices in Sweden rose as rapidly as those of day labour in agriculture during the nineteenth century. Thus, especially in the 1820s and 1850s, proletarian livelihood positions were increasingly well paid, which might go some way towards explaining the speed with which this social stratum expanded in Sweden during the nineteenth century.

However, this was not true of all parts of the country. The pattern of differences was rather surprising. Östergötland in general had a markedly lower wage level in agriculture than Sweden in general from the 1820s onwards whilst in neighbouring Kalmar county, real wages for farm labour actually fell by 14 per cent up to 1850.[48] The weak wage increase in the provinces of Kalmar, Kronoberg and Jönköping in south-east Sweden has been explained by a particularly rapid population growth and the predominance of small freehold farms, which meant that the demand for agricultural labour was consistently low.[49] The almost equally low wages in Östergötland, where numerous large farms and estates employed labour and where alternative employment in towns and industry was more plentiful, could perhaps be explained by a 'spill-over' of labour, migrating from surplus to demand areas. Of course, it may well be wrong to think in terms of a single labour market, with estates, farms belonging to persons of standing and peasant farms all competing in a common pool. A number of quite separate labour markets probably existed, and the flow of labour between the different kinds of farm seems to have been small: crofters on estates remained crofters, while farmers' sons stuck to farms as long as they could.[50] However, with an increased demand for proletarian labour, more of a common labour market was created, although we must remember that migration from one parish to another without permission was prohibited before the mid-nineteenth century.

Proletarianization and poverty It is easy to assume that a rapidly growing population, proletarianization, falling wages in some areas, and barriers to labour mobility must have caused an increase in poverty in nineteenth-century Sweden. This might be expected particularly in Eastern Sweden, where real wages fell and changes in the distribution of property were more significant than elsewhere. However, there were many provinces in Eastern Sweden which had significantly *lower* levels of poverty than their counterparts in the west: the provinces of Stockholm, Uppsala and Södermanland all had poverty rates of less than 20 per cent for people above eighteen years of age while in the west, Göteborg and Bohus, Älvsborg and Skaraborg provinces all had poverty rates above 23 per cent.[51] Östergötland and Skaraborg provinces, in the heartlands of the western and eastern demographic regions had almost identical levels of poverty in 1826, at 23.5 and 23.6 per cent respectively.

Poverty rates varied considerably over time. The 1840s experienced rising levels in the lake Mälaren basin but reduced ones in Östergötland and Western Sweden. The high levels of poverty continued during times of rising wages. In areas with numerous day labourers and cottagers, the poverty rate rose in bad harvest years, due to the way the Poor Law functioned. People who lived and worked full-time in their employer's households, such as hands and maids, were the responsibility of their employers and could not, by definition, be poor as long as their employers remained solvent.[52] The increase in poverty shown by the figures might, therefore, be rather illusory. The prevention of poor people without gainful employment from migrating outside their home parish without a guarantee of employment or a place in the household of a relative, tended to conserve poverty in certain areas (which of course was the intention of the legislation). People who were impoverished for more than a year or so tended to remain poor all their lives.[53]

126   *Peasantry to capitalism*

Developments in agricultural production projects

Changes in farm size and the labour force are linked through changes in the composition of farm output and the way it is produced. New crops or livestock could be innovated, or old ones combined in new ways, the proportion of the holding under cultivation could be expanded, techniques of production could be changed, or more value could be added to what was produced by working on it more intensively before it left the farm. In short, the production projects, the activities through which consumable products are created from farming, must have changed. Even in a limited region like our study area, we can anticipate that quite large differences would have existed between production projects in different physical environments, and that they grew with the wider differentiation of holdings and workforces in the eighteenth and nineteenth centuries. Estates were different from person of standing farms, which in turn differed from more traditional peasant farms, whilst peasants would have worked their farms in other ways than crofters and cottagers treated their smallholdings.

Traditional views of Swedish agriculture have stressed that obstacles to change, such as restrictions on farm subdivision and open-field village organization hampered not only the reclamation of land, but also the innovation of new implements, crops or animals. Apart from those who had farms outside traditional villages, farmers had to adapt to almost medieval patterns of land use. They ploughed together with the other farmers of the village, sowed and harvested the same crops and kept a limited number of animals together on the common, which was carefully guarded by the village community.[54] Recently, it has become clear that this picture of Swedish agriculture in stagnation is erroneous, and that it was not the case that legislative changes of the eighteenth century initiated change in a traditional and conservative peasantry. The changes in legislation were probably helpful, but agrarian production had started to change much earlier. When official village investigations were made, as in connection with enclosures when all the land of a village was mapped and measured, it was obvious that many farmers had already cultivated land on the common.[55] They also show, in the case of villages which had been mapped in the seventeenth century, that the reclamation of land within the open field system could have easily added another 50 per cent to the arable. It has even been suggested that the first (*storskifte*) enclosure came about, at least from the farmer's point of view, so that the reclamation process could continue, rather than begin.[56] In the same way, informal subdivisions of farms had apparently occurred below the permitted levels, but only when these levels were lowered could these informal subdivisions be admitted.[57]

Swedish farming in the eighteenth century was not very old-fashioned and static, suddenly exploding in an agricultural revolution over a few decades.

Rather, new implements had been put into use, more or fewer animals kept, and arable and meadow reclaimed gradually over a long period of time. Change was not something alien to the eighteenth and early nineteenth-century Swedish farmer – rustic as he may have seemed to foreign travellers and to members of the Swedish upper classes, used to a life in towns and manor houses.[58]

Land use

Hannerberg showed in the 1940s that the farmers of Central Sweden were already involved in arable reclamation by the seventeenth century,[59] but it is only in the last decade that his findings have been merged into a general picture of land-use development before the late eighteenth and nineteenth centuries. It has become clear that the strictly regulated cultivation of open-field villages left room for changes in the balance between arable and meadow, between the infield and the grazing, and between the number of animals and the area of arable land.[60] In fact, village common cultivation had been actively supported by the government as early as the sixteenth century.[61] Before the mid-eighteenth century, when the village domains became fully colonized, arable reclamation also took place outside villages on the hundred commons which had been saved for the purpose of firewood production and building materials. However, the clearing of forest soon became a source of conflict between farming and the iron industry's need for charcoal. Ironworks had exclusive rights to the charcoal produced in specified surrounding areas and it was feared that agricultural colonization of such areas would impoverish 'their' forests. In consequence, from the seventeenth century onwards the government tried to limit the number and extent of clearings outside traditional villages.[62] It was because of this policy that the establishment of crofts on peasant land was prohibited until 1755. The noise made by the government at the time, and all the regulations it issued, have contributed to the belief that little arable reclamation or other kinds of changes took place in the southern and central parts of Sweden before 1750.

Even if we now know that more land was brought under the ard than was previously thought, it is nonetheless true that open-field farming did present some obstacles to the colonization of land. In Sweden open fields were organized according to specific two-field or three-field régimes which could not easily be changed without adverse consequences: if a field with a higher potential for reclamation expanded out of proportion, the years in which the other field was used would be difficult to cope with. Moreover, some kind of consent among all the villagers was necessary if arable reclamation were to be possible at all. The numerous patches of arable in other places, on common land, were probably reflections of the lack of such consent and results of private reclamation efforts. Formally, the traditional field patterns were only dissolved through the reorganizations brought about by enclosures, but informally field organization could well have been considerably modified earlier. This was not formally documented because until the *storskifte* enclosure regulations were introduced in 1749, it was illegal to desert the medieval open-field organization.[63] Even in most *storskifte* plans, the old field pattern was kept, and the farmers were simply given fewer strips in the fields. All obstacles for farm reorganization disappeared only with the much more radical *enskifte* and *laga skifte* enclosure regulations in the nineteenth century.[64]

To depict land-use development only as a conflict between the old system and demands for reorganization and arable reclamation would be incorrect. All farms were not organized in large villages where communal agreement was necessary for reclamation, and all farms were not mainly grain-producing. Estate owners could certainly do what they liked on their manor farms, and on tenant farms if they owned all the farms of a village, but there were also other, more independent production units in the countryside. Eighteenth century maps show that many small hamlets, particularly on shieldland, only had exiguous open-field systems, where farms must have functioned much more independently than in large villages on the plains. Still, the general picture of development (not only in Sweden) is based on the latter. However, large areas were dominated by single farms and small hamlets, perhaps interspersed with a few estates. Such areas could develop their cultivation, reclaim land and reorganize holdings to a larger extent than areas where most farms had been in large villages since medieval times. Again, however, this picture is too simplified. To begin with, the fertility of the land varies widely and tends to be much higher on plainland, even if some fertile pockets exist within the rocky shieldland. Thus, if organizational complications were fewer and less severe for shieldland hamlets, single farms and estates, it was not necessarily the case that they could benefit from further arable reclamation. Furthermore, all farms did not produce grain as their main output. On the plains, livestock were kept mainly for traction and manure in a system geared to maximizing the production of grain, but on the shieldland, livestock were important not only as manure-producers, but to exchange for the grain of the plains and supply raw materials for working-up into goods for sale. On a livestock farm, the reclamation of further arable land would reduce the available meadow or grazing. Instead of reclaiming good meadow land and grazing into moderate arable, a livestock farmer who wanted to increase his output would be more likely to develop the livestock production rather than becoming an unsucessful grain farmer.

In fully colonized parts of Sweden, such as Östergötland and Västergötland, arable reclamation was not particularly great in the eighteenth century. Arable land increased in area by about 25 per cent at the expense of meadow and grazing in the Dala area of Västergötland over the course of the century.[65] The large villages were located on more fertile land, while single farms, which in most cases represented later colonizations, were generally on inferior land. About 60 per cent of the total arable acreage in the large villages in 1700 was controlled by estates in the form of tenant farms. Between 1700 and the 1770s, they grew by 14 per cent on average, while freeholders expanded their farms by more than a third. The arable on farms outside the large villages grew by 23 per cent from 1700 to the 1820s. Farms in villages added more than two-thirds of new arable over this longer period. Thus, even if outlying farms could more easily reclaim, their lower potential in terms of land quality, perhaps in combination with a concentration on livestock production, resulted in *less* reclamation. The arable also grew more rapidly after 1830, by 74 per cent in the large villages up to 1880 and by 51 per cent on the smaller farms and hamlets over the same period. Estates, apart from the manor farms, had to reorganize their tenant farms if they wanted higher degrees of reclamation. The expansion of arable on peasant initiative would probably have resulted in them having to pay higher rents, and was therefore unlikely.

There were similar arable growth rates on the opposite side of lake Vättern.[66] Extensions of arable land from 25 per cent to 100 per cent were recorded between 1640 and the 1770s. The area of arable in the shieldland parish of By grew more rapidly in the nineteenth century than in the eighteenth, particularly after the *laga skifte* enclosure, even though there had been a 50 per cent increase between 1750 and 1810,[67] whilst growth had been slow up to the latter year. Finally, between 1810 and the 1850s, the arable area grew by 184 per cent. Arable reclamation was more intensive in Närke province on the plains before the 1780s,[68] so that it seems that the expansion of the arable area on shieldland was later but more dramatic than in more fertile regions.

It is clear that in Sweden's central plains arable land was reclaimed steadily over at least the two centuries preceding 1800 within a framework of peasant open field villages. After the introduction of formally regulated enclosures, and in particular their later more radical forms, reclamation increased. Single farms and little hamlets, which were hardly affected by enclosures, also started to reclaim to a larger extent at that time. Despite the difficulties in the face of reclamation, villages on the plains reclaimed larger areas of arable than small hamlets and single farms, which were generally located on less fertile land.

Crops

The arable area of a farm does not equal the annually cultivated acreage. Before the introduction of fertilizers and root and leguminous crops, normally half or two-thirds of the arable area was sown annually, with periods of fallow and manuring in between cultivation to restore soil fertility. Unless other means for manuring were available (such as seaweed or turf), farmers needed a certain number of animals to provide manure. Rye and barley were the main grain crops.[69] From the sixteenth century onwards new rotations were brought into use in Europe to enable specialization and the gradual reduction of fallow, especially in more fertile regions. In Sweden the harsh climate and poor soils made this process slow. At the end of the eighteenth century two- or three-shift cultivation was still completely dominant in Central Sweden, with a marked difference between the west where three-shift was more common, and the east, dominated by two-shift.[70] Even though, formally, cultivation thus stuck to ancient forms, it is clear that some change had started by the mid-eighteenth century. Peas were grown on the fallow to restore nitrogen, and potatoes and fodder crops were becoming more common, grown on the reclaimed areas of arable.[71]

Quite marked differences in the range of crops grown had always existed in Sweden, not only between areas dominated by shieldland or by plains but on a larger scale. Western Sweden grew larger quantities of oats, while barley was more common on the plains in the east. Rye, the traditional winter crop, dominated on shieldland farms but was less common on the plains where more spring-sown crops were grown. Traditions in the cultivation of different crops were not that old, however. In prehistoric times, barley had dominated grain cultivation almost completely, while rye had accounted for about 8 per cent of total grain production in South and Central Sweden.[72] In the Middle Ages, rye gradually expanded its share, so that in the sixteenth century 26 per cent of the total grain harvest in Västergötland and 36 per cent in Östergötland was rye. In the counties around Stockholm, the share of rye was between 50 and 60 per cent in the 1570s. Oats were not grown at all in Östergötland at this time; they only accounted for 3 to 5 per cent of the total crop in Västergötland, and for between 10 and 25 per cent in Närke and Värmland further north. Thus, the patterns of the eighteenth century were in many cases less than 200 years old; arable reclamation had been coupled with changes in crop composition.

Within the regions of different crop composition, there were differences between farmers: crofters and peasants who grew grain for their own subsistence were not interested in the same crops as farmers who sold some of their output. In the same way, development over time on farms must have varied. Some farmers were open to the use of peas, clover and potatoes while others stuck to the system of their forefathers. The more enterprising farmers in terms of plot amalgamation and farm extension were probably more interested in changing cropping practices than those who farmed the same farm all their lives.

The introduction of new crops and rotations has been attributed to estates and 'well-informed' large landowners.[73] Only they could afford experiments on a large scale; if experiments were successful, ordinary farmers could benefit from this experience and introduce new crops and methods on their farms. The actual process of innovation diffusion was probably much more complex than this. The differences in scale between estates and peasant farms were so great that new crops and methods suitable to the former were impracticable on the latter. On peasant farms, with much smaller margins and a greater concern for family subsistence, risk had to be avoided. In consequence, development was much slower and not always in the same direction as on estates.

In Västergötland in the mid-eighteenth century oats accounted for more than 50 per cent of the grain produced on the plains while barley and mixed grain, each with about 20 per cent, made up most of the remainder. Rye (4 per cent), wheat (1 per cent) and peas (3 per cent) gave only marginal additions to the total. Most grain grown was spring-sown. Although potatoes were introduced to Sweden nearly a century before, their diffusion was very slow in most parts of the country.[74]

Changes in the mix of crops on the Västergötland plains were marked after the 1820s.[75] In the 1850s the three main crops of older days had been reduced to comprise 80 per cent of the total seedcorn quantity, while more wheat (3 per cent), rye (7 per cent), potatoes (7 per cent) and peas (4 per cent) were grown. Still, over one hundred years, the composition of the arable crops had changed only marginally. The major difference between 1750 and 1850 obviously did not lie in a massive change of crops, but rather in the cultivated area which grew by more than 90 per cent, and this is what must have had the main repercussions on other dimensions of agricultural production projects. Even if the proportion of oats was reduced by a few per cent, the total amount grown was virtually doubled.

Crop innovation was more common among the crofters and cottagers of Västergötland. In the eighteenth century, their crops had consisted of oats (56 per cent), mixed grain (28 per cent), barley (8 per cent) and rye (7 per cent). By the 1850s 16 per cent of the crofter seedcorn was potatoes, when they had almost completely replaced barley and mixed grain. The crofters and cottagers of the plains, who could not normally expand their cultivated area without encroachment onto the farmers' carefully guarded domains, sought methods and crops which could maximize subsistence production with uncosted labour. Farmers who could expand and who were by this time also partly dependent on the market and hired labour, had to make quite different decisions concerning their crops. What would sell best and give the best price, what could be grown most efficiently and give the largest harvests with the least possible expenses in terms of labour? Both kinds of farms had to avoid risk, but the strategy for minimizing it had to take different criteria into account.

These Västergötland figures can be compared with ones for Dahl hundred in western Östergötland,[76] which consisted almost completely of fertile plain-land. Oats were not important there; barley and mixed grain accounted for nearly 80 per cent of the crops, peas for 6 per cent, lentils for 1 per cent, wheat also for 1 per cent, potatoes for 5 per cent and rye for the remaining 10 per cent. The major difference between the two areas was that oats accounted for more than half the cropland in Västergötland, but were almost completely missing from the east. Colder winters and comparatively dry springs were probably why winter crops grew more extensively in Eastern Sweden, whilst the clay soils of Western Sweden were unsuited to them.[77] Barley thrives on calcareous soils and because it ripens quickly it could be grown if winters were late and cold.[78] Already in the sixteenth century, barley was almost completely dominant on the western plain of Östergötland, while rye comprised about half the harvest in all other parts of the country; indeed, in the south there are examples of harvests consisting only of rye.[79] The crop composition of the early nineteenth century contained a high proportion of mixed grain, something new, which continued to be grown, together with potatoes, up to the 1860s, while the share of barley shrank. In total, potatoes (13 per cent), rye (19 per cent), wheat (11 per cent), peas (5 per cent) and lentils (2 per cent) accounted for 50 per cent of the 'grain' crops. Change was much greater in western Östergötland than in Västergötland.[80]

On the shieldland, crops were quite different, reflecting climate and soil fertility. In Västergötland oats (79 per cent) were almost totally dominant, in combination with rye (16 per cent) while other crops were negligible in the latter half of the eighteenth century.[81] In the 1850s, when potatoes comprised no less than 43 per cent of the seedcorn, oats had gone down to 35 per cent and rye to 11 per cent. Barley and mixed grain were still only marginal crops. As the shieldland arable area had grown by 50 to 100 per cent over the same period, the total agricultural production must have grown too, albeit not by so much. On the shieldland potatoes became even more dominant among crofters and cottagers. By the 1850s they made up 50 per cent of the seedcorn, while oats, rye and mixed grain were further reduced. Potatoes seem to have become, here as elsewhere in Europe, the most important crop for local consumption, and especially so on shieldland, where the meagre soils were well suited to them.

A clear picture can thus be drawn of cropping changes in Sweden. In the west, oats dominated shieldland cultivation from the seventeenth century until the introduction of potatoes, while crops like rye, barley and mixed grain were of less importance. In the east, rye was of greater significance and oats virtually absent. On the plains in Western Sweden, oats and barley had early on been dominant. Over the eighteenth and nineteenth centuries barley was replaced by potatoes, rye, wheat, mixed grain and fodder crops. In the east, the plains cultivation was traditionally dominated by barley and rye, but mixed grain, wheat, potatoes, lentils and, later on, vetches were introduced, mainly at the expense of barley. The increased production of oats and other grains on larger farms might indicate increased market orientation, although it could equally well imply that more numerous traction animals were being used to reclaim arable for the production of potatoes, a typical subsistence crop. Crofts and cottages swiftly turned to potatoes, which accounted for about 20 per cent of their crop after 1850. On the shieldland, where crofts often could expand without barriers in the form of already existing freehold farmland, potatoes were more appropriate to environmental conditions and seem to have accounted for 50 per cent in the 1850s, with rye or oats declining. Here is a clear example of involution in the subsistence farming sector. However, little is known of the differences between what was grown on farms of different sizes. It is known that estates and large farms experimented a lot with new crops, and we can assume that larger farms had larger areas of fodder crops and grew more wheat, vetches and lentils, while smaller units would have had more potatoes.[82]

Farm implements
We will subdivide farm implements into those for preparing the seedbed and those for sowing, cutting and threshing grain. Important changes took place in each group in the eighteenth and nineteenth centuries. As in other aspects of agriculture, a spread of innovations is supposed to have taken place from estates and large farms, eventually to the rustic peasant farms on the shieldland who, as stalwart preservers of old habits, were late to start using such things as iron ploughs and mowers.[83] However, this is again to oversimplify. Small farms on shieldland needed quite different equipment from large granaries on the plains, crofts on the plains other things than the large farms nearby. Thus, development in implements was much more subtle than the processes described from England and the continent, not least because with more small farms and poorer soils, change would take less striking forms.

Seedbed preparation Ploughs and ards were the traditional implements for seedbed preparation, Eastern Sweden being dominated by different types of ards, Western Sweden by ploughs.[84] This corresponds to the cultivation of oats in three-shift rotations in Western Sweden and of rye in two-shifts in the east. Ploughs gradually became more common in Eastern Sweden in the late Middle Ages, but the use of ards was still dominant in the eighteenth century. The difference in effectiveness between a developed ard and the primitive ploughs in use at the time was in fact not great. On most farms using ards, crude wooden ploughs were used for breaking up meadows in the arable reclamation process after which ards took over. Such ploughs were reinforced with iron and made more effective in the early nineteenth century, representing a gradual development along traditional lines.[85] However, breaking with tradition, factory-made iron ploughs were introduced in Sweden in the 1820s. Ards were still used, as they were better for earthing-up potatoes and could also, especially if they contained iron, be used for reclaiming arable because they cut deeper than ploughs.[86]

In a part of estate-dominated Södermanland Province, the estates had new-design ploughs by the 1820s, when peasants used improved traditional ploughs only to a small extent, and ards were still their major implement of seedbed preparation.[87] In the 1860s, most farmers used improved traditional ploughs, and a few had even purchased ploughs of the new designs. Tenant farmers on estates were somewhat slower to change, probably due to smaller resources or a lack of incentive to develop cultivation. Not surprisingly, 50 per cent of the crofters still had ards as their major seedbed implement in the 1860s.

In Västergötland, where ploughs by tradition had a stronger position on the plains, iron ploughs based on traditional designs spread earlier than in Södermanland. In one area, 57 per cent of the farmers had such ploughs in the 1820s while in the 1850s 94 per cent had iron ploughs of different kinds.[88] In Dahl hundred in Östergötland, the timing of the introduction of modern ploughs seems to have been similar to Södermanland. On estates, iron ploughs became common in the 1820s, on large farms in the 1830s, on middle-sized farms in the 1840s and on small farms after 1860. The ploughs of small and middle-sized farms were developments of traditional designs, while those on large farms and estates were of an innovative kind.[89]

Apart from ploughs and ards, seedbed preparation could also require harrows, rollers and clodcrushers to cultivate and level the soil. Iron harrows became common in the sixteenth century in Sweden, but wooden ones were still used on light soils up to the nineteenth century.[90] The traditional harrow had a wooden frame and straight iron rods to scarify the soil. In the mid- to late eighteenth century, other kinds of harrows, normally more deeply penetrating, started to appear on the plains of Västergötland. By the 1820s, such harrows appeared in between 80 and 100 per cent of farm inventories. In other parts of the county, they only started to appear in the 1840s and 1850s and were least common in shieldland parishes.[91] In Södermanland, more elaborate harrows, like iron ploughs, came later in spite of the dominance of estates in agriculture. In the 1820s only estates had modern harrows to any extent, while by the 1830s about half of the ordinary farmers and tenants had acquired them. About 20 per cent of the crofters in the 1830s had modern harrows.[92] In Dahl, deeply cutting harrows only started to appear in the 1850s on the larger farms.[93] Again, the plains of Östergötland were later to change to modern implements than those further west.

Sowing, harvesting and threshing Sowing by hand dominated all through our study period on most farms in Dahl: only about a third of large farms had seed drills by the 1850s. Reaping, on the other hand, changed progressively from the seventeenth century. Sickles were gradually replaced by scythes, which had previously only been used in haymaking. In the mid-eighteenth century scythes were used in most plains areas of Sweden, but not so commonly on shieldland.[94] The reason for the change was that work is quicker with the scythe, although losses are greater and the degree of precision less.[95] In areas where fields were smaller, and labour also less expensive and more easily available, such as in the shieldland areas of Småland, the sickle remained the grain harvest tool into the nineteenth century, although all farms had the kind of scythe necessary. It has also been argued that the scythe, unlike the sickle, could not be used by women,[96] which could explain why sickles remained in use, although examples of female scything from the Baltic countries are common.[97] The sickle continued to be used on the plains after the scythe had replaced it in the harvest, to cut fallen grain and as a hook to pick up straw. It is possible to explain the change from sickle to scythe in another way. On the plains where barley dominated, straw was an important winter fodder by the 1800s. A scythe could cut the straw at the base without effort, whereas a person using a sickle had to bend over further to cut low. Thus, in areas where a lot of arable had been reclaimed early on, straw was probably of far greater importance as fodder, making the scythe the more appropriate harvesting tool.

Even with the scythe, harvest and haymaking were the most labour intensive parts of the annual production cycle. The family labour of the peasant was most stretched then, extra hands were frequently used, and crofters and cottagers had busy days on their landowners' farms. And the task had to be accomplished quickly because rain could mean disaster. Thus, there was an incentive for farmers to save both labour and time through mechanization, although mowers only started to appear in Östergötland in the 1880s, and the first reaper and binder in Sweden was imported to a large farm in that province in 1851.[98] The introduction of machinery would have been hampered by the field system in open field villages, but by the time that it was available few open field villages existed, and on estate manor farms this had been no problem anyway.

Horse-drawn rakes were introduced in the mid-nineteenth century and purchased by a few more substantial farmers in Dahl.[99] In the following decades, this implement started to appear on almost every kind of farm. Threshing grain was the final stage of the agricultural production process on the farm. Hand-threshing was labour intensive, and occupied the farm's menfolk for long hours on the barn threshing-floor in autumn and winter. The first mechanical threshing machines were large and needed a lot of manual labour as well as water- or horse-power.[100] The investment was much larger than for a modernized plough or a harrow (150–1,200 Rdr). In Västergötland the first mechanical threshing equipment appeared on large farms in the 1810s.[101] By the 1850s they were common on large farms, but hardly any ordinary farmers had yet introduced them. Smaller threshing machines were manufactured in Sweden from the 1850s and brought the investment down to more acceptable levels for smaller farms (50–160 Rdr). In parallel with the development of factory-produced threshing equipment, local craftsmen had started to build crude wooden threshers which needed three horses or more.[102] Only factory-made threshing-machines spread gradually from estates and large farms to smaller ones, while locally produced machines spread earlier and more rapidly among intermediate farms. In Dahl only estates and very large farms had threshing machines up to the 1850s, when they began to spread among smaller farms.[103] By the 1870s crofters had them.

Implements, production projects and labour   The major implement changes before the 1850s concerned seedbed preparation. A gradual switch had already occurred from ards to ploughs and from wood to iron before the nineteenth century. Harrows and rollers had also become common, and in reaping, sickles had been replaced by scythes. These changes occurred in parallel with arable reclamation and changes in crops, which were also slowly and steadily under way *before* the mid-eighteenth century.

The changes in implements were made to save labour, to increase speed in cultivation and harvesting, and also to get larger harvests from the same area. In combination, the time of ploughing, even with substantially extended fields, need not have increased. However, new crop rotations meant that every year a larger proportion of the arable was cultivated. This was compensated for by the use of more modern harrows after the first post-fallow ploughing, as harrowing used half as much time as another turn of ploughing.[104] The change from ards to ploughs implied a reduction in the amount of labour and traction used. Even though ards did not need more than one or two traction animals, the fields had to be arded seven or eight times each spring before sowing, while with an iron plough once was enough. Even though ploughing is slower than arding, which means that the amount of labour was not reduced eight times, this was the main time-saving from new implements during the first half of the nineteenth century.

Hay-making, harvesting and threshing, all vast consumers of labour and time, had only, through heavy investment, been mechanized on the large farms by 1860 – on small farms and crofts particularly on shieldland, this hardly occurred in the nineteenth century.[105] A reduction of threshing did occur on such units, though, because of the introduction of potatoes. At the same time, the calorific content of the produce rose by more than a third. Thus, even though the work of growing potatoes, including earthing-up and weeding, was more time-consuming than for other crops, the new crop meant a considerable amount of time saved overall.

Animal husbandry

According to Hannerberg, the number of livestock on Swedish farms was smaller in the eighteenth and early nineteenth century than it had been during the seventeenth.[106] This has been attributed to the growth in the grain production and the expansion of arable land. It could, of course, also have been a result of changes in implements, a growing population, and nitrogen-producing fodder crops like clover or peas being used to a larger extent. We have seen that such very general statements, with no regional differentiation or a categorization of farms into size groups, are inadequate.

On the plains of Västergötland, the number of traction animals grew until 1770, before being slowly reduced from the mid-1800s. The number of horses remained virtually constant, while the number of oxen and bullocks was reduced by half from four to two on average farms. The number of cows was also reduced by a third over a hundred years. These reductions could reflect both the introduction of new seedbed implements which demanded less work per acre, and an increase in the arable area coupled with population growth, which made animals more expensive to keep. The decrease in the number of animals on the Västergötland plains was even more marked among the crofters than the farmers, probably because their margins became smaller with an extension of farm cultivated areas. The crofters had all their large animals reduced by at least half.[107] Supposedly, all these reductions in animal numbers were a consequence of changes in the pattern of cultivation: when enclosures enabled further reclamation, the stress on arable cultivation became ever more marked for farmers, whilst the margins of crofters grew smaller and crofts were frequently relocated to smaller and poorer sites.[108]

Traditionally, the number of animals kept on farms was as large as possible in Sweden. They were fattened in summer and very poorly fed in winter, when they had to be given hay, straw and leaves indoors. Traction animals, which had to work all the year round, were the only exception to this, which meant very low winter milk production. The produce was used mainly locally, although butter and cheese were traditional market products. With growing towns and industrialization, the demand for dairy products grew, especially around large towns. In consequence, particularly estates but later on also smaller farms started to keep cows in a different way, aiming at all-year-round production.[109] Thus, there was a great difference between a cow of the eighteenth century and one of the nineteenth in terms of size and milk production. This makes it difficult to compare numbers of cows over long periods of time; a smaller number of 'modern' cows could well produce as much milk as the larger herds of traditional cows had.[110]

In Dahl there was a considerable difference in changes in animal stocks between different sizes of farm.[111] Estates and large farms had increasing herds of cows all through the nineteenth century, while smaller farms reduced their number, perhaps because both crofts and small farms were reduced in size during the first half of the nineteenth century. The average value of the cows grew much more rapidly on large farms, which thus not only grew in size and in consequence kept larger herds, but also deliberately developed livestock production more than traditional farmers.

The number of traction animals did not grow much in Östergötland. On small farms there was a tendency towards slightly fewer horses, while on the large farms horses were kept to the same extent over the nineteenth century, even though the number of oxen grew up to 1860. On smaller farms and crofts, the number of oxen was stable. Altogether, the importance of oxen grew during the first half of the nineteenth century, similar to Västergötland, although here the total number of animals did not fall by as much. Oxen became more popular with the modernization of traditional equipment, and only the introduction of factory-made implements reversed the trend back towards horses.

In sum, Hannerberg's general picture of falling animal numbers in Swedish agriculture seems to be verified by recent research. However, changes within the population of animals showed great variations: the number of oxen tended to grow on the plains, in response both to larger areas of arable and to improved implements which made oxen more economical. However, the demand for traction animals was reduced per unit of arable land by the new implements, which explains why there was little growth in the number of oxen despite the increase in arable land. A reduction in the number of cows, common in Västergötland and among smaller farms in Östergötland, could have been a consequence of shrinking areas of meadow and grazing, which were already small in the seventeenth century in Dahl.[112] The fact that herds of cows grew on large farms on the plains of Östergötland could be due to the fact that fodder crops were grown to a larger extent, and to the growing importance of market-orientated dairy production.

Farming systems
All aspects of the agricultural projects, including the workforce, were closely related to each other. The arable was extended, first slowly then more rapidly with the gradual dissolution of open fields and traditional farming systems. Hand in hand with this, crops were changing towards fodder and marketable grain and dairy produce, and new or developed implements were brought into use. Together with the reclamation of arable, this forced changes in the composition of livestock on farms. Population growth, irrespective of whether it was caused by increased demand for labour in agriculture or whether it was autonomous, made labour more readily available after the 1750s. With a growing labour demand generally, however, wages started to rise more consistently from the 1840s and this rendered it more profitable to try to save labour costs.

Many of the hallmarks of capitalist farming were clearly established during the early nineteenth century. The shrinkage of crofts and the replacement of crofters by married hands and cottagers gave a firm twist to the fuller proletarianization of labour on estates and the expanding farms of erstwhile peasants. However, many small farms survived, especially in the forested shieldlands. On these holdings, the introduction of potatoes in the absence of significant technical developments seems to have given a further shift towards subsistence autarky.

These conclusions seem clear, but they ignore homecraft production, which we might expect to be particularly significant in the shieldland regions which seem, otherwise, to have been in involution. Indeed, from what we have seen so far it is quite plausible to suggest that a professionalization of large farms was linked to increased craft production on small farms and in landless homes in a powerful 'hidden' economic process.

Rural by-employments: a hidden economy in the countryside?

Secondary production in peasant households, working up different kinds of agricultural produce or wood and metal had a long tradition in Sweden. Winters are long, cold and dark and only one harvest can be taken from the fields. Agricultural work in the winter was limited to animal husbandry, threshing and taking care of farm implements, which left time for other activities. It is not surprising that the Swedish government was keen to encourage homecrafts (*hemslöjd*) as supplements to peasants' income, believing that with farming they were 'the two sources from which all labour and earnings derive'.[113] These homecrafts can probably be accommodated within the conventional model of pre-industrial domestic mass production. But in Sweden the peasantry also had other traditions of non-agricultural production. Copper and iron ores were mined by peasants in various parts of the country, especially Bergslagen. The communal property of villagers, they were mined cooperatively, and metal was produced in individually or collectively owned works. Iron and copper were exported from the mining areas, not only within the country but also abroad. Government policy to try to make profits on more refined produce formed the basis for one of Sweden's major exports – bar iron produced in ironworks (*bruk*) in many areas of the country.[114] While at least part of the mining and smelting were done by peasants in iron-rich regions, bar iron was to a large extent produced by estate owners who during the sixteenth, seventeenth and eighteenth centuries built ironworks, employing twenty to forty people, where water-power, pig iron and charcoal were readily available.[115] It was also government policy to conserve forests in mining areas for blast furnaces and to encourage ironworks to locate elsewhere, even if this meant shipping ore and pig iron long distances. Blast furnaces and ironworks were established in Finland, northernmost Sweden and Småland, almost completely based on ore from Central Sweden.[116] Official policy was implemented through a system of privileges; it was not permitted to establish a works without sufficient supplies of ore, pig iron, water power and charcoal. This meant that only large estates could gain privileges, because they could command their tenant farmers and crofters to burn charcoal or gather lake-ore in the winter, and controlled areas large enough to provide the necessary quantities of water-power and charcoal.[117]

Thus, even though the part of iron production in peasant hands was shrinking by the eighteenth century, increased exports of bar iron meant important by-employments for peasants in many parts of Sweden. Charcoal burning for a normally sized ironworks with two or three hearths and a blast furnace took about 30 man-years of labour, and lake-ore gathering where it was available occupied the winter days of many peasants.[118] Keeping water systems running and hauling the massive quantities of charcoal, iron ore, pig iron and finished products were also important tasks for local peasants and crofters in the areas around the thousand or so ironworks established by the eighteenth century.[119]

Forestry was another important Swedish activity which is difficult to accommodate within a model of industrialization. Apart from the charcoal burning for ironworks which by the early nineteenth century was the economically most important part of forestry, tar and potash were significant products before the value of forests rose as a consequence of an international demand for timber and planks during the course of the nineteenth century.[120] Tar from Sweden (even more so from Finland) dominated the world export market for centuries.

Thus, in certain regions of Sweden, non-agricultural activities for export engaged peasants in their slack season. Many other industries were organized in large-scale plant (known as *bruk*) from the seventeenth century: large brass works like Gusum and Skultuna, as well as hundreds of small hand paper mills and numerous manufactories established over the course of the eighteenth century. All these had close connections with their agrarian surroundings, many put work out, and all offered other kinds of by-employments to rural households such as charcoal burning and haulage.[121] The regions where these activities were important were, self-evidently, the forested shieldlands, where water power was also abundant. It is suggested as well that such areas were important for the domestic production of textiles, woodcraft and metalwork: their often more difficult climate, niggardly soils and rugged terrain made agricultural expansion less profitable and gave more time for by-employ-ments.[122] However, pre-industrial mass production need not have been confined to the shieldlands. It is true that the Provincial Medical Officer of Mariestad, on the plain of Västergötland, observed that the arable-farming peasant 'having sufficiently of corn for his bread and aquavit ... considers it unnecessary to work and cloaks his negligence with the excuse that he can undertake nothing of practical value in the darkness'.[123] Shortage of wood for lighting and heating on the plains were more likely to lie behind this state of affairs than innate lassitude amongst arable farming peasants. And we know little about what went on in the cottages on the outlying lands of the farms. Just as crofters and other people in the forest needed by-employments in the winter, so did the growing numbers of day-labourers and their families on the plains. Even if the requirements for some branches of production existed only in forested areas, labour-intensive production could thrive on the plains.[124]

It is, however, extremely difficult to discover just how important by-employments were. It was not legally permissible, outside defined areas such as Bergslagen, to subsist from craft work outside agriculture in the countryside. Thus, non-agrarian occupations are rarely registered in official documents. Moreover, tax was mainly paid for landed property or per capita, irrespective of income or occupation, so we have no ledgers fully describing household economy in taxational terms. Knowledge of large portions of the home economy is based on inferences from fragmentary sources such as inventories, diaries and domestic accounts. As all the activities reflected in the ownership of tools and materials could also take place purely for home production and consumption, it is impossible to decide whether market production was occurring. All we can do is rely on evidence such as the quantities of raw materials and finished goods, together with debt and credit relationships with traders, to decide whether production for sale was occurring.

Parts of Norrland, Dalarne, southern Västergötland and certain areas in Småland were the most important regions for home craft production for the market.[125] In southern Dalarne, on the northern fringe of Bergslagen, smiths-with-land were indistinguishable from peasants-with-forges; together they comprised a large proportion of the population. In the middle of the eighteenth century, the peasants of By parish in this region produced 'horseshoes and harness, various types of axe, harrows, pitchforks, nails, cupboards, chests and boxes, as well as ... linen, spun and woven products'. The wealthier peasants in By 'generally owned many more metalworking tools and textile-manufacturing implements than did the smallholders ... [who] ... owned a greater number of woodworking tools ... which implies that [they] tended to work in wood, whilst the wealthier farmers pursued the craft of ironworking'.[126] Around the areas rich in iron ore, ironworks naturally dominated. In Lerbäck parish in southern Närke county, next to our study area, there were 600 active metal tack makers in 1825. The nail and tack industry of the parish was taxed at the value of 50,000 Rdr in 1843 and of 90,000–100,000 Rdr in the early 1850s.[127] In Godegård parish, on the opposite side of the Östergötland border, even larger numbers of the population had been engaged in nail production in the seventeenth and eighteenth centuries, although their trade had been out-competed by larger iron manufactories connected with the local ironworks by the early 1800s.[128] Metal craft and manufacture, dependent on raw materials which were not ubiquitous, dominated in Central Sweden, in areas next to the important iron mining Bergslagen. There were exceptions to this rule, mainly in Småland where there were both small iron mines and plenty of lake-ore which, together with mine ore from elsewhere provided raw material for numerous blast furnaces, which even though they were normally part of larger ironworks enterprises here, sold off pig iron locally for manufacture.

Throughout Sweden as a whole, textiles production was the most important market homecraft, at first mainly linen, though cotton grew in importance during the nineteenth century. Some flax was imported, but much was also home-grown by the peasants who retted, combed, spun and wove it.[129] Because it is very demanding of fertility, flax tended to be grown in areas where ample pastures supported large manure-producing herds, and textile production thus tended to be located in areas where arable land was heavily interspersed with forest grazing. It was particularly significant in the area of Sjuhäradsbygden in southern Västergötland. Flax and linen production also became important in the middle of the eighteenth century in Hälsingland and Ångermanland on the Gulf of Bothnia. Småland was the fourth region where market homecraft production had become very important by the end of the eighteenth century. Many parishes in the region were widely known for a particular line of craftwork, such as weavers' reeds from Markaryd; wool cards, carding wire, pins, needles and hooks from Gnosjö, or watches and clocks from Algutsboda.[130] Although the places mentioned were notable centres of market homecraft production, it was, in fact, common to all the forested shieldland areas of Sweden. Some such regions, too, produced itinerant craftsmen, like the tanners of the Malung district north of Bergslagen, who travelled round Sweden converting hides into leather for peasant farmers.[131]

Thus, shieldland regions were particularly heavily concerned in household craft by-employments. Although specialist craftsmen were only allowed to operate in small numbers in rural areas, the products of peasant homecraft poured out in rapidly increasing quantities in the second half of the eighteenth century in return for the corn of the plains, which reduced the demand for the craft products of the towns (and prevented evidence of manufacturing appearing in national economic statistics).

The growth of large-scale production in iron, textiles and other industries by no means eroded the importance of domestic industries. Indeed, they grew vigorously for much longer than in most European countries; even in the early twentieth century, 'among the few countries where domestic industries have succeeded in maintaining their position to any great extent, Sweden is, beyond possibility of contradiction, one of the principal'.[132] The increase in value of the output of handicrafts was, at fivefold, equal to that of manufacturing and mining over the late nineteenth and early twentieth centuries.[133] During our period, their rate of relative growth was undoubtedly much higher, continuing the acceleration witnessed by the late eighteenth century, because the stimuli to their earlier expansion either remained constant or increased. A very high proportion of the population remained involved in farming in an environment in which agricultural work was impossible in winter. Even in the late nineteenth century in Västergötland 'the summer was usually devoted to farming, berry picking and timber floating, while the winter was spent on forestry, carpentry, production of wooden crates, commerce in Christmas trees, and cottage industry in textiles' wooden clogs, barrels and ladles.[134]

More and more of the rural populace, too, had less and less or no land; indeed, it was only the growth of rural craft production that allowed such a large proportion of a growing population to remain in farming, especially because much of the population expansion occurred in forested areas and dependent livelihood positions on the plains. The increased production of iron and factory-made yarn increased the raw materials available to domestic industry. The government's attempts to stave off the impoverishment of the rapidly growing population intensified, and 'during the first half of the nineteenth century the handicrafts became the favourite subject of economic politics'.[135] In 1828, baking, brewing and butchery were removed from guild control; henceforth anyone could engage in these crafts for sale, though sale must still occur in towns, and in 1846 all Swedish householders, both male and female, were allowed by law to produce at home with their families any kind of goods whatsoever for sale.[136]

However, the 'impoverishment model' of proto-industrial development is by no means a complete explanation of this growth. Much domestic industry was still for self-consumption, rather than sale, and 'a parallel development of *sloyd* (sic) for private consumption and for sale seems to have developed during the first half of the nineteenth century'.[137] The production of yarn, cloth, clothing, tools, implements, furniture and so on continued, in fact, to be characteristic of rural Sweden right into the twentieth century, as the survival of local styles of dress and household goods until well after the end of our period testifies.[138] Even so, there is no doubt that production for sale amongst small peasant farmers, crofters and cottagers on the plains and most peasants on the shieldlands was responsible for most of the expansion of domestic production in the first half of the nineteenth century.

Even in Sjuhäradsbygden, where putting out to weavers by the proprietors of spinning mills was common, domestic producers generally kept contact with agriculture. 'In Sweden, the putting-out system in its pure form was less prevalent and most of the domestic production took place in family households where it was not the only source of income.'[139] The putters-out themselves continued to be involved in farming, and organized their often extensive manufacturing enterprises from their farmsteads. One peasant with whom Brace lodged on his journey through Sweden imported 2,000 bales of cotton a year directly from Mobile and New Orleans through Göteborg to supply the spinning mill in which he had a half share and 'about 1,500 spindles in fifteen different [parishes], and some 1,200 looms in families', to whom he paid wages of 3,000 Rdr a month. His shop, in a farm building, was full of patterned calicoes for sale to dealers and the public. Other farm buildings were full of harvested barley and rye; the stables contained eight to ten horses, and there were a bakehouse and brewhouse on the farmstead. All the farm produce raised was consumed by the peasant-manufacturer and his family.[140] The domestic workers of Sjuhäradsbygden also kept contact with the land, despite the rate of growth of the industry. The home weavers, whose families were larger than among the non-weavers in the same social groups, comprised about 41 per cent of the farmer households, 26 per cent of the crofters and 8 per cent of the landless, the latter a group which in total made up 32 per cent of the population. One reason for this marked social distribution of home weaving was probably that putters-out demanded security if yarn was to be handed out, which the landless could not provide. Whereas in an ordinary forest parish, the number of peasants would have increased, in Ahlberger's study parishes it did not, due to a concentration of capital into the hands of the putters-out who belonged to the peasant group. Thus, we have to consider not only the 'simple' agrarian production relations when analysing the patterns of livelihood positions in an area, but also the secondary ones which could turn upside down expected patterns of social structural change.[141]

Geographically, expansion of domestic production occurred in all the regions where it was traditionally important. The fine flax spinning and weaving industries of Ångermanland, introduced in the middle of the eighteenth century, produced goods worth 700,000 Rdr in 1857, when the coarse linen drill output of Hälsingland, further south on the Gulf of Bothnia, was valued at 500,000 Rdr. Amongst the newly introduced crafts of the early nineteenth century, pride of place must be given to the cotton spinning and weaving of Sjuhäradsbygden, much of which was still carried on outside factories during our period, and which employed 50,000 people in the 1870s.[142] However, there were many others, including the making of a distinctive kind of wooden basket in Göinge parish in Skåne and the advent of bulk shoemaking in Kumla parish in Närke, both of which began in the 1840s.[143]

Like specialist craft production, specialist dealing was restricted by law to the residents of towns before the late 1840s. Generally, peasant homecraft producers, members of their families, or their hands, must take their own products to the house-doors, fairs or towns where they hoped to sell them. In addition, permission to earn money from selling goods not produced by the seller could be obtained from Provincial Governors, but these permits covered only one journey of specified duration (usually six months), to sell specified goods (usually those produced in local peasant by-employment or works of some kind), in specified places. The only exception to this ban on rural dealing concerned the inhabitants of Sjuhäradsbygden around the towns of Borås and Ulricehamn in southern Västergötland, who were allowed to specialize as pedlars selling goods produced in by-employment within their home area throughout Sweden.[144] The existence of this conduit to distant markets was a major reason for the importance of homecrafts in this region.

However limited this system of sale by producers and itinerant pedlars may appear to have been, it was the basic medium of exchange through which the peasants of one ecotype exchanged their own specialist products for those of others through the fairs, across the length and breadth of Sweden. In Österlen (in the extreme south of the country) in the eighteenth century, 'small wooden articles, especially pieces of turnery such as bowls and spinning wheels, were bought from [peasant] artisans at the fairs. The same was true of casks, chests and baskets offered for sale by handicraftsmen from the northern forested part of the country.'[145] There were over 2,000 travelling pedlars in Sjuhäradsbygden in the 1850s, distributing wares from factories as well as domestic spinning wheels and looms. In most other regions, goods continued to be distributed by the producers themselves, or other members of their households who specialized in selling rather than making wares, and by pedlars with licences to travel to sell local goods granted by provincial governors. These journeyings to farm doors, markets and fairs throughout Sweden (and often abroad) were still the main conduit from the cottage firesides of even the remotest parts of Sweden to the eventual consumers of the goods made there. Most of the linen thread and cloth made for sale in Hälsingland and Ångermanland were taken to Russia by the peasants who made them.[146]

In Östergötland, which was not particularly noted for its domestic industries, large numbers of governors' passes were issued annually to people who wished to travel to sell goods made by themselves or their neighbours. In 1849 and 1850, respectively, over 500 and over 600 were granted. Some were taken out by urban merchants and craftsmen, others, testifying to the effects of the legislation of 1828, by rural women travelling to towns to cook and sell food at the markets. Most, however, were granted to hands, crofters, cottagers and maids to sell in distant places lace, buttons, nails, ironware, copperware, turned wooden articles and leather. The glassware of Rejmyre *bruk*, deep in the forested shieldland near the north-eastern boundary of Östergötland, was distributed far and wide in this way.[147]

Conclusion

Development in the Swedish countryside took a number of different paths. In agriculture itself, on the one hand, large farmers and estates developed in a capitalistic direction, producing grain and dairy products for the market. On the other hand, small farmers seem to have stuck to more traditional, self-provisioning agriculture, avidly innovating the potato on their shrinking holdings. The structure of livelihood positions changed in line with farm size, so that capitalist farms employed more labour, first crofters, hands and maids, then more day-labourers from the cottage population and, especially, the new category of married hands, whose wives must work in agriculture and as part-time maids. On estates, tenant farms providing corvées were replaced by crofters and married hands, together with cottager day labour.

In demographic terms, the proletarianization process seems initially to have involved the children of peasant freeholders and tenants, the social group which had the highest rate of reproduction and a shrinking number of available positions, who became crofters, hands and day labourers. The wages of proletarian labour seem, however, to have risen all through the last eighty years of the nineteenth century, in spite of population growth and a shrinking number of landed positions, although they varied widely over the country. There was also strong regional variation in the level of poverty which rose to high peaks in the late 1860s. In general, poverty seems to have been less marked on the plains, where labour was most proletarianized.

Smaller farms, which were more common in the shieldland areas, remained more peasant-like, using traditional methods and family labour, with little or no market production and a lesser degree of specialization. Was it here that the majority of by-employments existed, as the large farms sloughed off homecraft production and became professionalized? Or were homecrafts important on the large farms, helping to push the gap between them and the small farmers and landless ever wider?

It is clear that large farms, and in particular estates, had the means and interest to buy innovative equipment early on, new ploughs or threshing-machines, while smaller farms and farms on shieldland stuck to traditional implements. Smaller farms and farms on shieldland were, however, quick to accept the potato when it was introduced on a large scale at the turn of the century, reflecting their character as subsistence farms in comparison with large farms where potatoes were of less importance. In general, grain production grew on the larger farms, with intensified cultivation and increased arable areas, wheat increased its share together with the mixed grain and in Western Sweden oats, while the proportion of barley was generally reduced. This was also reflected in a shrinking number of animals in general and a switch from horses to oxen as traction animals during the first half of the nineteenth century. In Östergötland, however, cows were kept in larger numbers on the plains, probably in association with the new rotations including fodder crops.

Work on other parts of Sweden leads us to expect that the changes in farm size which we have shown to have occurred in western Östergötland would have been associated with some of the paths of change in peasant society that we described in chapter 1. New crops, implements and farming systems using increasing amounts of labour which was further and further proletarianized did occur in some other plainland areas, which were obviously witnessing the emergence of agrarian capitalism on estates and some expanding erstwhile peasant holdings during the early nineteenth century. As seems inevitable, pauperism came with it, although levels of poverty were lower on the plains than on shieldland. At the same time, the still numerous small farms of the plains seem to have been following the path of involution. In contrast, the continuing relative importance of small farms on the shieldland seems likely to have been associated with increasing homecraft production, and perhaps fully fledged proto-industrialization.

These three paths of change might, of course, have been intertwined. Perhaps the shrinking farms of the peasant rump on the plains provided wage labour to their expanding neighbours, even, indeed, that the availability of wage work on large farms is what allowed the latter to survive despite their shrunken areas. Perhaps, too, at the same time, craft products from the shieldland provided wares that the large farms of the plains were no longer inclined to make for themselves, and the proletarian households there did not have the resources or creditworthiness to produce. The shieldlands might, furthermore, have provided land for colonization by people who were dispossessed on the plain, but baulked at proletarianization. Craft work, pastoral farming and potatoes could provide such colonists with enough to consume and exchange for the growing grain surpluses of the areas from which they had come.

Hypotheses such as these can be inferred from work on other parts of Sweden. We will now go on to test them as far as we can by looking at how labour forces and production projects actually changed during the early nineteenth century in the different environments of our region. The information about flows of produce and people between different ecotypes which is also necessary to provide reliable answers is examined in later chapters.

Agrarian livelihood positions and production structures in western Östergötland, 1810–60

On peasant farms

The fall in the number of farms as they expanded in size in parts of Aska and Dahl brought a reduction in the number of positions for peasants and an increase in the number and character of subservient livelihood positions in agriculture. We will first discuss these changes, then deal with changes in the composition of the households of peasant and other families engaged in agriculture.

There were large changes in the composition of livelihood positions in Aska and Dahl, albeit perhaps not always in directions which would be expected (see figure 4.1). The number of cottages, and consequently the cottager population, grew in all three subregions. Surprisingly, this occurred to the greatest extent in Dahl, where peasant agriculture became most capitalistic. The number of crofts shrank in every subregion and most in Dahl, where the number was more than halved. Both estates and peasantry incorporated croft land into their farms and thus changed the status of the erstwhile croft buildings into cottages or evicted their inhabitants and had the croft pulled down.

It is reasonable to assume that this was why the number of crofts fell fastest in Dahl, more slowly in southern Aska and virtually not at all further north. Figure 4.1 shows that the major drop in the number of crofts in Dahl came at the same time as the enclosures – after 1830. Between 1810 and 1820 croft

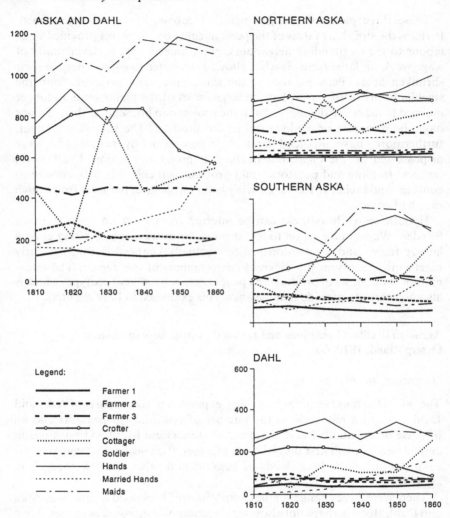

Figure 4.1. Changes in the structure of livelihood positions in Aska and Dahl, 1810–60.

numbers were still growing. From the diagram, we can also see that the cottager population only started to grow rapidly from the 1850s, which, of itself, implies a further intensification of agriculture *after* the enclosures, requiring more and more seasonal labour.

Table 4.1 and figure 4.2 show that the proportion of crofts on large farms was higher than on small ones in 1810 while in 1860, the smaller farms in size groups II and III had a higher proportion than would be expected from their share of the total value of farmland. In Dahl estate crofts were reduced from 59

Table 4.1. *Numbers of crofts and cottages on estates and farms of different sizes in 1810 and 1860*[a]

	Estate		Farm I		Farm II		Farm III	
	1810	1860	1810	1860	1810	1860	1810	1860
Dahl								
Crofts	59	17	63	41	48	15	20	13
Cottages	9	21	31	104	42	55	16	23
S. Aska								
Crofts	53	26	30	60	35	43	39	43
Cottages	11	25	47	100	57	107	57	106
N. Aska								
Crofts	117	99	18	65	35	51	119	59
Cottages	19	43	5	11	10	49	81	79
Total								
Crofts	229	142	111	166	118	109	178	115
Cottages	39	89	83	215	109	211	154	208

Note:
[a] There were also 76 crofts and 40 cottages in 1810 and 34 crofts and 114 cottages in 1860 on plots in Aska and Dahl which were worked by farms located outside the region.

Source: Mantalslängder, taxeringslängder. Landsarkivet, Vadstena.

to 17 over the fifty years, while the peasant crofts shrank in number from 131 to 69; on category I farms crofts were reduced in number in spite of the expansion of the proportion of the total farmland worked by category I farms from 50 per cent in 1810 to 80 per cent in 1860. On the other hand, in 1810 about half of the cottages were located on large farms, which in 1860 had about two-thirds of the cottages. This could be because croft land was being taken over by farms and the croft buildings becoming cottages, at the old location or in new, less favourable cottage 'slum' areas into which many erstwhile village crofts were moved.[148] But cottages became more numerous than could be explained in this way, and it is also clear that the increasing cottager population was not viewed unfavourably by peasants. The cottages often contained labour which could be hired on a day to day basis, they did not mean large obligations and they did not have to be allocated precious farmland like the crofts. Thus, on large farms where the aims were market-orientated, crofts were disposed of while cottages were kept.

A number of reasons can be suggested why the smaller farms kept proportionally more of their crofts. To begin with, smaller farms could get less

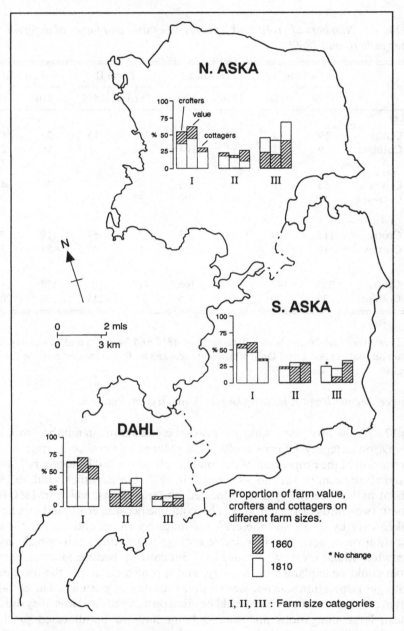

Figure 4.2. Farm values, crofts and cottages per farm size category in the subregions of Aska and Dahl in 1810 and 1860.

favourably located land in the *laga skifte* enclosure, more marginal in relation to the old village, where crofts and outlying land had traditionally been situated. Hence, many crofts could be transferred to moved small farms on the village periphery. Secondly, smaller farms might have had other, less market-orientated aims in their farming. As traditional family holdings, their farmers were perhaps less keen to use every marginal patch in production, and blood relationships with the crofters might have meant more to them than to large farmers. Whatever the exact reason, the survival of crofts on small farms and the growth of cottages on large ones, especially when taken together with the disproportionate increase in the number of married hands on large farms, shows clearly that the wage nexus was penetrating the economy of large farms much further than that of small ones.

Herrestad Storgård, belonging to Lars Larsson and then to his son Per Johan, can be used to illustrate how one of these larger farms developed its labour force. In 1860, it comprised two plots in Herrestad village and one in Väversunda, in total worth 7,000 Rdr. One of the farm's Herrestad plots continued to support a grenadier's croft, together with a cottage containing a destitute family. On the other Herrestad plot, which was larger and held the farmstead, there was another grenadier's croft. On the Väversunda plot, no smallholdings survived at all. The labour force of Storgården consisted of full-time employees, comprising one single and three married hands at the Herrestad farmstead and a single and a married hand on the Väversunda plot. The latter seems to have operated quite independently of the home farm in Herrestad during most of the year. The composition of Larsson's labour force was quite typical of the larger farms in the Dahl subregion in 1860, consisting mainly of full-time employees. The labour force working on the same plots in 1810 (all of which did not then belong to the Larsson farm) contained fewer hands, no married hands, and a couple of crofts in addition to those of the soldiers.

Due to its location on the most fertile part of the plains where little marginal land remained by 1810, Herrestad parish had always had fewer crofts than other parts of Dahl, where changes during our period were more dramatic. Whereas Herrestad parish had only seven crofts in total in 1810, of which one remained in 1860, Väversunda, which was a smaller parish but contained a large area of marginal land next to Mount Omberg, had twenty-four crofts in 1810. Ten were left in 1860, most having been moved to less favourable locations in the *laga skifte* enclosure of 1840.[149] Herrestad's twelve cottages of 1810 increased only moderately to sixteen in 1860, so that its total of nineteen crofts and cottages had shrunk to seventeen in 1860. Crofters were not replaced by seasonal wage labour from cottages, but by full-time hands and maids.

The peasant farms of Dahl thus ranged from family holdings where there was still room for crofts, to large multi-plot farms with full-time labour forces of hands, where all land was utilized by the farms themselves and there was only space left for cottages which could provide seasonal labour as well as a small rent. The number of married hands grew throughout our period, especially quickly from 1850. As the number of single hands did not go down over the period, and the number of cottages, in which we can assume that seasonal labour lived, increased rapidly from 1850, it is probable that the total labour force in peasant agriculture and in particular on the large farms, grew markedly. It was almost entirely on the large farms where the married hands were employed in 1860. These large farms instead had a lower number of single hands than the proportion of their total tended value would have indicated. The same was true of maids; even though their total number on large farms grew, it did so by less than that of the proportion of tended value of large farms. It is simpler to organize a trim labour force on a large unit where three hands are actually needed, than on a small one where perhaps only half a hand is of use. Even so, the larger farms in Dahl were obviously replacing their single hands with married hands who not only provided one man's labour but also a certain proportion of female work. As in the case of the cottagers and crofters, the differences between parishes were significant, reflecting the distribution of large farms. Thus, Herrestad parish which had a number of substantial farms in 1810 had forty single hands distributed evenly over the farm size categories. In 1860 a few of the large farms employed married hands, in total ten. The number of single hands had gone down slightly to thirty-two, of which eighteen were employed by large farms. A randomly selected category I farmer in Herrestad in 1810 with a farm worth 5,000 Rdr had two crofters, one cottager, three single hands and three maids in his employment. His counterpart in 1860 had one cottager, three married hands as well as one single hand, three maids and five people living-in on the farm.

The changes among the smaller category III farms were less significant. It was not uncommon to find farms of this category which had no employed labour at all in 1810 or in 1860. In Örberga, where there were twenty-eight category III farms in 1810, only one in four employed a hand and equally few a maid, while one in seven had a croft on its land. In 1860, the fifteen category III farms that remained had seven hands as well as seven maids, four of the hands were married. One farm in three had a cottage on its property. Even among what might be thought of as 'family' farms, therefore, intensification and change in the labour force occurred. It seems that changes in Dahl were similar in kind among all farms, but that they were more drastic on larger farms.

In the other two subregions, the change in structure of livelihood positions on peasant farms came later, in parallel with the development of compound farms. In southern Aska where category I farms controlled 46 per cent of the property values in 1810 and 61 per cent in 1860, there was a growth in the number of crofts up to 1830 and their decrease started only in the 1840s. Over the entire period, the croft number shrank only by 10 per cent. In northern

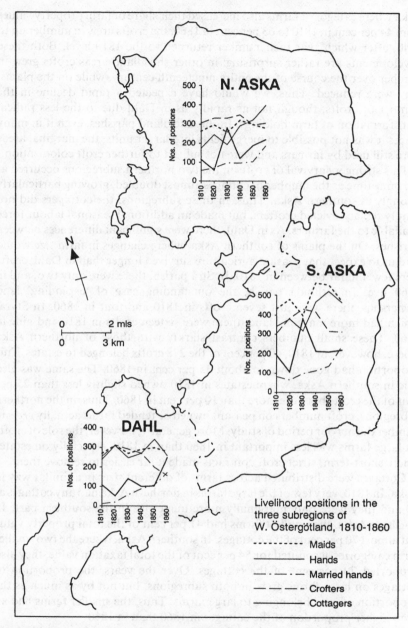

N. ASKA

Nos. of positions

500
400
300
200
100
0

1810 1820 1830 1840 1850 1860

S. ASKA

Nos. of positions

500
400
300
200
100
0

1810 1820 1830 1840 1850 1860

DAHL

Nos. of positions

400
300
200
100
0

1810 1820 1830 1840 1850 1860

Livelihood positions in the
three subregions of
W. Östergötland, 1810–1860

- - - - - - - Maids
───────── Hands
─·─·─·─ Married hands
─ ─ ─ ─ Crofters
───────── Cottagers

Figure 4.3. The changing structure of livelihood positions in the subregions of Aska and Dahl, 1810–60.

Aska where category I farms also increased their share of farm property values from 44 per cent in 1810 to 63 per cent in 1860, the crofts grew in number up to 1840, after which year their number returned to the 1810 level. Both these developments are rather surprising: in other shieldland areas crofts grew in number over the course of the entire nineteenth century, while on the plains, they were reduced. Thus, we would have expected a rapid decline in the number of crofts, though not as rapidly as in Dahl due to the less radical transformation of farm holdings. In the shieldland parishes, even if in many places it was not possible to merge fields into larger units, the marginal areas were still used by farmers and were not subject to further croft colonization.

This stronger survival of crofts in the two northern subregions occurred at the same time as the number of cottages almost doubled, growing particularly strongly in southern Aska. Thus, in these subregions, the cottagers did not simply replace evicted crofters, but made an additional seasonal labour force available to the farmers. As in Dahl, there were significant differences between parishes. On the plains of southern Aska, where changes in farm size would lead us to expect that peasant agriculture survived longer than in Dahl, crofts were few and far between in 1810. In Strå parish, there were only two, and in 1860 five. In Orlunda parish, the outstanding case of 'exploding' land ownership, there were just seven crofts in 1810 and four in 1860. In Styra, which had more marginal land, there were sixteen crofts in 1810 and nine in 1860. These small numbers contrast starkly with those of northern Aska where, however, in 1810, 40 per cent of the 278 crofts belonged to estates. This proportion had gone down to about 35 per cent in 1860. The same was also true in southern Aska, where estates in 1810 owned slightly less than 25 per cent of the crofts but little more than 10 per cent in 1860. Thus, in the northern subregions, croft numbers on peasant owned or tended farms actually *grew* in number during our period of study. More generally, however, the role of crofts on large farms was less important in 1860 than in 1810, especially on estates where short-term, strict croft contracts made it far easier to remove them.

Cottages were distributed across farms of different sizes in a similar way in Aska. In 1810, very few of the large farms in northern Aska had any cottages at all and they were only marginally more numerous in the southern part. In northern Aska, the smallest farms had 41 per cent of the total property value but almost 70 per cent of the cottages. In southern Aska, where the two smaller farm categories accounted for 54 per cent of the total taxation value, they also supported 70 per cent of the cottages. Over the years, the proportion of cottages on large farms grew in both subregions, but not by as much as the proportion of land belonging to large farms. Thus, the smaller farms had an even higher proportion of the cottages in 1860 than in 1810.

It was clear that in Dahl the large farms replaced crofters with labour recruited from cottages but also that the input of labour was extended to a certain degree through the employment of married farmhands. These and their wives, to some extent replaced the ordinary hands and maids. In Dahl,

Table 4.2. *Numbers of hands and maids employed on farms in each size category in each subregion in 1810 and 1860*[a]

	Estate		Farm I		Farm II		Farm III	
	1810	1860	1810	1860	1810	1860	1810	1860
Dahl								
Hands	32	15	84	139	75	66	34	15
Married hands	5	35	0	79	5	10	0	9
Maids	26	19	91	129	89	76	33	17
S. Aska								
Hands	38	28	90	159	112	104	81	75
Married hands	14	78	8	54	24	18	18	16
Maids	50	43	106	153	128	113	106	77
N. Aska								
Hands	36	18	18	35	12	46	79	82
Married hands	33	90	0	0	3	29	14	35
Maids	54	39	2	31	14	36	100	88
Totals								
Hands	106	61	192	333	199	216	194	172
Married hands	52	203	8	133	32	57	32	60
Maids	130	101	199	313	231	225	239	182

Note:
[a] There were also 54 hands, 32 married hands and 121 maids in 1810 and 345 hands, 160 married hands and 200 maids in 1860 employed on plots in Aska and Dahl which were farmed from outside the region.

Source: Mantalslängder, taxeringslängder. Landsarkivet, Vadstena.

married farmhands were almost exclusively employed on large farms. In the two Aska subregions, where the concentration of plots occurred later and not as strongly, a less radical transformation of the farm production structure could be expected. In 1810, the married farmhands were few in both northern and southern Aska, even though not as few as in Dahl hundred. More than half of the married farmhands in northern Aska were, however, employed on estates (see table 4.2). In southern Aska, the concentration of married hands on large farms was less marked, and smaller farms actually had a slightly higher proportion than their share of the farmland value would suggest. Even though in the north farms remained small, they did not completely conform to the model of family labour supplemented only by neighbours' sons or daughters.

The dominance of large farmers as employers of proletarian labour became much more marked over the period. In southern Aska, 80 per cent of the married farmhands were employed on the large farms, which tended 61 per cent of the farmland, in 1860. The smaller farm categories each had around 10 per cent of the married hands. On the shield, the large farms were less prone to employ married farmhands than on the plains, even though the number grew over the period and the smaller farms employed more than their share of the total farmland would indicate. This was even more notably the case for ordinary hands and maids. When we consider the total hand labour force of the different subregions, it becomes apparent that the northern farms employed *more* people per amount of farmland than those of the plains in either southern Aska or Dahl. This must have been either because the pastoral farming of the shield was more labour-intensive, or because more supplementary employment existed outside farming on the shield than on the plain.

We will illustrate these trends with some examples. Övre Götala hamlet, on the plain in southern Aska, had six farms in 1810, one of which also farmed a plot in nearby Styra. There were no large farms, four were in category II and two in the smallest group. There were five single hands and eight maids, three crofts and six cottages. Labourers, who probably worked on the farms in peak season, occupied two cottages, but the other four housed aged and poor people, three of them destitutes. Few of the peasant households contained more than the families, though one unmarried farmer kept a family of six living-ins. The total village population was fifty-seven people including eleven minors. In 1860, Övre Götala had seven farms, one of which owned a plot in another village. One Götala plot was owned and tended from a neighbouring village. Three farms now belonged to category II and four to category III. The total labour force consisted of just four hands, one of them married, and seven maids. Two farms had living-in families. All crofts had gone and the cottages were still intact in number. Thus, unlike in Herrestad, the small and intermediary farmers in Övre Götala had come to rely more heavily on family or seasonal labour and hands and maids were fewer. The cottages had similar people to those of 1810; two housed destitutes, two aged people, and two day labourers. The total population of the hamlet now was eighty-five people of which thirty-two were minors. On the basis of the livelihood positions they supported, we can regard the farms of Övre Götala as surviving peasant family holdings.

One of the 'exploding' *hemman* of Orlunda and Strå parishes, also in southern Aska, comprised Orlunda *utjordar*.[150] In 1810 they were tended as one farm by a peasant and his family and contained six cottages, two occupied by widows and the other four by apparently productive people. In 1860 Orlunda *utjordar* contained twenty-nine smallholdings, each worth only between 30 and 290 Rdr. Most of them were tended by their owners and can be characterized as very small peasant farms. Their labour consisted entirely of the peasant families. Three had cottages with destitutes living in them. The *utjordar*'s population increased from thirty-five people, of whom ten were minors, to ninety-one, including thirty-six minors, but no productive labour at all.

In our examples, the existing crofts disappeared while the cottages remained virtually the same. This differs from the subregional statistics. Obviously, in other parishes in southern Aska, croft numbers must have increased slightly. Fivelstad parish was more typical of southern Aska in general. It had only six crofts and thirteen cottages in 1810; all the crofts and half of the cottages were on small farms, which accounted for less than one third of the parish land taxation value. In 1860 there were twenty crofts and forty-five cottages, thirteen of the former and twenty-nine of the latter on small farms. Clearly, some areas still had margins to be utilized by smallholdings, and farmers who still preferred to increase the number of rent-paying crofters rather than incorporate croft land into their own farms.

There were many crofts and cottages in Björka/Björketorp in northern Aska in 1810. All the seven farms were in size category III; all had crofts and cottages, but in total they employed only four hands and five maids. The seven farms, fourteen crofts, seven cottages and grenadier's croft of the hamlet existed on land worth in total only 6,000 Rdr, less than many single farms on the plain. Some of the crofts were sold off to become independent farms at enclosure, and farms were subdivided so that their number – all of them still small – had more than doubled by 1860. The number of crofts was the same and the number of cottages shrank by only one, whilst the number of single hands and maids rose to nine and seven, respectively. Although the farms had been subdivided and multiplied, they must have reclaimed arable land, thus maintaining the productive area per farm. The large increase in total population from 102 to 235 on farms which still had the character of traditional peasant units, with the family as the major source of labour supplemented by a hand, maid and croft, together with the heavy subdivision of holdings, might imply that large areas of forest grazing or opportunities for secondary production were available in this shieldland hamlet.

In conclusion, larger peasant farms, especially on the plains, tended to get rid of their crofts and to some extent their single hands and maids, replacing them with more fully proletarianized cottagers and married hands living on their farms. The smaller farms in categories II and III on the plains remained fairly similar over the period, basically as peasant family farms with a maid or a hand employed on the larger ones. In between, there were many variations: some intermediate farms stuck to their crofts and to family labour, while others turned in the same direction as the larger farms. In northern Aska on the shield, where estates still dominated in 1860 and small peasant farms were much more numerous than on the plains, changes were less distinct. However, the process here was superficially no different in kind: cottages grew in number, crofts declined after increasing until 1840 and married hands, unlike traditional unmarried hands and maids, became more common, although less so than on the plain. However, workforces in the north tended to be larger per unit of land taxation, either because farms in the forest had a stronger emphasis on animal husbandry, and hence a greater need for labour employed all year, or because of an increasing reliance on part-time production projects outside agriculture.

On estates and farms of person of standing

Two kinds of estates were defined in chapter 3: traditional manorial *säteri*, and holdings of 25,000 Rdr or more, which were normally larger. Land worth 334,000 Rdr, which was 57 per cent of the total belonging to estates, was tended as manor farms, the rest by tenant farms or as parts of other estates. Estates defined in both ways could belong to the new kind of non-noble owner, a person of standing who was less prone to stick to particular plots and obviously considered land as income-generating capital. Because it is possible that the traditional noble estate owners behaved differently, we must examine the development of the workforce on both kinds of estate. Estates on shieldland might also have developed along different lines from those on the plains because of the different resources at their disposal.

In 1810 few estates were fully fledged versions of either the *Gutsherrschaft* or *Grundherrschaft* models. Most employed a mixture of tenant farmers, crofters and hands and the differences between noble and non-noble estates were negligible in this respect. Bona in Västra Ny employed one crofter and eight hands on a manor farm valued at only 2,250 Rdr. This workforce and labour from fourteen tenant farmers could easily tend the manor farm, especially as the tenant farms supported nineteen crofts. On Kungs-Starby, by Vadstena, on the other hand, there were no tenant farmers at all, and thirty-seven crofters and fifteen hands tended the largest manor farm in the region, worth about 50,000 Rdr in 1810. Estate agriculture in our region seems not to have been particularly advanced along capitalist lines by 1810: proletarians comprised little more than half of the total labour force. The noble estates, in particular the larger ones, had more proletarian labour than those of persons of standing. Only two non-noble estates had any married hands at all, and of the noble estates only Bona, Medevi and Karlsby had more than five employed married hands each. All three were on the shield in northern Aska. Northern Aska estates generally had more than one crofter per 1,000 Rdr value of land, whereas in southern Aska they had about one third and Dahl about half of a crofter for the same value of land. The number of hands per 1,000 Rdr of land (single as well as married) was twice as large in northern as in southern Aska or Dahl. The generous amounts of meadow, grazing and forest on the shieldland might explain differences, irrespective of whether estates were noble or not, because animal husbandry needed more labour all through the year, as well as more women for milking and attending to the animals. The physical impossibility of creating large manor farms on the shield must be why estates on the shield kept, and even increased the number of, their crofts. On the plains, large manor farms could be formed, and seasonal labour could much more easily be combined with a nucleus of permanent workers in large-scale production. The most significant difference, however, was that on the shield estate activities were not restricted to agriculture. Their mills, sawmills and ironworks needed provisions, labour and haulage from the ordinary workforce of the estate.

Generally in Sweden crofts and cottages were kept under strict control on estates, and only permitted to be established if more labour was actually needed, but it seems that less strict control was kept of cottages on estates in our region. Estates supported almost forty in 1810, mostly in northern Aska, mainly housing aged widows or male destitutes. This is another indication that more part-time secondary activities occurred there. In general, more estate land was tended directly as part of manor farms in 1860 than in 1810, more so in southern Aska and Dahl. Whilst this was occurring, the nobility were declining markedly in importance. Whereas in 1810 the region contained a dozen noble estates, only five were owned and tended by noblemen in 1860. All this meant that the value of land under noble estate agriculture was reduced from 390,000 to 220,000 Rdr, while the land of non-noble estates increased in value from from 184,000 to 320,000 Rdr. In 1810, the manor farms of noble estates were worth 244,000 Rdr, in 1860 130,000 Rdr, while non-noble manor farms increased in value from 105,000 to 266,000 Rdr over the same period. Total manor farm agriculture increased by 15 per cent. What did this entail for livelihood positions on estates?

The non-noble estates in general had fewer tenant farms than traditional estates like Ulfåsa and Bona, which still had substantial tenant farmer demesnes. In other respects, the differences between structures of livelihood positions on noble and non-noble estates were small. In proportion, noble estates had a few more crofters and married hands, and the non-noble estates had a few more single hands and maids than one would expect. This probably reflects the dominance among the person of standing estates of manor farms without any outlying properties. The number of tenant farms on estates went down between 1810 and 1860, while in parallel the number of estate crofters was reduced from 229 to 142. The replacement of estate owners and managers in daily agricultural activities by foremen also indicates more capitalist organization, with proletarian labour under the control of employed supervisors. The married hand group quadrupled to 203 on estates while ordinary single hands and maids were reduced by between 30 per cent and 40 per cent. The 288 hands, maids and *statare* of 1810 were replaced by 365, a majority of whom were *statare*, in 1860. This meant an increase in the total labour force on the manors despite the reduction of crofters.

The cottager population of estates grew, as did the living-in group on manor farms. Part of the reason for this increase in the less productive groups is that estate owners permitted aged crofters and former tenants to stay on in a cottage or a dwelling alongside the manorhouse instead of evicting them. However, whereas in 1810 virtually every cottager was an aged widow or a destitute, in 1860 a cottage frequently contained an able-bodied man and wife of productive age, sometimes with other productive people living-in. Thus, not only did the cottages become more numerous, they also came to contain more labour capable of seasonal farm work.

In the south, development on estates was radical. In Dahl, estate tenant farms did not decline in number, probably because they were already down to a low figure in 1810. However, crofters diminished from fifty-nine to seventeen and the number of single hands was halved from thirty-two to fifteen. The increase of married hands from five to thirty-five could not compensate for the reduction in the number of crofters, hands and maids. The total value of land tended as manor farms was reduced in Dahl by almost 25 per cent, compared with an increase of 15 per cent in the region overall. This is part of the reason for the reduction of labour on Dahl estates, plus the fact that seasonal harvest labour could be hired from cottages on peasant land. In 1810, Kungs-Starby royal estate on the plain was tended by the nobleman major Vult von Steijern, with the assistance of thirty-seven crofters, eleven hands, four married hands and eight maids. In 1860, when Kungs-Starby was part of the Lallerstedt estate, the number of crofts had been reduced to thirteen. Instead, most of the work was done by thirty married hands, assisted by twelve hands and ten maids. It had obviously become more profitable for estate owners on the plains to employ married hands, supplemented by seasonal labour at peak times. Labour had become more easily available and it was no longer necessary to provide crofts and land to tie them to the estate, and new agricultural methods with large areas of arable or larger fields had been introduced. The upshot was much fuller proletarianization of labour on estates on the plains.

In the northern part of our study area, where the crofts mostly lay on isolated patches of till and clay in the forest, it was unprofitable to switch completely to a married hands system. Again, the majority of estate cottages were in northern Aska where, in general, the reduction of the traditional labour force had been far less radical than on the plains. The number of tenant farms on estates in northern Aska fell by only one, while crofts were reduced by little more than 10 per cent and the increase of married hands was much less than in the south. This was in part because of the physical restrictions on transforming estate agriculture on the shield, but also because whatever else besides farming went on estates in the north consumed more labour than the agriculture in the south. Bona estate in Västra Ny parish was owned by the Grill family all through our period. It gradually extended manor farm agriculture, from two plots and 2,250 Rdr worth of land in 1810 to six plots and 13,610 Rdr in 1860, and its crofts from ten in 1810 to sixteen in 1860. The number of married hands increased too, from seven to seventeen, while unmarried hands and maids provided only a marginal reinforcement. This combination of crofters and married hands was suitable for an estate like Bona, where arable land was in widely scattered patches and animal husbandry in the forest must have been important. In spring and summer, when fields needed attention, the crofters' corvée duties could be used for ploughing or harvesting while in the winter, when the married hands could manage on their own attending to the cattle, crofter labour could be used in non-agricultural production like charcoal burning or transport.

In conclusion, the estates of the area transformed their labour force quite rapidly in the nineteenth century, usually substantially faster than the peasant farms. Generally, they had fewer cottages than peasant farms in proportion to their taxation value throughout the period. In 1810, few estates had any proletarian labour but relied heavily on crofters and tenants for running the manor farms, which were often quite small. In 1860, however, manor farms were larger, and most estates were mainly worked by proletarian married hands. It is difficult to distinguish significant differences between the traditional noble estates and those which were owned and tended by non-nobles. There were, however, significant differences between estates on the plains and those on the shield, which to a larger extent kept their crofters.

Labour, wages and poverty

In general the waged labour force grew, while some peasant freehold farms expanded and tenant farmers on estates and crofters on estates and peasant farms shrank in number. Thus, if crofters, former tenants and some of the previous freeholders wanted to stay in the region they had to become hands or cottagers. Unlike married hands, cottagers did not live and work full-time on a particular farm or estate. They could work for different farmers at different times, or be occupied in handicrafts, haulage, forestry or milling. Thus, we cannot be sure that the growing number of cottagers meant an increased supply of agricultural day-labour or increased production in the 'hidden' economy. In consequence, a transfer of a crofter or farmer to this group might reflect a change from the primary to the secondary sector as well as a change from a landed to a landless position.

This makes it difficult to analyse labour supply and demand on farms in our region. Without doubt, the demand for agricultural labour grew. Without doubt, too, the supply in the form of erstwhile crofters and tenant farmers grew. To these agricultural and social structural changes we have to add in-migration and natural population growth. The rapid population growth in Sweden as a whole during the eighteenth and nineteenth centuries must, in

Table 4.3. *Adult wages (in Rdr) in 1833 and 1858. (The 1833 figures are estimates made in 1858; the 1858 figures have been deflated by two-thirds.)*

	Östergötland		Aska		Dahl[a]	
	1833	1858	1833	1858	1833	1858
Farmhand	167	250	165	275	150	250
Winter day rate	0.42	0.67	0.44	0.80	0.41	0.67
Summer day rate	0.67	1.00				
Maid	120	175	120	175	100	166
Factory and craft worker[b]	60–70	156–166				

Notes:
[a] Average of figures for Dahl and neighbouring Lysing hundred.
[b] Probably excluding payments in kind, which would raise the totals slightly above those for farm work.
Source: 1858 finanskommittés handlingar, Riksarkivet, Stockholm.

principle, have meant that the rural labour supply grew because population growth occurred mainly in the countryside. Labour should therefore have become relatively cheaper over time. However, with a complicated changing labour market, and the fact that the population did not grow much at all in some areas, especially in Eastern Sweden, we cannot be certain of such a supply situation. Even within our limited study region, there were subregions which were quite different from each other. It might have become easier for farms to hire labour over time due to an extended supply; but it could equally well have become more expensive, either because of competition from other sectors of the economy, or because the available labour force did not grow as quickly as labour demand.

In 1858 the annual wages for a hand in Östergötland, including payment in kind, was 250 Rdr (deflated for devaluation) or about the taxation value of a smallholding. For Dahl hundred, hands' wages were the same as the county average, while they were 10 per cent higher in Aska,[151] as were male casual day labour wages. Day wages for female labour were the same in the two parts of our region (see table 4.3). These figures seem to show that the general demand for male wage labour was higher in Aska than in Dahl in the 1850s. In northern Aska about twice as much labour was employed per unit of taxable estate land. Different forms of agriculture or other production were obviously important determinants of farm wages; otherwise the shieldland farms could not have employed such large labour forces.

Wages for hands in Östergötland rose by 40 per cent between the early 1830s and 1858 and by a third for day-labour, suggesting an increase in relative demand for the former group, most probably a reflection of the demand for

Table 4.4. *The proportion of the adult population who were destitute in the subregions of Aska and Dahl, 1810–60*

	1810	1820	1830	1840	1850	1860
Dahl	10.8	10.7	11.3	7.4	4.4	9.8
Southern Aska	14.7	12.9	14.7	10.8	13.0	11.2
Northern Aska	9.5	19.5	7.6	6.2	9.2	6.0
Regional total	12.2	14.3	11.6	11.5	9.7	9.3

Source: Mantalslängder, Landsarkivet, Vadstena.

married hands. These were an attractive prospect for farmers, who paid only one-third of a maid's wages extra for the hand *and* his wife.[152] The rise in wages for annually employed agricultural workers might also indicate an increase in demand for full-time labour from other sectors. The wage trend in Dahl was typical of the county while payment for day labour in Aska almost doubled over the twenty-five years to 1858.

For maids, daily wages more than doubled in Dahl and again reflected the difference in demand between the two hundreds. It was more expensive to employ casual labour than single hands if the wages of the former were multiplied by 300 – the normal working year in terms of days laboured – and married hands and day-labourers had higher wages than single hands. Thus, the way in which agriculture changed was expressed in a significant way in wage rates. During the enclosure period in Dahl, the 1830s and 1840s, the demand for day labour was even higher than later when the projects taking place on farms were limited to agriculture and not to house-moving, ditch-digging and arable land reclamation.

Did this restructuring of rural livelihoods and the way they were remunerated mean an increase in poverty? It is usually difficult to get reliable measures of the development of poverty in Sweden.[153] People who were recognized by local officials as needing support from the community were labelled as destitute in the taxation registers, but economic historians have used a wider definition, including everyone exempt from all personal taxes.[154] According to this statistic, poverty levels of about 20 to 23 per cent have been found in Östergötland during the 1820s, when the Lake Mälaren basin had lower levels. However, this method yields totals which are 5 to 12 per cent higher than the numbers actually labelled as destitute in the taxation registers.[155] We have used only the latter, because we consider it more important to discover whether poverty grew or was reduced according to a consistent criterion, rather than to compare the level of poverty in our regions with others. Table 4.4 shows that it did *not* increase. The highest level of poverty was 19.5 per cent in the shieldland of northern Aska in 1820, although the figure for that subregion in 1860 was only 6 per cent. Dahl, where proletarianization was proceeding fastest, had the lowest level, oscillating between 11.3 per cent in 1830 and 4.4 per cent in 1850, while southern Aska, where many characteristics of peasantry survived, had a higher but more stable level of poverty, ranging between 10 per cent and 15 per cent. Northern Aska had large proportions of poor in only two of the years studied: in 1820 with 19.5 per cent and in 1840 with 15.5 per cent. In between, its figures were similar to those for Dahl.

It is curious that the subregion with the largest changes in the composition of livelihood positions had the lowest level of poverty. Moreover, areas like Dahl with large numbers of cottagers and day-labourers were those in which poverty grew most in other parts of Sweden in bad years.[156] At the same time, however, we might assume that the labour force would be under closer control by the farmers in Dahl (who also paid lower wages) than anywhere else. Certainly, the cottagers and living-ins of that subregion were generally of a more productive age group, and therefore less prone to destitution, than in the other parts, where they still tended mainly to consist of the aged and infirm. A comparison of the average ages of the destitute in the subregions would probably show that destitutes in Dahl were older than elsewhere, because labour was demanded from people in active ages if they were to be allowed to stay in dwellings on the land of capitalist farmers.

Demographic development

Complex demographic analysis is outside the aims of the study, and we have simply calculated how families in different environments and social groups changed in size. Family size is defined as what was recorded in taxation registers of particular years, rather than completed family size. This will tell us enough for our purposes about how changes in farm production structures were related to demographic change.

There were more small farms and crofts in the northern shieldland, and fewer small and more large farms and more married hands in the southern part of the plain. Over time, the difference grew more marked. Table 4.5 shows that this was also true of the size of farmers' families: on average they contained more children in northern Aska and fewer in Dahl. The number of children per family under eighteen years of age grew everywhere, but by most in the north. On average, the farmers of our region had more children in the mid-nineteenth century than those of Dala in Västergötland, while fifty years earlier they had had the same number. Not surprisingly, Dala, which resembled plainland Dahl, was most dissimilar from northern Aska. It is also clear that larger farmers had larger families in all our subregions towards the end of the period. In 1810, this had not been the case in northern Aska, while in the south there was least change in this respect. According to studies of other Swedish farmer populations from the eighteenth and nineteenth centuries, we have no reason

Table 4.5. *Average numbers of children per family in rural social groups in Aska and Dahl in 1810 and 1860*

	Dahl			S. Aska			N. Aska		
	N	Ch1[a]	Ch2[b]	N	Ch1	Ch2	N	Ch1	Ch2
1810									
Farmer I	48	0.48	1.52	70	0.57	1.58	9	0.33	1.33
Farmer II	90	0.40	1.61	134	0.28	1.78	14	0.78	1.71
Farmer III	82	0.18	1.50	222	0.26	1.32	140	0.49	1.68
Crofter	206	0.04	1.51	217	0.09	1.34	278	0.19	1.62
Soldier	50	0.00	0.92	89	0.04	0.95	33	0.00	1.27
Cottager	109	0.09	0.87	193	0.07	0.55	117	0.09	0.42
Living-in	93	0.08	0.35	211	0.11	0.51	105	0.00	0.10
Married hand	10	0.00	0.00	92	0.07	1.12	58	0.05	1.32
Maid	271	0.00	0.01	433	0.00	0.01	253	0.00	0.04
1860									
Farmer I	48	0.35	2.89	63	0.51	2.81	21	0.33	3.43
Farmer II	54	0.52	2.13	111	0.52	2.55	30	0.43	2.50
Farmer III	61	0.26	2.00	233	0.36	1.85	121	0.45	2.30
Crofter	88	0.16	1.97	206	0.21	1.74	274	0.16	2.11
Soldier	68	0.12	2.28	107	0.09	1.90	32	0.06	2.72
Cottager	248	0.17	0.96	399	0.14	0.61	220	0.17	0.94
Living-in	142	0.18	1.28	449	0.25	1.05	367	0.23	1.24
Married hand	168	0.02	1.97	241	0.02	1.99	186	0.08	1.65
Maid	281	0.00	0.02	472	0.00	0.00	280	0.00	0.02

Notes:
[a] Children of eighteen years and over living with parents.
[b] Children under the age of eighteen living with parents.
Source: Mantalslängder, Landsarkivet, Vadstena.

to believe that family formation was out of control of those involved:[157] if families grew bigger, this was deliberate.

The reason could range from a need for more family labour on an expanded farm to the ability to support more children because of increased production. If a farmer who had inherited or built up a large holding chose to have more children than average, this can be seen as a reflection of the growing surplus on expanding large farms as they took over the land of small ones. On the other hand, if a small farmer on the shield had more children, in spite of a constant arable area, that could either mean that additional family labour was preferable to expensive employees, or that the farm had become involved in non-agrarian production. The increase occurred even among peasants with shrinking farms, who according to the principles of peasantry ought to have reduced their family size. In fact, there was a general increase in the number of children over the period in *all* households in our area, even among the landless. This perhaps reflected a general national trend at the time resulting from 'peace, inoculation and potatoes'.

Such general tendencies cannot, however, explain deviations from a common rate of growth; the stronger increases in certain of our social groups. Thus, farmers in category I must have had reason to let their families grow so that by 1860 they were more than double the size of those in 1810 in northern Aska, and almost double the size in southern Aska and Dahl. We have no reason to believe that the need for family labour was larger on the most capitalistic farms which employed more labour than the others. Instead, it might have been because the traditional criteria of family formation were no longer relevant, despite the continuance of partible inheritance, and that other considerations had become more important. Certainly, all the children of the larger farmers could not be provided with large holdings of their own on their parents' death: the flocks of children must soon have outnumbered the maximum number of large farms the region could accommodate. Thus, the large numbers of children on larger farms must have represented a broken connection between property, production and family size. This in turn meant that other sectors of the economy had been opened to capital accumulated in agriculture.

In 1810, other social groups had as many children as the farmers, most of whom, irrespective of wealth, had about 1.5 children at home. Up to 1860 crofters increased their families by about the same proportion as the smaller farmers, that is, by 0.5 children per household. For crofters, as well as peasants, it was logical to have a certain number of children to help out with agriculture instead of hiring labour, particularly those on estates with severe corvée duties. This could also explain the admittedly small differences between subregions. In the north estate crofters were more common while in the south, crofts were more normally situated on peasant land. Thus, in Ekebyborna parish, in southern Aska, with many estates, the average number of crofter children was 2.1 in 1860, compared with an overall average family size of 1.7 in the subregion. The size of crofters' families was growing whilst the area of crofts was shrinking, so that large numbers of children would have to be fed on incomes from farm work outside the croft, or from home production for the market.[158]

The growth of married hands' family size was no smaller than in other social groups, in total about 0.5 per family in southern Aska and in Dahl. Finally, cottagers had less than one child per family on average. The number was slightly higher in northern Aska and in Dahl than in southern Aska. The married hands and cottagers comprised the major part of the proletarian workforce in agriculture. These were the people with only labour to sell, who might be expected to have large numbers of children to sustain the family economy. We can see some signs of this. Although hands and cottagers had fewer children than all other groups averaged out over our whole study period, by the end of it hands on the plains had families almost as large as those of crofters and small farmers. The cottagers seemed to become a more productive group during the early nineteenth century. Many cottages came to house day labour rather than the aged and infirm, hence the increase of children amongst cottagers.

The transformation of agriculture brought about changes in the proportions of the different livelihood positions in the countryside, as well as in their numbers and the resources available to them. Demographic behaviour seems to have mirrored these changes. However, this was not invariably the case. For example, we might expect small families amongst crofters in Dahl, whose number was reduced by almost two-thirds after 1830 and whose positions were deteriorating as the enclosure process reduced the size of many crofts. In northern Aska the number of crofts was stable over the study period; the productive areas of crofts were larger because they were frequently estate crofts; there was potential for a certain amount of arable reclamation, and off-farm work was available. However, crofters in Dahl had almost the same number of children in 1860 as those in northern Aska. It is clear that we cannot explain differences or similarities in family size between social groups or over time only from considering the development of their livelihood positions.

To summarize, in Aska and Dahl farmers' families were larger than those of the landless, although this situation only developed during the nineteenth century. As elsewhere in Sweden, proletarian groups had smaller numbers of children, but by 1860 the families of married hands had grown almost to equal those of crofters and small farmers on the plains. The other change which stands out is that large farmers bred more children than any other group by 1860, most markedly on the shield. It seems clear that both these changes were connected with the transformation of agriculture, the second a response to increasing wealth, the first to fuller proletarianization. Other findings are less easy to explain. We might have expected some social groups to have had fewer children by 1860 in consequence of deteriorating living conditions, but this did not occur. We might also have expected cottagers to have had larger families by 1860 than they actually did.

It is not possible to see where the growing groups were recruited from by examining family sizes. To do that, both the degree of reproduction and the levels of migration must be known. Work on other parts of Sweden suggests that both farmers and crofters bred 'surplus' populations: that there were not sufficient positions within their own social groups to provide a living for their offspring. Thus, if the region had been a closed purely agrarian one, many children of farmers and crofters would have been proletarianized into the landless groups. However, the region was closed neither to out-migration nor to movement into other sectors of the economy. This is indicated by the differing rates of population growth in the subregions. Table 4.6 shows that northern Aska's rural population grew by 77 per cent between 1810 and 1860,

Table 4.6. *Populations of the rural parishes and subregions of Aska and Dahl in 1810, 1830 and 1860*

	1810	1830	1860
Nässja	201	283	256
Örberga	494	653	909
St Per (part)	256	232	251
Strå (part)	322	369	429
Rogslösa	640	922	1,102
Herrestad	324	347	436
Hov (part)	71	60	102
Källstad	249	266	300
Väversunda	281	326	374
Dahl total	2,838	3,458	4,159
Vinnerstad	464	704	916
Ekebyborna (part)	267	400	500
Ask	200	246	292
Västra Stenby	779	994	1,172
Hagebyhöga	568	662	713
Varv	463	597	671
Fivelstad	430	634	732
Styra	333	356	429
St Per (part)	231	232	336
Orlunda	317	439	589
Allhelgona (part)	396	527	499
Strå (part)	114	194	228
Hov (part)	102	130	140
Bjälbo (part)	43	56	104
Southern Aska total	4,707	6,171	7,321
Västra Ny	1,228	1,510	1,989
Motala	1,420	2,553	2,822
Kristberg (part)	388	579	553
Northern Aska total	3,036	4,642	5,364
Grand total	10,581	14,271	16,844

Source: *Mantalslängder*, Landsarkivet, Vadstena.

while that of Dahl only grew by 46 per cent. Southern Aska also experienced moderate growth. Even if most of the family size averages were larger in northern Aska by 1860, they were not so in 1810 and the differences in family sizes give no reason to expect that the actual growth should be nearly double that of Dahl. Instead, the differing population growth rates suggest that there was considerable migration within our region, or into and out of it.

Naturally, due to the changes in the composition of the livelihood positions, the composition of the total population into social groups also changed. Additional changes due to different rates of reproduction in the groups were hardly more than marginal relative to, for example, the number of hands in agriculture more than doubling in northern Aska, or the number of cottages doubling in southern Aska. Even if family size was larger among the farmers than among the hands or cottagers, the fact that their positions were reduced in number meant that their proportion of the total population shrank. Dahl reduced its total farmer population by one hundred people, or 10 per cent of the total population, in spite of an average number of children exceeding two per farm. Combined with the differences in total population growth among the subregions, our observations on the analysis of family sizes become even clearer; a large transfer of people occurred, either in the form of proletarian-ization from the farmer group to the landless, or to other sectors of the economy, or to other regions. In the last case, a net inflow of proletarians must also have taken place in order to build the large cadres of cottagers, hands and maids which dominated the population in 1860. A flow of population towards the shieldland within the region was also possible, but if it occurred it had to be combined with proletarianization. These surmises are all logical, but only an analysis of the *flows* of people through the positions of the area can explain how the changes in the agricultural sector affected the lives of the inhabitants of our study area. This is the purpose of chapter 7.

Conclusion

From a superficial point of view, the most striking features of agrarian western Östergötland during the first sixty years of the nineteenth century were that population grew by 80 per cent whilst the number of farms and estates fell. The number of landless people was growing most rapidly, even if some of the population growth was produced by larger farmers' families. At the same time, changes were occurring in the structure of the landless population as it became more proletarianized. These changes were more marked in the south where the large farms were growing more rapidly and a seemingly capitalist agriculture was most developed. The number of crofters was more than halved. They were replaced by cottagers, who more than doubled in number and by married hands, who increased even more rapidly. On the shield and the northern plains crofts were at most only marginally reduced, but cottages still doubled in number. Day-labourers and married hands also became more numerous. Relative to taxation value, the labour force grew more in the Aska subregions than in Dahl, which might suggest that labour was being used outside agriculture or that agricultural production projects were becoming more labour intensive.

Estates developed along two lines. The traditional *säteri* retained numerous farms located close to the manor as a source of labour, while more recently created estates had large manor farms, independent production units in other parishes and, above all, more crofters and married hands. In general, the labour forces on estates were larger in the north than in the south, as on peasant farms. The trim labour forces of Dahl were paid slightly lower wages, and less frequently ended up in destitution. The growth of agrarian capitalism, which was most intensive in Dahl, did not increase poverty. Towards the end of the period, the married hands had almost as many children as crofters, but there was little sign of attempts to compensate for a lack of resources through increasing household labour by breeding large numbers of children.

How were these changes related to changes in the production projects of the region's farms? How did their productive resources change, not only in terms of taxation values but in the specific terms of amounts of arable and meadow land? Did the innovation of new crops, rotations and farm implements affect output and labour demands? Were any changes restricted to farms of certain types, or were they general? Finally, how important was non-agricultural production in different parts of the region?

Agricultural projects in western Östergötland

Land resources

In order to discover if the expanding farms were innovating new methods as well as employing more proletarian labour, we need to look in detail at agricultural projects on a local level. It is not possible to do this in the same way as with ownership or the structure of livelihood positions on farms, for which there are sources which cover everyone in every farm or estate, down to the smallest child in the poorest cottage on the smallest farm. There are no easily available statistics on agricultural projects. Instead, we must rely on fragmentary sources such as inventories and *ad hoc* surveys relating to individual people or villages, which cannot give us either time series or complete cross-sections.

We have seen that the traditional view of the role of enclosures in Sweden was that they dissolved village communities and open fields, and made massive arable reclamation possible. Recent research has shown that the eighteenth-century (*storskifte*) enclosures in our region had varying effects on the areas of land cultivated. In southern Aska, some reclamation did occur after the *storskifte* had been carried out, but the pace of reclamation did not necessarily quicken.[159] Portions of outlying land still existed to be reclaimed, even on the plain, but where it did not enclosure could have little effect on the cultivated area. In Väversunda parish in Dahl, for example, arable reclamation virtually came to a standstill. The only visible consequence of the first enclosure there was a reorganization of the common fields into wider strips. This is probably why reclamation was concentrated on outlying land in the parish: it was simpler not to involve all villagers in a major reclamation project linked with the old fields. Also in Väversunda, and the parishes bordering it, Lake Tåkern yielded up cultivable land when its level was lowered during the eighteenth and nineteenth centuries. Few of the lakes in Central and Southern Sweden were left untouched in this way. In total more than 1,600 acres of arable was improved or created by lowering lake levels in Aska between the early 1830s and 1858, and 2,400 in Dahl and Lysing hundreds, a large part which came from the lowering of Lake Tåkern.[160] Thus, lakes were not unimportant as sources of new arable land, even if many projects failed or yielded only permanently waterlogged meadows. Because the initial investment was high, those who benefitted were most often large farmers, smaller farmers being limited to reclamation from meadow and grazing land, which could be made progressively with the investment distributed over time.

Small farms and split-up fields with many parcels of land meant that it was difficult for farmers to accomplish drainage projects on their own. After the nineteenth-century *laga skifte* enclosures farmers' land was organized into one or a few parcels and drainage was easier to carry through. Still, it was a costly procedure and thus took place first and most commonly on estates and large farms. In total, between 1833 and 1858, only 308 acres were drained through covered ditches in Aska hundred (1.5 per cent of the total area) while in Lysing and Dahl hundreds, 941 acres were drained (about 5 per cent of the total).[161] At the same time, 17,000 acres were drained in the county of Malmöhus.[162]

A number of measures were taken to improve the fertility of agricultural land. Manuring has already been mentioned, and new fertilizers such as marl and Peruvian guano were introduced during the 1840s and 1850s. Artificial fertilizers like phosphate from new iron-making processes became common only after our period of study. But soil fertility improvements do not only take the form of adding fertilizer to the topsoil. New crops, or old ones used in new rotations, could also improve fertility: peas, clover and vetches all help to raise the nitrogen content of the soil. So could waterlogging the land. Even though this had a long history in Sweden and elsewhere in Western Europe, it was only in the early nineteenth century that it began to be widely used. The principle was simple: flooding the fields during winter raised the temperature of the soil, added fertilizing matter, and watered the land. The method was clearly best adapted to farms comprising one large continuous area. In our region, only 79 acres were adapted for waterlogging in Aska hundred from the early 1830s to

1858, and 55 acres in Lysing and Dahl hundreds. In climatically more favoured and estate-dominated Malmöhus, about 70,000 acres were flooded in this way during the latter part of the nineteenth century.

Only after enclosure can significant differences in agricultural projects between farmers in the larger villages be expected. But estates where the manor farms were not bound into village open fields, could change their production without any consideration of neighbours. Many of the person-of-standing farms and also some peasant holdings were either single settlements or parts of small ones where the organization into open fields can hardly have been inhibiting. In the same way, but on a different scale, the crofters' small patches of land were independent from the agricultural activities of others. There were also regional differences between farms in restrictions on cropping. Only Dahl and parts of southern Aska consisted mainly of open-field villages and, even there person-of-standing single farms and estates were sprinkled about. In the north, most settlements were single and the inhibitions on innovation depended more on such things as the demand for charcoal from ironworks or the restrictiveness of the natural environment.

In Dahl single farms and small hamlets reclaimed arable land earlier and to a larger extent than some of the large villages like Väversunda, where the village community had to compromise.[163] The single farms and small hamlets were to a larger extent in the hands of persons of standing, which might also have been why they experienced earlier and more extensive reclamation. In Väversunda village, virtually no arable land reclamation took place between the late eighteenth century and the *laga skifte* in 1838–41. Even though members of the Sjöstéen-Enander person-of-standing family owned all the larger farms of the village, they were unwilling or unable to extend their arable land before the final enclosure. Until then, the only way to extend agriculture in the large villages was through buying or tending more plots.[164] On the other hand, Kungs-Starby estate expanded its arable from less than 200 acres in 1800 to between 800 and 900 acres in 1870, and the single farm of Östra Kedevad in Strå parish grew from 80 acres arable in 1775 to 250 in 1860.[165] However, whereas Väversunda's arable grew by only a few acres between its enclosures, that of Herrestad village grew during the same period by 29 per cent. Up to 1876, when the first map of the entire hundred was drawn, a further growth in arable of 26 per cent occurred in Herrestad and in total, over a hundred years, the arable area had grown by 62 per cent. Thus, large villages need not always have been hampered by conflict between different categories of landowners, and could develop steadily and quite early, even if like Herrestad they were completely dominated by peasants throughout the period. For a multiple plot owner like Lars Larsson of Herrestad, arable reclamation within the traditional field boundaries meant that his arable, assuming the constant value of his farmland, would increase by the same proportion as that of the village as a whole (29 per cent up to 1853). The enclosures also meant that Lars Larsson had his plots of land concentrated in fewer locations. Even so, after the *laga skifte* of 1853 Lars Larsson still had more than a dozen strips in Herrestad alone, although now they were organized independently of other farms into separate integral fields.

On the plainland of Dahl, where it was possible to plough almost all land, the main limitation on arable reclamation was the need to keep cattle for manure, of which the larger farms of our region were chronically short.[166] The introduction of marl and guano, and new crop rotations with fodder roots and grains meant that arable land could be extended further at the expense of meadow and any surviving forest grazing. The proportion of land under arable varied between villages, hamlets and single farms and estates. On domains where large-scale conversion to arable had been undertaken early, such as Kungs-Starby, the potential for extension was less than in villages such as Herrestad and Väversunda, where little had happened in the first decades of the nineteenth century. The Dahl part of Strå parish had the highest proportion of its total area under arable in the hundred in 1825 at 65 per cent, which had risen to almost 100 per cent by 1875. Strå village's plots were in the hands of a few large farms quite early on, and there were some single farms or small hamlets where land ownership was also highly concentrated. Thus it can be expected that the parishes of Dahl hundred which had a lower degree of owner/tenant concentrations also had a higher degree of late arable reclamation, from the 1840s. Nässja, which had the least concentrated land ownership in Dahl, also had the biggest conversion to arable in our period, the arable proportion rising from 35 per cent in 1800 to 80 per cent in 1860.[167]

The Dahl parishes bordering Lake Tåkern, like Väversunda and Herrestad, were actually physically expanded in area by lowering the level of the lake. Svälinge village in Herrestad gained 150 acres in this way in the 1840s, all of it going to the farmers who had paid for the drainage scheme.[168] All had category I holdings; even though some Svälinge plots were owned by single plot owners, they were by the 1840s tended by people like Anders Larsson in Broby or Anders Persson in Östra Starby, both of whom were among the biggest farmers of the area. The quality of the land by the lake was at first not particularly good, and further investment was necessary to bring it into full arable cultivation.

In our other subregions, the physical restrictions on reclamation were much stronger, especially on the shieldland of northern Aska. Despite the slightly lower fertility of southern Aska, plots were of about the same sizes as in Dahl. In Övre Götala village in Styra parish, most of the arable yielded six times the seedcorn, slightly less than in the Dahl plains, and only 7 per cent of the land gave three times the seedcorn. Here arable reclamation had been significant between the first and second enclosures; the arable area grew by 25 per cent over a century to a total of 247 acres in 1865. The remaining potential for reclamation was 130 acres, leaving only about 5 acres of outlying land. A farm in Övre Götala assessed at 1,800 Rdr in 1860 contained 45 acres of arable and a reclamation potential of more than 20 acres as well as an additional 10 acres of meadow. A hundred years before, such a farm had comprised 36 acres of arable. The situation in southern Aska was not, therefore, entirely different from that in many villages in Dahl hundred; most of what could be put under the plough had been by 1860. There was little room for further reclamation or for many cattle. Already the 1762 *storskifte* enclosure deed from Övre Götala mentioned that 'due to the lack of manure, only one-tenth of the arable can be sown with rye'.[169] The variations in acreage and fertility over the sample villages discussed so far were not very large. From 20 acres per 1,000 Rdr of value in Herrestad to almost 28 acres per 1,000 Rdr in Övre Götala is inside the limits within which the added dimension of fertility makes sense. Bondorlunda, one of the 'exploding' and most heavily subdivided villages, had very low figures of arable per 1,000 Rdr. Farms there had less productive potential per 1,000 Rdr worth of land, and far less per actual farm, than elsewhere.

These examples show that during our period of study the productive arable areas in most villages on the plains grew by at least 25 per cent and in many cases, where much potentially useful land had previously only been used as meadow or grazing, by 50 per cent or more. It seems reasonable to conclude that at the beginning of the period, 1,000 Rdr's worth of farmland in Dahl and southern Aska included between 15 and 20 acres of arable, and at the end between 20 and 30, with correspondingly less meadow and grazing land.

In northern Aska on the shield, grazing and forest were of much greater importance, both in terms of area and as sources of output. In Björka/ Björketorp village there were 91 acres of arable in 1846, representing 16.5 acres for a 1,200 Rdr farm. It was of low quality, yielding only 2.5 to 3 times the seedcorn planted. However, the small, poor arable resources of a 1,200 Rdr farm were supplemented by 70 acres of meadow, presumably of low quality, and 230 acres of forest and grazing. The area around the village was full of crofts, some of which had by the time of the enclosure been given a formal *mantal* and tax assessment as the crofter had bought his croft. These crofts-cum-farms had between 1 and 6 acres of arable, similar to the small farms of Bondorlunda on the plains. But with it came quite large areas of meadow and forest for the shield croft. For example, Jan Carl Jansson's plot of just one-fortieth of a *mantal*, assessed at a mere 120 Rdr in the taxation register for 1860, had 2.25 acres of arable, 12 acres of meadow and 37 acres of forest. Thus, the low formal assessment of such a croft, similar in value to a very small holding on the plains, did not mirror its real economic potential.

There were large differences between the resources available per 1,000 Rdr value of farmland in different parts of the region. In the south, where large villages were common, a lot of highly fertile arable belonged to each farm per unit of taxation value, while there were shortages of meadow, grazing and forest. Further north, the fertility was low and arable acreages small, but forest

Plate 2. A shieldland croft.

grazing was a ubiquitous resource, giving a much higher capacity for animal husbandry. It is curious to see how large differences in the size of farms of the same taxation value were over the region: clearly, an old-fashioned, purely agricultural perspective was being taken by the tax assessors. The large forest areas belonging to northern farms were not reflected in the taxation values. It is only when farmers are looked upon as pure producers of grain on arable fields that the shieldland and intermediary zone farmers automatically appear to have been unfavourably endowed with resources. In the next few sections, we will show that this was not correct even within agriculture. Many of the ways to improve land fertility and reclaim arable depended on the scale of the project. The larger farms and estates could develop along quite different lines from the smaller units where fewer options were open and where capital was scarcer. This means that the profit potential was higher on large farms, and could partly explain the structure of farm size and how it was changing.

Crops

What was grown on farms and smallholdings was not entered in documents such as enclosure deeds or taxation registers. It is not possible to get a complete picture of the crops of a parish or village at a given date before the agricultural censuses of the late nineteenth century. Our description of cropping is based

Table 4.7. *Crops listed in the estate inventories of District Judge Duberg of Lindenäs, Motala (1844) (A), and War Commissioner Lallerstedt of Kungs-Starby, St Per (1851) (B)*

	No. of Barrels	
	A	B
Barley (of various kinds)	23	234
Lentils	24	
Mixed grain, black	106	435
Mixed grain, white	21	
Oats	238	83
Peas	19	121
Potatoes	390	42
Rye (seedcorn)	85	43
Rye (other)	273	
Vetches	12	
Wheat	8	32
Various second rate grain	87	75
Malted mixed grain	8	

Source: Bouppteckningar, Aska och Dahl Landsarkivet, Vadstena.

on a sample of inventories from different parts of the region and different categories of farm. It is not intended to give a complete picture, but simply seeks to discover whether new crops were being introduced. The task is complicated by the fact that not all inventories specify crops or seedcorn, and some farms had obviously sold off their grain when the inventory was taken.

Table 4.7 shows that large estates must have had varied and complicated crop rotations in the 1840s and 1850s, and that there were clear differences between Dahl and northern Aska. Far fewer potatoes were grown on the plainland farms of Kungs-Starby than on those belonging to Duberg further north, and much more barley and mixed grain, while vetches signify the innovation of modern crop rotations on the plain. Even so, traditional crops of the area such as peas and lentils were still grown. Large peasant farms had fewer fancy crops like vetches, and crops were grown in different proportions than on the estates. Thus, Gustaf Lundberg of Jernevid, in Nässja parish in Dahl, who died in February 1820, left behind crops almost completely dominated by barley and mixed grain. Only 20 per cent of his production, according to the inventory, comprised other crops. Olof Andersson of nearby Herrestad Norrgård, who died in January 1818, had also mainly grown barley and mixed grain. His brother Gustaf, who died in 1836, still had a lot of barley,

Table 4.8. *Annual harvest in Östergötland in 1858 (in barrels)*

	Aska		Dahl and Lysing	
Wheat	3,549	3.0%	3,601	2.6%
Rye	15,959	13.0%	17,528	12.7%
Barley	5,732	4.8%	21,824	15.8%
Oats	6,706	5.6%	2,490	1.7%
Mixed grain	28,503	23.7%	46,420	33.6%
Peas	3,820	2.5%	5,575	3.9%
Potatoes	55,356	46.0%	32,509	24.4%
Other crops	1,303	1.0%	8,476	6.2%

Source: 1858 finanskommittés handlingar, Riksarkivet, Stockholm.

but rye and wheat constituted larger proportions than previously. Surprisingly, none of these large farmers in Dahl grew any potatoes. When Anders Samuelsson, a nephew of Olof and Gustaf Andersson, died in Herrestad in 1850, the composition of his crops was still very much like his uncles', despite the completed enclosure in the village. Barley and mixed grain dominated, there were no potatoes, and only 2 barrels of wheat and 9 of rye out of a total of 107. Sven Swärd, of Kasta, Örberga, another large farmer of Dahl had, however, at the time of his death in the 1850s, grown small proportions of potatoes, peas, lentils and swedes.

The composition of crops on the smaller farm seems not to have been similar. Thus, Jonas Larsson, of Kasta, Örberga, a category III farmer who died in January 1822, left behind mainly barley and mixed grain. By the 1850s this had changed. Some 36 barrels of crops remained on a category III farm in Ullevid, Örberga in March. More than half was barley, but some potatoes kept at this time of the year indicates that they had become a subsistence crop of importance. Apart from potatoes, none of the modern crops appeared in the inventories of small farmers of the plains. They were obviously part of complex crop rotations only practised on estates.

Table 4.8 shows that rye dominated cultivation on the shield in 1858, while on the plains of Aska, as in Dahl, farms grew mainly barley and mixed grain. About half the land on small and intermediate farms in northern Aska had been under autumn-sown rye in the 1820s, and we have found no indication at all of potatoes being grown, so that bread must have formed an important part of the diet. In the 1850s, however, large quantities of potatoes were grown by everybody on the shield, replacing both some winter rye and spring crops of barley, oats and mixed grain. A small farmer in Västra Ny who died early in March 1857 left 16 barrels of crops; 10 contained potatoes and 5.5 rye. The wealthier category II farmer Fredrik Kjellman of Kavlebäck in Västra Ny, who died in July 1860, had sown 6.5 barrels of rye, 8 of spring grain (oats and

Table 4.9. *Average crop composition of three inventories of farmers in northern Aska in the 1820s (in barrels)*

Winter rye (seedcorn)	1.5
Winter rye	3.5
Rye from slash and burn fields (*fallråg*)	0.8

Source: *Bouppteckningar, Aska härad*, Landsarkivet, Vadstena.

barley) and 35 of potatoes. As animal herding was important in the north, some of the spring crops were probably fodder.

Crofters and cottagers grew various kinds of grain, peas and lentils in the 1820s. Grenadier Kratz of Örberga, who died in the winter of 1823, left behind a mixture of crops comprising 42 per cent barley, 21 per cent lentils and mixed grain, with peas and rye making up the remainder. For a crofter from the same parish who died thirty-four years later, potatoes accounted for 85 per cent of his total crop. The crops stored at an Orlunda croft in 1859 were more varied: 40 per cent potatoes, 10 per cent peas, 18 per cent barley of two different varieties and 32 per cent mixed grain. On the shield rye was more important as a grain crop than barley or mixed grain among both peasant farmers (see Table 4.9) and smallholders in the 1820s, accounting for about half the arable acreage. In the 1850s potatoes were almost completely dominant. A crofter in Västra Ny, who died after having sown his fields in 1851, had planted 60 per cent of his land under potatoes, 20 per cent under rye and another 20 per cent under flax and mixed grain. A neighbour who died a few years later had grown 95 per cent potatoes and only 5 per cent rye. Thus, potatoes became massively important for crofters on the shield by the end of our period.

To sum up, it appears that the major change in western Östergötland's crops was the introduction of potatoes, which occurred quite late all over the region, probably in the 1830s. Vetches and other fodder crops were of less importance, only grown on larger farms using complicated rotations. The differences between the subregions became more marked over the study period as potatoes became completely dominant on the shield, while the grain harvest remained intact on the plains of both southern Aska and Dahl, except among crofters and smaller farmers, for whom potatoes also became important.

Farm implements

Of the Dahl farm inventories studied for the 1820s, only two out of thirteen contained ploughs, which seem to have been cheap wooden ones, worth less than one *Riksdaler* each. None of the larger farmers of that time ploughed with anything but ards. The situation was similar in Aska, where in the 1820s two out of five farmers had cheap wooden ploughs, both on the shield. Not even persons of standing seem to have had more elaborate ploughs at that time: of the two vicars whose inventories are included in our 1820s sample, only one had a cheap plough, worth 0.5 Rdr. Of the twenty-seven crofters' and cottagers' inventories examined, only that of a Kungs-Starby crofter contained a plough.

By the 1850s the plough had been accepted on larger farms throughout the region. Three of the four category I farmers in the sample had ploughs, two of them modern, all-iron versions. Among the thirteen farms of category II we studied, eleven had ploughs of different kinds, four either specified by brand (for example 'Nävekvarnsplog') or worth more than 10 Rdr, indicating that they were improved ploughs. In the small farm category, nine out of twenty-six farms had ploughs but only one of these was of the expensive, all-iron kind. However, only one of the five persons of standing's inventories examined contained a specified brand of plough, and only two had ploughs of high value. No significant differences between the farmers of the three subregions could be distinguished. Almost all farmers of categories I and II had some kind of plough in the 1850s while less than half of the small farmers had, irrespective of whether they owned or leased their farm or if it was located on the shield or the plains. Of forty-eight crofters' inventories from the 1850s, only two mentioned ploughs and then only of the cheap, wooden kind. As in the 1820s, numerous crofters and soldiers had no ploughing equipment at all. None of the nine Dahl crofters' inventories had ards compared with nineteen out of twenty-four in northern Aska, implying that crofts were larger in the north and that their counterparts in Dahl were gradually becoming more like cottages with the expansion of large-scale farming at the expense of croft land. Over the period, the value of ards went up faster than inflation, indicating that this traditional implement was being improved. It was frequently equipped with iron parts by the 1850s. This reflected growing potato cultivation: ards are more effective than ploughs for earthing up potatoes.

Traditional harrows valued at 2 Rdr or less each dominated in the 1820s, amongst persons of standing and peasants. Of twenty-one farmers' inventories sampled from the 1820s, only five do not mention harrows, all from the plains. As we mentioned earlier, harrows could be replaced by ards in preparing the seedbed at the expense of greater labour input. Only seven of twenty-six crofters had harrows, most of them estate crofters with larger holdings. No harrow was described in such a way as to indicate that it was a modern, deeply cutting one. The situation was no different in the 1850s. Of the forty-four farmers' inventories examined for that decade, only five mentioned deep-cutting harrows. Most farms had traditional ones, but some of the smaller ones still stuck to their ards and had very few other implements. All five estates and person-of-standing farms had harrows, four had modern iron ones. The crofts were again less well stocked in the 1850s. None of the nine studied from Dahl hundred had a harrow, although, nineteen of the total of forty-eight from the entire region did, one even a deep-cutting iron one. Only six of twenty-four crofters' inventories from the northern subregion were without harrows in the 1850s – another reflection of the difference in the scale and nature of the production projects of crofters on the plains and those on the shield by the end of our period.

Rollers were specified in twenty-one out of forty-four 1850s inventories, clod-crushers in eight. All estates and person-of-standing farms had them, while only three of forty-seven crofters had a roller and just one a clod-crusher. The last were all from the plains; it seems as if harrows were sufficient on the shield. Compared with the 1820s, when a clod-crusher only apperared in one person of standing's inventory, this was a common item of equipment in the 1850s, but rollers were about as common in the 1820s as in the 1850s.

The major development during our period was thus in the ploughs and harrows which no doubt led to substantial improvements in the yields on the farms of the plains, where they were mainly innovated. However, not much complicated harvesting equipment was introduced. Scythes seem to have taken over from sickles in most parts of the region by the beginning of our period, reducing the time taken to cut grain by half. Sickles still existed on the farms but were probably used only for cutting fallen straw.[170] Mowers only appeared in Dahl during the 1880s although the first mower in Sweden was imported to central Östergötland in 1851.[171] Threshing machines were introduced in Dahl during the first decades of the nineteenth century and, in spite of being substantial investments, they had become common on the large farms by the 1850s. Four of the five estate and person-of-standing inventories in our sample from 1850s contained threshing machines; Judge Duberg of Lindenäs had two on his extensive estates, one worth 200 and the other 150 Rdr. Among the forty-four farmers, five in category I had threshing machines. Valued at between 50 and 200 Rdr, they were by far the most expensive pieces of inventoried farm equipment. The large investment was obviously considered worthwhile. Previously, threshing had been both time-consuming and difficult. Machines could also thresh relatively recently harvested grain, whereas hand threshing had to wait until the grain was properly dry. It has been claimed that the massive acceptance of machine threshing in the 1850s coincided with rising wages. Certainly, its diffusion could be a reason for both the apparent lack of pressure on farm labour supplies on the Dahl plains, despite reclamation and intensification, and for the rapid growth of seasonal cottager labour, which would be attractive to farmers when winter threshing was no longer available for hands.

In 1858 the average value of farm implements in Dahl was 750 Rdr per *mantal* and in Aska about 1,000 Rdr, which for the smaller farms would give averages of between 100 and 250 Rdr. The increases in value of farm implements between the early 1830s and the late 1850s were assessed at 100 per cent in Aska and 150 per cent in Dahl. This increase could not be due only to new ploughs and harrows plus inflation, but must have signified larger *quantities* of modern farming equipment and threshing machines on the large farms. Different categories of farm invested in different ways and on different scales. By 1860 large farms had equipment, apart from wagons and status-conferring vehicles, much more than proportional to the difference in tax value between them and smaller farms, and this difference grew continuously. In Dahl, the value of farm implements on small farms little more than doubled between the 1810s and the 1860s, while on large farms it went up twenty-eight times! This reflects a number of things. Large farms grew bigger while the average small farm was reduced in size. Geared to production for the market, they found investment which increased production and decreased unit labour costs worthwhile, while subsistence farmers were less tempted by either of these motives. Finally, there was probably an element of boastfulness in the attitude of successful large farmers. The break point between large farms which invested heavily in equipment and small farms which were more modest seems to have been somewhere in the lowest part of category II in Dahl. As the intermediate farms of this category were rapidly disappearing, the differences in levels of investment between capitalist large farms and smaller subsistence farms became very stark.

Livestock

It has been claimed that agriculture on the Östergötland plains was in many ways old-fashioned and neglected. In particular, this was supposed to be true of animal husbandry. Too few animals were kept to manure the fields properly, and hence yields were much smaller than the potential.[172] The surveyor who carried out the first *storskifte* enclosure in Herrestad village in 1768 wrote that

there has been so little occasion so far to accomplish a correct arable cultivation, so little knowledge and ambition can be seen, only the old habit is followed. And such a wonderful soil that exists here could become the most fertile arable land, if consideration and thrift were used and a correct *storskifte* enclosure could be accomplished, but instead the weeds are the most prosperous vegetation and the good grain is destroyed by the lack of ditches, which are considered unnecessary and a waste of land. The meadow does not correspond to the arable ... grazing is scarce and meagre and hardly sufficient for the traction animals.

The number of cows, bullocks and oxen increased between the early 1830s and the late 1850s in both Aska and Dahl (see table 4.10) despite the extension of arable at the expense of meadow and grazing land. The number of horses was reduced by 10 per cent in Aska but remained constant in Dahl. Horses were only half as common as oxen; there were nearly twice as many cows as oxen in Aska, and oxen increased more in relative numbers on the large and

Table 4.10. *Animal stocks in Aska and Dahl in 1833 and 1858 (1833 figures estimated in 1858)*

| | Aska | | Dahl and Lysing | |
	1833	1858	1833	1858
Horses	1,145	1,045	1,590	1,590
Oxen, bulls etc.	2,101	2,488	2,878	3,369
Cows	3,490	4,235	3,640	4,151
Heifers	779	951	1,336	1,477
Sheep	3,135	3,275	5,639	5,639
Pigs	2,087	2,386	2,790	2,790

Source: 1858 finanskommittés handlingar, Riksarkivet, Stockholm.

intermediate farms of Dahl than on smaller farms there and in Aska generally. Intermediate farms increased their stock of oxen later than large farms.

Inventoried cow populations on the plains were surprisingly small in the 1820s. Large farms usually had more traction animals than cows; half the inventories list as many oxen as cows. The situation was similar among the intermediate and small farms; all had both horses and oxen, about as many of each, and their total number was as a rule greater than that of the cows (see table 4.10). We only looked at half a dozen inventories from small farms in Aska, but the results are easy to interpret: the plainland farms there had similar animal populations to farms in Dahl, whereas on the shield the number of cows, although small, was about one unit larger than horses and oxen together. Quite large numbers of sheep were kept on the more substantial farms.

The number of animals grew on large farms and estates during the 1810s and 1820s after which it was kept almost constant up to the 1880s. The intermediate farms, conversely, reduced their number of animals in the 1820s and 1830s and only increased them in the 1860s. The *value* of individual animals increased on large farms from the early nineteenth century, on intermediate farms from the 1830s and on small farms only from the 1870s. Large farmers and estate owners were more aware of and could develop quality among their animals, while the smaller farms still had cows, oxen and horses of low value. Crofts had minute stocks of animals. Only the estate crofts of Kungs-Starby kept horses, and oxen were listed in only two of the croft inventories examined for the plains. The normal animal stock of a croft on the plain seems to have been one cow, a few sheep and, in about half the cases, a pig or two. On the shield, crofters had equally few traction animals but the average number of cows was higher than on the plains.

The two person-of-standing farms included in the 1820s inventory sample were both from the plain, and contained animals in similar proportions to the large peasant farms there. They had as many or more traction animals as cows; up to the 1850s increases in oxen were modest and confined to larger farms, where expansion required more traction animals. However, the range of farm sizes in category I makes generalization difficult. Anders Larsson of Broby, in Strå parish in Dahl had seventy-six oxen on his 18,475 Rdr farm, while a more normal number on a large farm of 4,000 Rdr value was between three and ten. Fewer cows than oxen were kept on the larger farms of the plains. On small and intermediate farms, the average number of cows per farm was 1.5 and 5 respectively, again smaller than the number of traction animals. The number of cows on small farms seems to have fallen from the 1820s and risen on intermediate farms, reflecting the change in the average value of land per farm in the two categories. Similarly, the number of horses and oxen fell slightly on small farms and was unchanged on intermediate ones. The number of cows grew on large farms, in line with their expansion in size, bringing an increase in the number of cows in both Aska and Dahl.

There were significant differences between farms of our three size categories on the plains and on the shield. They were largest between the smallest farms. Shieldland farms in the smallest size group had an average of 4.5 cows in the 1850s, their counterparts on the plains only 1.5. In the intermediate group, the shield farms had seven cows on average, those of the plains five. This must have been related to urban growth along Motala Ström after the 1830s, which increased the local market for milk, cream, butter and cheese for the forested lands of northern Aska, to which smaller farms seem to have responded most strongly.

Sheep flocks were also larger in the forest, where on average small farms had six sheep compared with two on the plains. Sheep seem to have been kept mainly for self-provisioning on the plains; equipment for processing wool to finished cloth was found in most of the inventories. The larger flocks on the shieldland might suggest wool and textile production for sale. However, the contracts of hands and maids guaranteed them various items of new clothing each year, so that cloth production for self-consumption on a farm could be substantial. Sheep declined in number during our study period. In Dahl large farms had an average flock of sixteen sheep in the 1820s, but fourteen some decades later, while small farms went from eight to two sheep. Similarly, in northern Aska, the small farms of the 1850s had just over six sheep on average where there had been ten in the 1820s. The reduction was most dramatic on small farms on the plains, probably for similar reasons as for cows: small farms shrank as larger farms expanded and specialized in grain. The reduction in northern Aska is perhaps more surprising. Here, no shift towards grain growing occurred and the small farms were still the most numerous category by 1860. We have seen that some of the meadow and grazing in these parts was reclaimed to arable land. This might have reduced the room for more marginal production such as keeping sheep. Another reason for a reduced number of sheep could be that the market demand for Swedish wool was small. Due to its crude character, it could not be used in the machines in the textile mills which instead imported wool from abroad.[173] The fact that more and more cotton garments became available at low prices, and could be used as part-payment to the farm employees who had previously been supplied with home-made woollen clothes, might also have been important.[174]

At the start of the nineteenth century most farms had one or two pigs, probably mainly for their own meat. As in Västergötland,[175] the number of pigs in Dahl was reduced but the value of pigs per farm grew in consequence of improved breeding.[176] It is difficult to see any consistent trends in the inventories in our sample, because most pigs were butchered in the autumn. In summer or early autumn, farms had many more pigs than in the winter and early spring. Small farms on the plains seem to have had about 3 pigs each in the 1850s, compared with 3.5–4 in the 1820s, while their counterparts on the shield also kept 3 pigs where 4 had grazed three decades earlier. The large plains farms could have as many as 30 pigs, but more commonly about 10 in the 1850s and 7–8 in the 1820s. The intermediate farms in the 1820s had about 5 pigs on the plains and 6 on the shield, and in the 1850s 7 in both subregions. Few cottages had pigs according to the inventories, while half the crofters kept them. On average, crofters on the plains had slightly less than one pig each while on the shield each crofter had about three-quarters of a pig in the 1820s. The figure was reduced by 20 per cent up to the 1850s for crofters in both subregions.

Changes in the stocks of pigs, like those of cows, thus reflected changes in the farm size distribution and the fact that pigs were mainly kept for subsistence. Production of pork for the market can only have been possible on the few large farms with herds of twenty or more, although these may have been destined to satisfy the demand for piglets to be fattened over summer and autumn for winter food in smaller farms and cottages, which rarely kept fully grown pigs, and could therefore not breed their own stock. Industrial and urban workers also kept piglets, and were probably in the market for piglets rather than pork.

The few persons of standing's inventories we examined testify to their interest in market production. For example, Dean Ek of Rogslösa parish in Dahl, for whom an inventory was taken when his wife died in 1823, had six horses, sixteen oxen, twelve cows, forty-eight sheep and twenty-three pigs, even though his farm was assessed at no more than 5,000 Rdr. The relative numbers of Dean Ek's animals were typical for Dahl at that time: more traction animals than cows, quite numerous sheep and pigs. In comparison, Bergsäter estate, on the fringe of the shield in Motala parish, kept four horses, sixteen oxen, forty-five cows, fifteen sheep and twenty-two pigs at the death of its owner in the early 1850s. The balance there was quite different, with a strong dominance of cows, few sheep and numerous pigs. The value of Bergsäter in 1850 was 8,000 Rdr, not much bigger than Rogslösa vicarage but a much larger potential for keeping cattle in the wide areas of outlying forest land.

The livestock of agricultural giants such as District Judge Duberg of Lindenäs, Motala, and War Commissioner Lallerstedt of Kungs-Starby was similar in kind although much larger. At Duberg's death in the 1840s his agricultural property was concentrated in the parishes of Motala and Västra Stenby, but within this structure, production was specialized. Cows were kept on eleven plots, and animals of any kind on thirteen plots out of fourteen. But the main agricultural units for livestock were Lindenäs, Norra Freberga and Bromma, while at Lilla Hals only four oxen were kept as traction animals for the stone quarry there and for ploughing. The total balance reflects the fact that Duberg was involved in non-agrarian production at his stone quarry and mills on Motala Ström. He had no less than ninety-two oxen and nine horses, but only fifty-three cows and forty-seven heifers. The large number of heifers may hint at an expansion of cattle rearing, in response either to increased demand for meat or growing needs for manure and dairy products on the estate. War Commissioner Lallerstedt had fewer non-agrarian interests, and obviously felt more keenly about the fertility of his fields, which were mainly concentrated in two parishes next to Vadstena. He kept 175 oxen, 211 cows and 21 horses there, a concentration on animal husbandry which is somewhat surprising for the plains. Even though many of the inhabitants of Vadstena kept cows and pigs on townland farms of their own, the market for dairy products for a farm next to the town must have been thriving.

In summary: first, the number of livestock grew between 1810 and 1860, less in Dahl than in Aska. Secondly, oxen were more and more used for haulage, while the number of horses fell. Thirdly, there were large differences between the subregions' different sizes of farm. Small farms and crofts on the plains, which shrank in number and size, had smaller animal stocks, while their counterparts on the shield remained largely the same. Increased animal stocks on the large and intermediate farms must have been the reason for the small total increases in the number of cows (15 per cent), oxen (8 per cent), sheep (5 per cent in Aska) and pigs (10 per cent in Aska) between the 1820s and 1850s.

Conclusion: the development of agricultural production

The cultivated area grew throughout Aska and Dahl during the first half of the nineteenth century, but more so in the north where there was still a lot of spare forested land. On the plains, the cultivated area expanded mainly at the expense of meadow. In consequence, grain production grew in most parts of the area, and this was boosted further by the introduction of new techniques, fertilizers, crops and rotations on estates and large farms. In chapter 3, we

Table 4.11. *Net production for different yields and seedcorn quantities in some Aska parishes in 1858 (in barrels)*

	Total seedcorn	Yield per acre	Average yield
Wheat	459	0.75	7.7
Rye	2,742	0.75	5.8
Barley	919	1.40	6.2
Oats	1,480	1.16	4.5
Mixed grain	5,720	1.12	5.0
Potatoes	9,425	5.15	5.9

Source: *1858 års finanskommittés handlingar*, Riksarkivet, Stockholm.

showed that the agricultural land over almost the entire region became progressively concentrated into larger farms and estates. This inevitably meant that more wage labour was involved in agriculture in 1860 than in 1810, and we have seen that the number of married hands and cottager day-labourers grew much faster than would have been needed to replace erstwhile farmers and crofters. Farm implements were much the same over the period on the smaller farms, while the large units improved their equipment substantially, which reflected the different objectives and production capacities of farms of different sizes. Large farms could employ a lot of labour, and had resources for investment in the reclamation of arable land and the purchase of plots from small peasants in trouble after enclosure, and they innovated new equipment and crop rotations.

It is not admissible to assume that grain production, for example, grew by the same percentage as the arable land of a farm was increased by reclamation. The entire complex of output components has to be considered, which makes calculations of farm production difficult: many attempts have been made, but few have resulted in anything more than vaguely realistic.[177] The combination of factors which determined total production can be expressed in measures such as seedcorn per acre, seed/yield ratio, hay produce per acre of meadow, plus the amounts of milk, meat, wool and hides produced per unit area per annum. It is not difficult to calculate the levels of production of grain per acre for larger areas for which general yield measures have been published, as for the entire hundreds of Dahl or Aska in the 1858 Finance Committee report. However, it is difficult to distribute such measures over farms of varying sizes with different numbers of animals, using different implements and labour forces on different soils in different locations relative to markets. In consequence, we have not tried to elaborate complex mathematical models, as the values we can put into them are inadequate to produce anything useful. Table 4.11 gives figures of net production for a number of different yields and

Table 4.12. *Average yields for different crops in Aska and Dahl in 1858*

	Aska	Dahl and Lysing
Wheat	7.7	6.6
Rye	5.8	6.6
Barley	6.2	7.6
Oats	4.5	5.0
Mixed grain	5.0	4.9
Peas	5.3	6.6
Potatoes	5.9	6.1

Source: *1858 års finanskommittés handlingar*, Riksarkivet, Stockholm.

seedcorn quantities. It is easy to see from this how big differences could develop from yields of, for example, 5 or 7, which could easily exist between neighbouring farms during our period of study. Table 4.12 gives average normal yields for different crops separately for Aska and Dahl plus Lysing hundred. Because Aska and Lysing both contained plain and shield the actual differences between Dahl and the shieldland parts of Aska were bigger than the table suggests.

In the *storskifte* enclosure documents, the fertility of fields was normally given in terms of yield. A field yielding six times the seedcorn was naturally considered more valuable than one yielding four times. These figures cannot be used to calculate yields of specific crops because yields varied a lot between crops, but they do illustrate differences in potential between the plain and the shieldland. Thus, in Dahl it was not uncommon to have fields which yielded eight times, while on the shield it was rare to find yield assessments higher than five, and in certain hamlets maximum yields were only three times the seedcorn. It is obvious from this that calculations based only on average hundred or county yield data are quite useless if we want to discover how production changed according to farm size or ecotype. Instead, we have used some examples of individual farms to assess how grain production developed over our period. We have assumed that soil fertility affected specific crop yields and have therefore accounted for this in the calculations by giving soil fertility in the most fertile parts of Dahl a value of 1.0, and that of the worst patches in the north 0.3. Even these coarse measures reveal significant differences in the development of production (see table 4.13). To this should be added changes in yield over time due to innovations in manuring, implements levels and crop rotations.

Throughout the region, the expansion of arable land normally meant a

188   Peasantry to capitalism

Table 4.13. *The development of productivity in two villages in western Östergötland from the mid-eighteenth to the mid-nineteenth century*

	Cultivated area (in hectares)	Soil fertility	Yield	Rye production (in barrels)
Svälinge				
1760 (two shift)	65	0.75	6.6	326
1850 (crop rotation)	82	0.75	6.6	495
Övra Lid				
1776 (three shift)	6	0.50	5.8	24
1857 (three shift)	18	0.50	5.8	72

Sources: Enclosure deeds, ÖLM, Linköping; *1858 års finanskommittés handlingar*, Riksarkivet, Stockholm.

shrinkage of meadow, which reduced fodder production unless new meadows were opened up or fodder was grown on the arable. As little new meadow was created, thus only the fields could produce extra animal fodder. New crop rotations certainly included crops which could be used for this, such as vetches, peas, clover and mixed grain. But the reclamation of arable land in most parts of the region brought reduced herds on smaller farms, while large farms maintained similar or slightly larger animal stocks on their greatly expanded areas. Does this imply that the large farms began to grow fodder on the arable? At the same time, why were supposedly more old-fashioned small farmers reclaiming so much arable instead of sticking to old methods with a 'sound' balance between arable, meadow and grazing? Although small farms experimented for ways to escape from the traditional two-shift cultivation of the plains and three-shift of the shield, they did not do so as extravagantly as larger farms and estates, so that yields probably increased more quickly on large farms. But some expansion of output must have been achieved by the smaller farmers who reclaimed land. On the less fertile shieldland and the areas next to it, animal husbandry could still grow despite arable expansion because marginal areas could be used for meadow and grazing.

The differences between large and small farms grew over our study period, so that by the end of it large and intermediate farms on the plains were producing substantial surpluses for the market while small farms were mainly subsistent. On the shield, market production must mainly have been of animals, and even small farms had cattle herds of a size indicating market production. With expanding population on the shield, surpluses from animal husbandry may well have been needed for the purchase of grain from the plains, but only farms of category I and II could have produced significant surpluses.

Table 4.14. *Increases in seedcorn quantities in Aska and Dahl, 1833–58*

	Seedcorn quantity in 1858		Increase, 1833–58	
	Aska	Dahl and Lysing	Aska	Dahl and Lysing
Wheat	459	542	30%	12%
Rye	2,742	2,674	22%	15%
Barley	919	2,863	25%	11%
Oats	1,480	498	50%	25%
Mixed grain	5,720	9,534	26%	12%
Peas	573	846	25%	6%
Potatoes	9,425	5,353	55%	12%

Source: 1858 års finanskommittés handlingar, Riksarkivet, Stockholm.

How big were surpluses produced, and by how much did they grow on different kinds of farm? Table 4.14 shows that production expanded as the seedcorn quantities in Aska increased by between 55 per cent and 22 per cent, depending on crop, and in Dahl and Lysing by between 6 per cent and 25 per cent. However, grain production did not grow at double the rate in Aska that it did in Dahl. Just as important as total increases in grain production is their distribution over the spectrum of farm sizes. The largest farms were those which had the more complex crop rotations in the 1850s and most other agrarian production units, irrespective of whether on the plains or the shield, had stuck to traditional crops together with potatoes. Potatoes, which benefit from lighter soils, were grown to a larger extent in Aska than in Dahl. However, as about 80 per cent of the farm land in Dahl and Lysing hundreds was reported as either under three-shift cultivation or a complex crop rotation (about 10 per cent) while the traditional two-shift only accounted for 20 per cent, most of the large and intermediate farms had intensified their production as well as expanded in size and reclaimed arable land. Petersson calculated the surplus of large farms in Dahl at about two and a half times the subsistence of their inhabitants, while her intermediate class of farms (some of which would count as large in our classification) using modern crop rotations could reach 1.5–2 times subsistence level. Smaller farms would have had difficulties even to reach subsistence level before 1860.[178] A category I farm in Dahl with a taxation value of 6–7,000 Rdr, cultivating 100 acres of arable, could produce a surplus of 500 barrels or more, which in monetary terms, depending on the mix of crops grown, was 2–3,000 Rdr. With lower levels of fertility, large farms on the shield would not have managed that amount. Smaller farms on the shield and the plain would have found it difficult to reach surpluses even of a few barrels per annum.

There were forty-eight category I farms in Dahl in 1860, each with a calculated surplus, including animal products, equivalent to about 1,200 barrels of rye. On top of that, the fifty-four intermediate farms each had a surplus of about 300 barrels. These figures assume that modern crop rotations were used, whereas in reality only some of the larger farms and virtually none of the intermediate ones did so. Therefore, the actual surplus must have been much smaller – probably about 500 barrels for each large farm and 50 for each intermediate one. Small farms were supposed not to produce any surplus at all. If one-third of the large farms used crop rotations in 1860, the total surplus for Dahl would have been equivalent to almost 35,000 barrels of rye. Considering that total annual grain production in the hundred was between 50,000 and 60,000 barrels,[179] this seems too high. More realistically, the total surplus for farms in 1860 was probably equivalent to between 20,000 and 30,000 barrels of rye. This was worth about 250,000 Rdr, or 162,500 Rdr in terms of the deflated money values used to standardize farm sizes in 1860, to the farmers.

The creation of large farms and innovation of new methods caused the surplus to rise considerably from that of the early 1830s, for which we can calculate[180] at the equivalent of 11,500 barrels of rye. This figure, representing about 120,000 Rdr, is even more approximate than that for 1860 because the Committee report contained only assessments of the 1833 figures. However, a doubling of the surplus does not seem unrealistic for Dahl, considering the large changes that took place, and considering the amount of increase in the export of agricultural produce from Vadstena and other loading places in Dahl between March to May and September to October of 1835 and 1856. If these months are representative, rye exports were paramount, growing from 602 to 2,152 barrels out of totals, including wheat, barley, oats, peas and potatoes, of 750 and 4,690 barrels. Unless other crops were exported in inverse proportions in other months, the structure of these long-distance exports is curiously out of line with the shrinkage of rye acreages and the expansion of mixed grain and potatoes. Potato exports during the months concerned fell from 50 to 15 barrels.[181]

In Aska, the dominance of large farms was not so marked, which naturally affected the level of surplus. So did lower yields on the shieldland, although here we have had to use average yield figures for the entire hundred, which did not differ much from those of Dahl. However, the large areas of meadow and grazing on the shieldland must to some extent have compensated for smaller arable areas and yields. As virtually no farmers in Aska had introduced crop rotations by 1860, the calculated surplus of 27,000 barrels of rye in 1860 is quite low compared with total grain production of 63,000 plus another 55,000 barrels of potatoes. The total surplus in Aska, according to these crude calculations, was thus about the same as in Dahl even though Aska had 33 per cent more arable.[182]

In the early 1830s, the calculated surplus for Aska was 11,600 barrels, less than half the figure for 1860. Thus, in this subregion too, changes in farm structure, labour organization and methods of cultivation had brought a substantial increase. In reality, it might have been bigger because our calculations do not fully account for changes in the number of cattle. However, a round figure equivalent to 50–55,000 barrels of rye, actually made up mainly of milk, meat, butter, potatoes and barley, meant a cash income of about 550,000 Rdr per annum on top of subsistence, compared with 220,000 Rdr in the early 1830s. This represents about one-sixth of the total taxation value of the area – a considerable capital for investment.

Non-agricultural farm production

Introduction

The calculations of output and surplus in the previous section do not necessarily reflect the total sums of money earned by farmers because they exclude secondary production on farms. What was the scale of by-employments and other means of adding value through processing output on the farm: how big was the 'hidden economy' which lurked in the huge proportional share attributed to agriculture in official statistics of both taxation and population?

Apart from itemizing on-farm distilling in years when it was taxed, the taxation registers provide no evidence of part-time economic activities in the households of the region. For information on domestic manufactures themselves we must turn to inventories, of which we examined 47 from the 1820s and 207 from the 1860s for evidence of craft work. They were sampled from all social groups and each subregion, but there are too few to allow cross-tabulation by social group and parish.

On-farm processing of crops and milk

In pre-market economies, brewing and distilling alcoholic drink were palatable ways of preserving surplus grain and other vegetable farm products for future consumption. In a market economy, they add considerable value to crops at the same time as rendering them imperishable. For the first of these reasons, on-farm 'distillation for household use' had a very long history in Sweden, and many eighteenth-century restrictions on its extent were removed at the turn of the nineteenth century. In consequence, 'it became ... a national peril',[183] although the provincial governor of Blekinge considered that distilling was 'so to speak the real purpose of agriculture' in the 1820s,[184] and in Ljuder parish in Småland in 1846 'the price of grain was so low that the peasants must distill their produce in order to farm profitably' at all.[185] The rapid spread of potatoes was due in part to their value in distilling, with the mash serving as livestock fodder.

Figure 4.4 shows that distilling was widespread in our region in 1840, when farmers were allowed to distil alcohol in quantities proportionate to the size of

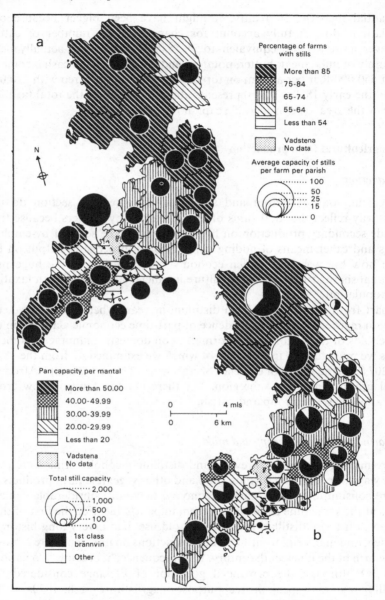

Figure 4.4. Distilling in the parishes of Aska and Dahl in 1850: (a) The proportion of farms in each parish with stills and the average capacity of those stills; (b) The distillation pan capacities of each parish, the proportion of stills producing first class liquor, and the ratio of still capacity to mantal value. Still pan capacity in *kannor* (1 *kanna* = 2.62 litres). *Source: Taxeringslängder* Landsarkivet, Vadstena.

their holdings and taxed upon the capacities of their stills.[186] The destination of the potato and mixed grain surpluses of the region is now quite apparent! The value of liquor tax owing exceeded the total assessed land tax in all but six of the twenty-seven parishes in 1840, by a maximum of 5.35 times in the northern parish of Kristberg and by at least a factor of 2 in three other parishes, two of them also on the shield. The upper map of figure 4.4 shows that the proportion of farms with stills was also highest in the forested north. The lower map shows that still capacity per *mantal*, and therefore the amount of distilling relative to the area of arable land, was generally highest in that subregion, too, although there was a broken belt of parishes running from Nässja to Orlunda where the ratio was only slightly lower. The reasons for this were probably that potatoes were much more commonly grown in the north, and estates, which accounted for a much greater proportion of the land in the shield, were particularly heavily involved in distilling. Nearly all the large stills in the highest taxation bracket were on plots belonging to estates. On the other hand, the peasant farms of the north were in general much smaller than those on the plains of the south, and so, therefore, were their still capacities. This is why average still size per farm did not vary much from parish to parish.

However, the size of stills varied very little across the vast majority of farms. Of over 400 stills in the region, only 22 had a capacity of 50 *kannor*, roughly double the average. Only seven of these large stills belonged to peasant farmers, all of whom had accumulated very large holdings on the arable plain.[187] The rest belonged to estate owners. War Commissioner Lallerstedt had stills of 90 *kannor* capacity at both Kungs-Starby in St Per and Börstad in Orlunda, and Assessors Jansson and Dahlberg's 138 *kannor* still on Södra Freberga estate in Västra Stenby parish was the largest in the region. The other thirteen large estate stills were in the northern shieldland. The relationship between still capacity and size of holding meant that there was a close correspondence between the distilling capacity of farms and their labour forces, so that potential on-farm consumption was closely correlated with the amounts of liquor distilled. It is impossible to tell from taxation records how much liquor was destined for sale. However, rough estimates of still outputs from information given in inventories[188] suggest that a very large proportion of the region's liquor must have been produced for export. At the national annual rate of consumption of 4 quarts per head at the end of the 1840s, the stock in the Lallerstedt warehouse alone could have supplied the consumption of the whole region's population in 1852. Even at the much higher annual average consumption rate of 44 quarts per head in 1830, this one source could have supplied about one-tenth of the whole region's annual demand. If we assume that the Lallerstedt warehouse contained the output of the three stills listed in his inventory (where there were no stocks of liquor); that the quantities listed for the three distilleries were annual outputs, and that the still sizes of these farms were the same in 1840 as when the inventories were taken a decade or so later, then we can calculate an average output of 77 *kannor* per unit of taxed still volume. This would mean an annual output of 1,925 *kannor* for the average-sized still of the region, enough to supply nearly 1,000 people at the annual average consumption of the late 1840s, or 100 at the 1830 rate. The region's total output would have been well over 1,000,000 *kannor* a year, enough for a population more than five times that of the region even at the very high consumption rate of 1830. At about 2 *kannor* per Rdr, it was worth over half a million Rdr on the market. The annual tax value of all the agricultural land in the region in 1840 was 2.1 million Rdr, so that, according to our calculations, distilling was bringing a notional return on agrarian capital of about 25 per cent!

More prosaically, War Commissioner Lallerstedt, Judge Duberg, Assessors Jansson and Dahlberg, and Anders Larsson must have been selling the vast majority of the huge quantities of liquor they distilled, and getting large return for it. Even Anders Kastbom's much more modest holding at Västgöteby in Motala parish (valued at 800 Rdr in the taxation registers of 1850, but at 4,000 Rdr in his inventory of 1852) supplied considerable quantities of liquor for sale. He was owed 182 Rdr for liquor by a customer from Börrum, near Söderköping at the oppposite end of Östergötland. Perhaps this connection was not fortuitous. Shipments of liquor along the canal increased dramatically. In the four months' accounts we examined for 1835, two ships carried liquor from Motala to Stockholm and one to Göteborg, one from Vadstena to Stockholm. Total shipments were 5,729 *kannor*. Over the same months in 1856 (mainly in May), thirteen ships loaded liquor in our region at Motala, Vadstena and Borghamn (Rogslösa), bound for Stockholm, Göteborg, Västerås, Jönköping, Kristinehamn and Askersund. They shipped 93,082 *kannor*, worth 42,663 Rdr at 22 Skilling per *kanna*. Two thirds was shipped to Stockholm from Vadstena, the only source in the region of liquor for the capital. This perhaps helps to explain the large size of the Vadstena Crown Distillery, the Lallerstedt and Larsson distilleries, and the high still capacity to *mantal* ratios of Nässja, Strå and Orlunda, in the vicinity of the town.

Considerable value was, therefore, added to grain and potato crops through on-farm distilling, and liquor as well as unprocessed rye was exported from the region to many parts of Sweden by the middle 1850s. A comparison of the estimates of income from distilling in the 1840s with our earlier calculations of the value of raw produce surpluses in the 1830s and late 1850s suggests that 25 per cent value-added could be earned from distilling, which gives some idea of the investment opportunities available to farmers and shows that distilling was not a desperate measure forced onto farmers by low grain prices. However, the restrictions imposed on on-farm distilling in 1855 removed this opportunity, and thenceforth distilling was concentrated in plant under the control of the authorities. Although these were often the stills located on estates and large farms, the possibilites for ordinary farmers to profit from distilling were greatly reduced when their role was restricted to supplying grain and potatoes to stills operated by others.

Judging from the inventories, on-farm brewing was conducted on a much smaller scale. Like stills, brewhouses occur only in inventories of very large farms, but they were obviously much smaller affairs.[189] One or two farmers had small quantities of malt when their inventories were taken, but the only substantial stock was that held in the warehouse of War Commissioner Lallerstedt at Kungs-Starby, where there were 1,450 *Skeppund* (283 metric tons), valued at 725 Rdr. There is no evidence in the Göta Canal accounts we examined that farm-produced beer or malt were exported from the region, and it seems that brewing was generally carried out only to supply on-farm consumption.

The normal herd of up to three cows per small farm was insufficient to supply more than household needs for milk, butter and cheese. Moreover, as with brewing and distilling, dairies and cheese- and butter-making equipment are only listed in the inventories of the largest farms. However, it is evident that large farmers who were heavily involved in distilling also had high stocking densities. Whether these herds were used mainly to supply meat, milk or manure, we cannot say. Large quantities of hides and meat were stored in the Kungs-Starby warehouse, and the Lallerstedt farms at Egeby and Kungs-Starby had cheese dairies (worth only about one-tenth as much as the distilleries), while Anders Larsson's dairy at Strå contained about 400 kilos of cheese, worth 133 Rdr, at his death. Small quantities of butter and cheese were shipped out of the region along the Göta Canal in 1835 and 1856. Thus, it seems that some large farms were producing milk for conversion to cheese and butter on a commercial scale on the farm, which might well have been related to the use of spent mash as cattle fodder. Even so, cheese and butter making for sale consumed far less investment and generated far less income than distilling.

Whether or not distilling, brewing and butter and cheese making are rural 'crafts' as that term is generally understood is a moot point. Certainly, they cannot have been part of a proto-industrial process, in which on-farm production for putters-out led to the development of factory production of the same goods. But there is no doubt that the first in particular yielded considerable income to estates and large peasant farms.

Homecrafts

The remaining household manufacturing activities to be considered sit much more comfortably within the categorizations normally used in the study of rural industries. All could be practised by non-landholders as well as farmers, and three of them at least have been widely identified as the progenitors of later large-scale production elsewhere.

Table 4.15. *Percentages of inventories containing evidence of rural homecrafts in the 1860s*

	None	Spinning	Weaving	Leather	Wood	Metal
Parish						
Västra Ny	19	54	45	7	62	15
Orlunda	36	45	45	2	6	0
Örberga	19	56	51	10	54	19
Social group						
Estates	0	100	100	50	0	50
Peasants	12	58	58	3	42	12
Crofters	32	34	38	4	51	11
Hands	21	29	29	0	40	7
Others	19	48	46	2	46	4
Unspecified Women	31	54	35	3	23	5
Total	29	49	43	6	37	7

Well over two-thirds of the inventories sampled contain evidence of homecrafts of some kind, in the guise of tools and equipment or stocks of raw materials or finished products (see table 4.15). Given the emphasis in synoptic literature on the relationship between pastoral farming and rural industries, it is not surprising that the plainland households of Orlunda apparently practised less craft work than those of forested Västra Ny. By the same token, however, it *is* surprising that inventories from plainland Örberga show more evidence of craft work than those from Västra Ny. In terms of socio-economic groups, the peasant farmers had the lowest proportions without craft work, which again, given received wisdom, is to be expected. However, a majority of households in all groups practised crafts of some kind, and the size of the proportion for hands is remarkable, given that they generally lived in the households of others, working at the behest of a master, subject to constraints on behaviour which did not apply to other social groups.

Örberga had the highest proportion (84 per cent) engaged in textile production. As with crafts as a whole, it spanned all social groups, from estate owners, through peasant farmers and their hands, to people who were ascribed a full-time occupation unconnected with either the land or textiles (such as sexton or boat skipper), or no occupation at all. Although the proportion with textile equipment or materials was higher amongst the peasant farmers than other groups, this was not markedly so. It is impossible to discover from this evidence how much craft production was geared solely to supplying the needs of the households in which it was practised. However, it is possible to draw some tentative conclusions from the types and quantities of goods and equipment listed. Types of equipment were very numerous, and most people had quite a variety, but they were not valuable. Tools such as wool cards, flax hackles and combs, bobbins, reels and ball-winders were frequently listed with spinning wheels. The former were rarely worth much more than 1 Rdr, in whatever quantity, and spinning wheels varied from the two valued at 32 Skilling (Sk) to two at 5 Rdr 32 Sk. We do not know whether these variations reflect differences in the sophistication of wheels which spun thread of different thicknesses and degrees and types of twist. The same inexplicable variety is evident in the case of looms, which were normally valued together with their accessories.[190] Looms with many treadles and harnesses to weave complex fabrics and those for broad cloth must have been more valuable than ones to weave crude narrow cloths, but the inventories do not record whether this is the reason for the wide variations in loom values. Very few people, even farmers with weaving rooms in their houses, had more than two looms, and the value of the biggest stocks of textile equipment rarely exceeded 10 Rdr, a far cry from the cost of plant for distilling, or even butter and cheese making.

The percentages of households with spinning wheels or looms are given on table 4.15. Many people did not have both, implying that some households bought in yarn and cloth made by others. Because the absolute numbers on which they are based are small, small variations between the percentages might not be significant. However, it seems that spinning was more common than weaving in all parishes except Orlunda and amongst all socio-economic groups except crofters – especially, perhaps not surprisingly, amongst single women.

Hemp and pig's hair yarn were listed as well as dyed and undyed wool, linen and cotton thread, and they were combined in mixed cloths as well as woven alone. The baffling variety of weights, values and unit prices of yarns, as well as the variety of spinning wheel values, suggests that yarns of many different degrees and types of twist and fineness were spun, even from the same fibres. The quite strong spatial patterns of different types of cloth imply that they were made commercially. The percentage frequencies of the common types of fibre, yarn and cloth are listed on table 4.16. Linen was made nearly everywhere in the region, with cloth rather than yarn being particularly prevalent in Örberga, though to a lesser extent also in Västra Ny, and in the inventories of crofters. Flax was certainly grown locally: according to a number of inventoried standing crops it was sown together with mixed corn on the arable, and cargoes of flax seeds were exported from Motala on the Göta Canal in 1856. Indeed, flax growing was given particular government encouragement in the region from the middle of the eighteenth century to ensure supplies for the damask linen factory at Vadstena. One of the reasons for the shortfall of thread supplies to the factory[191] might have been the same as that for its eventual failure: the commonness of home linen weaving, which was done in a quarter of the households of the region.

Table 4.16. *Percentages of inventories with various types of fibre or yarn and thread*

	Wool		Linen		Cotton	
	Fibre/yarn	cloth	Fibre/yarn	cloth	Fibre/yarn	cloth
Parish						
Västra Ny	29	8	25	34	6	2
Orlunda	15	2	19	13	2	4
Örberga	25	13	22	43	5	5
Social group						
Estates	100	0	100	50	50	50
Peasants	30	9	30	26	2	9
Crofters	21	11	15	32	2	0
Hands	14	7	21	26	5	5
Others	24	13	31	21	10	4
Unspecified Women	15	5	24	28	7	5
Total	20	7	21	26	5	5

Some of the cloths listed were wefts still on the loom or otherwise unfinished, and the majority of finished cloths were simply described as *lintyg*. However, some descriptions were more specific. *Lärft* (shirt-quality linen) appeared in two Örberga and five Västra Ny inventories, whilst coarse tow cloth (*blångarnsväv*) was listed in all three parishes. 'Half linen' and 'mixed linen' cloths were also occasionally itemized, as was *grövre dräll* (huckaback, a figured cloth used to make such things as towels), also sometimes of half linen. Most of the yarn listed was plain or white; some was washed, some dyed black, yellow or blue, the last presumably to be made into the 'blue and white cloth' mentioned in some inventories. The yarn woven with linen into mixed cloths was probably cotton, which might also have been the raw material of the five women from Örberga whose inventories contained lacemaking equipment. Cotton fibre and yarn were apparently widely and cheaply available in Östergötland by the 1820s,[192] and very large shipments of raw cotton passed through along the canal between Göteborg and Norrköping.

Wool, and the yarn and cloth made from it, was second most common to linen, although it was apparently a long way behind in terms of cloth, if not of raw fibre or thread. The raw material must have been ubiquitous, if nowhere in super-abundance. As we saw earlier, almost all farm inventories list a few sheep. Outside Örberga, the cloth produced was nearly always simply decribed as 'woollen', only Västgöteby on the shield specifying *vadmal*. In Örberga, however, four out of eight inventories with woollen cloth specify this type, one a number of different kinds of it, and half-*vadmal* and half-wool cloths were also recorded there. *Vadmal* is an ancient form of heavy cloth, used for saddle cloths, harness linings, work aprons, draught-proof curtains and so on.[193] It must have been fulled because one of the watermills in the adjacent parish of Strå was described as a *vadmalstamp* (fulling-mill) in the tax register of 1840. Usually, both the cloth and yarn were undyed, but there were examples of dyed wool yarn and *vadmal* cloth. However coarse and old-fashioned, *vadmal* was more valuable than linen and cotton cloth. It was worth from half to one *Riksdaler* per ell, whereas the most valuable figured damask and *lärft* were both 15 Sk per ell and cotton 12 Sk, although the most expensive cloth of all was valued at 3 Rdr 16 Sk per ell. Destined for use as trousers, it must have been some kind of worsted or fine woollen.

It is impossible to say how much of this wide variety of cloths was made for sale, but the narrow concentration of lace and vadmal in Örberga implies that they were produced commercially. The fact that four out of six references to *lärft* were in Västra Ny might suggest the same thing. In the *passjournaler* of 1849 and 1850, permits were granted to crofters and maids from parishes near Vadstena to sell lace made in the town in distant places.[194] Thus, the normal means of peasant marketing were in operation in the lace trade, and lace must also have been sold at Vadstena's three annual fairs in February, May and October. Lace production seems to have spread through the countryside around Vadstena during our period. Rural inventories from the 1820s contain no lace cushions, but in the 1850s landless households in both Örberga and Fivelstad were making lace, as the craft developed in counterpoint to the concentration of land in agriculture.[195]

Only 26 of the 182 inventories with cloth had more than 10 ells (about 6 metres) of any one type: the vast majority of households contained no more than might be needed for their own use (although cloth put out to a person would not belong to them, and would not therefore appear in their inventory). In these 26 inventories, the overall dominance of linen and the importance of woollens in Örberga are marked. Cotton was much more commonly held in larger quantities, too, perhaps suggesting that it was more often produced in larger volumes, and therefore mainly for sale. Peasant farmers and women were the main holders of large cloth stocks. The latter were unusually wealthy, but with one exception the peasants had category III farms. Large farmers such as Anders Larsson of Broby, Lars Larsson of Herrestad and Johan Nilsson of Lycketorp, Orlunda, owned equipment, yarn and cloth (the last even had a weaving room), but none of them possessed 10 ells of a single type of cloth. The largest quantities of all exceeded 200 ells, which could not possibly have been wholly produced and consumed by the household in which they were held. But there are few clues as to how these cloths might have been acquired and disposed of. Given that the inventories are completely reliable, however, it is impossible that all twenty-six people with more than 10 ells could have produced the cloth they possessed because eleven of them did not have looms. On the other hand, Mrs Ölander of Piltorp in Orlunda, with the second largest stocks of cloth, had two looms with accessories in the East Room of her house, stored 828 Rdr's worth of fibre and yarn of all possible types in the attic over her kitchen, and kept her cloth stocks separately elsewhere in the house. In 1850 the large holding of Piltorp employed only three maids and three hands, so it is unlikely that Mrs Ölander was running some kind of proto-factory. The purchase and sale of yarn and cloth from and to spinners and weavers would explain the conjunction of a large stock of materials and a small household labour force. So, of course, would putting-out. Five other inventories contain quantities of thread worth more than 10 Rdr, all in households with little or no cloth: there were local people from whom others could acquire more yarn than they could produce for themselves.

Although it is impossible to know whether household linen and clothing itemized in inventories were destined for sale, some contain so many items of a particular kind to suggest that they must have been. Seventeen pairs of curtains of various sorts, large quantities of serviettes and other household linen, amounting in total value to 519 Rdr, were itemized in the inventory of Mrs Ölander of Orlunda, as well as large quantities of yarn and cloth. A single woman of Orlunda had 74 Rdr's worth of household linen, though the balance of her inventory was only Rdr 349. Johan Källman of Stavlösa, Orlunda, who was owed for 21 ells of cotton cloth at his death, possessed tablecloths, linen and *blångarn* sheets, pillowcases and table napkins in much greater quantities than one would expect to adorn the house of his small farm. Isak Jönsson, an Orlunda hand with 12 ells of cloth, had no loom and his stock of cloth, though relatively small, was the most valuable by far per ell. He had eight racks of trousers made from *vadmal* and other types of cloth, one silk, three half-silk and three other waistcoats, and the very large quantity of 66 Rdr's worth of bed clothes and household linen. It is inconceivable that they were all destined for his own use. In Västra Ny on the shield one crofter had eighteen shirts; another had fifteen sheets, eighteen table napkins, nine coats, ten waistcoats and twenty-eight shirts; a single woman had nineteen sheets, twenty-four towels and fifteen table napkins, whilst Peter Carlsson of Nedra Lid, whose holdings of linen cloth were the second largest in Västra Ny, possessed fifteen sheets, eighteen hand towels, six pillow cases and eighteen table napkins. Stocks such as these could have been the sources of the 1.5 *skeppund* of household linen (*duktyg*) loaded at Motala for transit along the Göta Canal to Göteborg in 1856.

The evidence of commercial production is circumstantial, and the vast majority of the people with textile-making equipment and materials were probably working up locally produced flax and wool for their own use. Nonetheless, it is certain that some of the people who made cloth, particularly linen generally and *vadmal* in Örberga, sold it, and that some people made up cloth into textile goods for sale. Even though all social groups were involved at large and small scales, farmers with smallholdings and wealthy women seem to have been most heavily involved in producing or dealing in large quantities of cloth, crofters and hands in making it into articles for sale. However, the value of the goods concerned was very small compared with the sums involved in distilling liquor. The supplementation of livelihoods by textiles production for the market, or the saving of expenditure by home production, made a significant contribution to the household budgets of only the poorest social groups.

Large numbers of other kinds of tools and materials were itemized in inventories, but it is even more difficult to discover the extent to which they were used in production for home consumption or sale. In the case of some tools, such as turning-lathes, it is not even possible to decide the type of craft in which they were used. Hides and skins were itemized in twelve inventories. Apart from War Commissioner Lallerstedt, whose Kungs-Starby warehouse contained fourteen cow hides and sixty calf skins worth 133 Rdr, leather work was carried out on the same scale, in terms of cash values, as most textile production. Fur pelts were listed in only one inventory, that of a crofter from Västra Ny, and although some farmers had tanned and untanned hides, most of the skins were from calves and sheep. It is possible that the farmers were simply processing the skins of their own animals for their own use. However, this cannot have been true in all cases, and it seems that the preparation of animal skins, even more than textiles work, was the prerogative of crofters and cottagers, and amongst them, of women. Three of the possessors of skins were single women, who had seven sheepskins, four unprepared sheepskins, and five sheepskins and a hide, respectively; another, who had three-quarters of a tanned hide, half a curried hide, four unprepared calfskins, six prepared calfskins and one unprepared sheepskin, was the wife of Örberga's sexton. The only mention of quantities of bark is in the inventory of an Örberga grenadier. Perhaps the dirt and smell of tanning meant that farmers were loath to do it. It is likely that skins were obtained from local farmers, and the possession of parts of tanned hides may suggest that small pieces were sold to neighbours as they required them. However, an Örberga shoemaker owed 4 Rdr to tanner Ulander of Vadstena, presumably for leather, and an Örberga crofter owed even larger debts to tanners Ulander and Norling of Vadstena, but his inventory contained no hides or skins.

Table 4.15 shows that woodworking was the second most common craft to textiles, evidenced by the itemization of tools such as saws, chisels, planes and axes, belonging to women as commonly as men. These may well have been used wholly for preparing wood for use about the house and farm, even when held in large quantities. It is probable, too, that some of the saws and hewing axes were used, especially in forested Västra Ny, for part-time work felling timber which, in the form of planks and boards, was one of the main cargoes shipped along the Göta Canal from Motala, and increased from 2,329 cubic metres in April, May, September and October 1835 to 5,585 cubic metres over the same months of 1856. Stocks of timber were listed in a few inventories, most of them from Orlunda, where one farmer had 40 Rdr's worth of 'good timber after the *laga skifte* enclosure of Stavlösa'. Johan Nilsson of Lycketorp's 150 Rdr's worth of birch, boards, planks, 'many poles and staves' may represent his share of this windfall opportunity to improve his large farm and homestead. So might the stock of eleven waggons and seven pairs of sledges owned by an Orlunda farmer. The only other stock of wooden goods in greater than usual quantities were the six unfinished stools of a 'joiner and sculptor' in Västra Ny. Turning-lathes were listed in far greater quantities in Västra Ny than elsewhere. They might have been used to convert the abundant timber of the parish into saleable articles, but because they were often listed together with pieces of iron we classified them as iron-working tools. In that case they perhaps indicate that the crofters and cottagers who mainly owned them were taking in turnery work from forges and ironworks in the parish. Skills in household metal working were certainly present in the region: an Örberga farmer had a set of clockmaking tools. A few farmers had other 'smiths' tools', and the largest of all had well-equipped smithies with hearths. The Lallerstedt estates's smithy and its tools were worth 50 Rdr; the Larssons' smithy at Broby which had '10 stamps, 3 hammers &c' and was probably powered by one of their watermills, was worth 30 Rdr. Perhaps it was connected with the diffusion of iron farm implements.

A number of other activities supplemented farm and wage income. Orlunda was the most heavily subdivided parish in our region, with the greatest plethora of tiny arable holdings on the plain. Perhaps in consequence, the farmers, crofters and cottagers there seemed to do everything imaginable to eke out the maximum return from the natural environment. Judged by the number of young fruit trees listed in their inventories, it seems likely that the sexton of Orlunda and his wife were growing them for sale, whilst the crofters, cottagers and farmers of the parish had unusually large numbers of goats, geese, ducks, chickens and bee hives.

Conclusion: was there a hidden economy?

Agriculture was supplemented in many ways by most rural people in our region, ranging from distilling through clothmaking to leatherwork, wood and metal turning and clockmaking. Much of the output must have been for consumption in the households where it was produced, as was typical of peasant societies. However, some crafts were practised on a scale which must have meant the sale of produce. In these activities, people used whatever resources were readily available. Differences in the resources available between ecotypes showed through in the by-employments practised. However there was no sharp distinction between plain and forest parishes in terms of the the most important by-employment, textiles.

Many households practised a number of crafts simultaneously. Small farmers and wealthy women seem to have been predominant in textile production or dealing on a large scale. But cottagers and crofters, whether married or single, also engaged in textile production on a scale which suggests more than self-provisioning, and leather production, woodworking and turning were mainly their preserves. Moreover, the inventory balances of crofters and cottagers were often larger than those of peasants with very small holdings. The hand and soldier who were wealthy enough to lend money at interest to small farmers in Orlunda were heavily involved in animal keeping and cloth making, the latter in wood dealing as well. Perhaps by-employments on a large scale over a wide enough range of activities were capable both of supplementing the produce of a very small arable holding sufficiently to provide a family livelihood, and also of helping to bridge the gap between landlessness and small peasant proprietorship. Certainly, the practice of homecrafts blurred the boundary between small peasant farmers, crofters and cottagers. Even so, the most lucrative by-employment by far, distilling, was by law restricted to landholders, and the scale on which it was permitted was linked to farm size. In this case, at least, the boundary between the landed and the landless was a sharp one, and the function which described the relationship between area of arable and scale of livelihood had a steep gradient.

Conclusion

The agricultural population increased by about 80 per cent between 1810 and 1860, while the resources available to hired farm labour were reduced. This was particularly so in Dahl, where more than half the crofts disappeared, the number of cottages more than doubled, and the number of married hands grew even more rapidly. This radical transformation occurred mainly on large rapidly expanding farms; on small farms, which must have produced mainly for subsistence, crofts and non-productive people remained numerous. Dahl had fewer people employed per unit of arable than the other two subregions, suggesting that the large farmers kept their labour-forces as trim as possible, and perhaps also that opportunities for non-farm work were fewer on the open plains. By 1860, the workforces of large Dahl farms consisted mainly of hands, maids and married hands, who must have been joined by cottagers in the most intensive agricultural seasons. The workforce per unit of agricultural land was substantially larger in northern and southern Aska, and many more crofts survived; in the shieldland of northern Aska, there were as many in 1860 as in 1810. At the same time, married hands also became quite common on the larger farms of Aska, but there were many more examples in Aska than in Dahl in 1860 of substantial farms reminiscent of 1810, staffed with crofters and single hands and maids. This suggests that the rationalization and capitalization of farming was slower there, and/or that the workforce was at least partly occupied in pastoral and non-agricultural production.

Two kinds of estate can be distinguished. The traditional *säteri*, which did not change much in terms of land ownership, also stuck to a more old-fashioned form of labour organization. Although some of their farms were merged, they still supported large demesnes and numerous crofters in 1860. The more recently formed estates, which were much less stable in plot composition, did not have a demesne, but numerous farms spread over several parishes, and like the large peasant farms of Dahl they employed mainly married hands and supported few crofts and cottages.

The higher degree of proletarianization in Dahl might be expected to have produced greater poverty, but it did not. There was more poverty in the northernmost part of the study area. Neither did the greater proletarianization of Dahl lead to rapid population growth as wage-workers produced more children to compensate with labour for their lack of resources. To an extent, this did occur. The average number of children per cottager family grew in all three subregions as cottages came to be occupied by seasonal labour, rather than the aged and infirm, but normally they had fewer children than farmers. Among non-farmers, the crofters of northern Aska had most children in 1810 and 1860, but married hands and crofters in Dahl, who had many fewer 1810, came close to matching them with about four children per family in 1860. The number of children per landless family also increased in southern Aska, but not by as much as in Dahl. Thus, even though the high degree of proletarianization was not reflected in particularly high poor rates, it was so in a demographic sense. Where this increased population went, we do not yet know.

Changes in farm size and labour supplies were associated with changes in production projects. Most peasant farmland was enclosed during our period. Not only were plots reorganized into more nearly integral farms, but within the plots the opportunity for reclamation grew as they were wrested free from open field organization. In the entire region, the amount of arable grew substantially at the expense of other land-uses. In Dahl, most of what could be used as arable already was by 1810, and the conversion of meadow to arable was under way. Some Dahl villages also reclaimed land from Lake Tåkern, others from the marshland next to the lake, in the process doubling their fields. Several villages and hamlets on the shield also doubled their arable area, but most increased it by about 50 per cent from the late eighteenth century and up to the 1850s. Yields were much higher on the plains, where a 50 per cent increase in arable meant a much larger addition to total production than a similar growth on the shield. At the same time, the potential for animal production was higher on the shield's large areas of forest grazing.

The growth in arable land was connected with changes in crops, in the kind of and number of animals kept, and in farm equipment. On the plains, potatoes were not grown much on the larger farms, which stuck to grain. However, all farmers on the shield, and smallholders on the plains quickly adopted potatoes, which became their major crop. Together with grain, potatoes were used in the farm distilleries which pumped out vast quantities of liquor from both the plain and the shield.

The development of livestock was closely related to the size, location and changing arable practices of farms. On the plains, large farms increased their traction animals, cattle and pigs as they expanded while smaller farms reduced their livestock as they shrank in size. The largest farms and estates developed big herds of cattle, to some extent to produce enough manure to fertilize their extended fields. The growth in livestock numbers was smaller in Dahl than in Aska. Greater grazing potential in the north meant that an extension in the arable area did not necessarily significantly reduce meadow and forest areas. On the plains, this was inevitable unless more fodder crops were grown on the new arable. This was not generally the case before 1860 except on large estates.

All over the region, scythes replaced sickles, iron ploughs replaced wooden ards, improved harrows became common, and on large farms threshing-machines came into use. The first and last in particular must have been heavily implicated in the changes of the labour force. Smaller farms had stocks of implements in 1860 which were little different from those of 1810. There were few differences between the subregions, but smaller farms on the southern plains do seem generally to have been less well equipped than those further north.

All social groups everywhere were involved in homecrafts, but it is hard to believe that any but distilling brought significant income to the large farmers. The expansion of homecrafts among peasants with very small farms and crofters, was probably related to the shrinking size of their farms and crofts, amongst cottagers and married hands, to the increasing seasonality of farm labour.

Our findings have clear implications in terms of the general models of peasant societies discussed in chapter 1. Non-traditional estates and large freehold farms were following a capitalist path of development, using the market more and more to allocate both land and labour to those forms of production which yielded the highest monetary returns. Swedish traditions and legal codes affected both the exact nature of the proletarianization process and the particular significance of distilling in the farm economy, although the latter was also appropriate to the location of our region relative to the markets of Stockholm and Göteborg along the Göta Canal. The new-style estates were somewhat ahead of the peasant farms, but this does not indicate that large landowners were forcing development onto a reluctant peasantry, as suggested by the model of agrarian capitalist development. Free peasants followed of their own accord where persons of standing led. What Kautsky and Lenin averred to be an inevitable consequence of commercial production occurred: the gap between large and small farms grew as the former expanded, the latter shrank, and intermediate farms became far less numerous. However, differences between the rural economies of various ecotypes were perhaps less marked than general literature leads us to expect. Estates and large farms were heavily involved in distilling and the manufacture of dairy products on both the shield and the plain; homecrafts were practised by small farmers, crofters and cottagers everywhere. Indeed, although crafts based on forest products were inevitably concentrated on the northern shield, Örberga, the most important parish for textile production, was located deep in the southern plain. Similarly, small farmers and crofters everywhere were conservative in technique, but quick to take wholeheartedly to the potato. On the other hand, the commercial agricultural economies of the estates and large farms of the subregions diverged as the plains became more specialized in grain production and the shield turned to potatoes and animal husbandry.

Complementarities thus seem to have developed during our period between the commercial products of different ecotypes, and between those of farms of different sizes in the same ecotype. All produced more than enough liquor, but the grain deficit of the shield was matched by surplus on the plain; falling livestock production on the plain by an increase on the shield; falling craft output on large farms by an increase in that of the shield generally and of small farmers, crofters and cottagers on the plain. These reciprocities are striking, but they do not necessarily demonstrate the existence of autonomous hidden rural economic development. The changes on the shield might have owed at least as much to local non-agrarian development; the large farmers of the plain exported produce far and wide, and they might have done so to acquire urban 'Z goods' rather than the homespun rural wares of small farmers, crofters and cottagers.

5
Industrial and urban livelihoods

Setting the scene

Non-agricultural livelihoods in Sweden before the nineteenth century

According to taxation and parish records,[1] only 20 per cent of the Swedish population was occupied outside agriculture in 1750. This was still so in 1850, when 8 per cent worked in mining, manufacturing and handicrafts, and 2 per cent in commerce and transport. Mining and manufacturing were rural industries based on scattered bodies of ore, forest charcoal and water power. Urban handicrafts and commerce were of little significance in aggregate statistical terms. Even in 1800, towns accounted for less than 10 per cent of the Swedish population (compared with 31 per cent in England and Wales in 1801 and 19 per cent in 1700); only Stockholm, with 76,000 inhabitants, and Göteborg with 13,000, housed over 10,000 people in 1800, although Norrköping was not much smaller.[2]

However, Postan's caution against the lure of national statistical aggregates[3] is as appropriate to Sweden as anywhere else. Because it was seasonal, much of the work in industry disappeared into the farming category of occupational tabulations. In fact, from medieval times industry was by far the major earner of foreign exchange and helped to support population where farming alone could not, especially in the forested shieldlands. Towns, too, were more densely spread in some regions than others, and exerted a qualitative influence far beyond the size of their populations. Nearly all full-time handicraft work, retail and wholesale dealing were confined by law to places defined as urban for administrative reasons.[4] Thus, all full-time livelihood positions in much of the secondary and the whole tertiary sector were in principle urban. The formal definition of urban places excluded some large concentrations of population which were functionally towns.[5]

Metal and other manufacturing industries

In the seventeenth century Sweden almost monopolized European copper output from Stora Kopparberget at Falun (see figure 3.1) and the mines near

Åtvidaberg in Östergötland. Wealth derived from it and from the silver of Sala was the basis of Sweden's political prominence and military power in the following century. Copper mining declined from the middle seventeenth century, and the relative value of copper exports slipped to 30 per cent by 1700 and 10 per cent by 1720. One reason for this was that Sweden became Europe's pre-eminent iron-exporter in the 150 years before 1800. All Swedish iron until the sixteenth century was *osmund*, recovered as a lump directly from small furnaces. It was of high quality and in great demand throughout Europe, especially around the Baltic and in Germany. Because output was low and wasteful of raw materials, the government encouraged the innovation of blast furnaces of the French type, with two bellows working in tandem, in the sixteenth century. To keep as much processing as possible in the country, the introduction of hammer-forging techniques to convert pig into bar iron was also encouraged. The Crown tried to keep forging independent from smelting and as large and well-capitalized as possible. Many forges ended up in the hands of Swedish noblemen and Walloon immigrants.[6] The Walloon process, with separate finery and chafery hearths, was introduced in the seventeenth century, mainly to forge low phosphorus haematite ores from Dannemora, 30 km from the Baltic in Uppland, into *Öregrund* iron, which was the most prized raw material for steel throughout Europe.[7] Swedish iron output increased fivefold between 1600 and 1720, when bar iron accounted for 75 per cent of exports by value and Sweden produced more than one-third of world bar iron output, supplying between 80 per cent and 90 per cent of England's iron imports and 40 per cent of its total iron consumption.[8] This strong position in international markets continued through the eighteenth century, but a fall in world prices in the 1730s and early 1740s had important effects on the Swedish industry for the following century.

Because of its importance to military supply, foreign exchange earnings and tax yields, the iron industry was heavily regulated by the government from the sixteenth century, especially after the creation of a special department of state, the Board of Mining and Metallurgical Industries (*Bergskollegium*) in the second quarter of the seventeenth century. All forges had to trademark their bars to allow checks on the quantity and quality of their iron in the exporting ports and inland trading centres. Exports of *osmund* iron, pig iron and ore were completely forbidden.[9] In the early 1730s, the ironmasters responded to falling export prices by increasing their outputs, raising fears of forest depletion and further price falls. Increased output by individual furnaces and forges required charcoal to be hauled over longer distances, causing unit costs to rise and forcing the industry to spread from its traditional base in Bergslagen north of Lake Mälaren into forested regions further south, such as Godegård, Tjällmo and Västra Ny in northern Östergötland. The government decided in 1747 to limit iron production to conserve forest resources, force up prices and keep individual furnaces and forges small and supplied with charcoal as cheaply as possible.[10] The opening of new ironworks was forbidden and a maximum permissible output was allocated to each existing forge. Exports of ore, pig iron and bloomery iron being forbidden, these measures also limited their outputs, and therefore total charcoal consumption. Total forged iron output was fixed at 46,650 tons annually, the quantity produced in the year preceding the restriction, and an allocation was made to each forge, with favourable consideration for ironworks in newer areas of production where pressure on charcoal supplies was least. Forges could only exceed their allocation to make up shortfalls of previous years.[11] The fixed annual output of bar iron, plus small quantities of pig used in casting, consumed 65,000 tons of pig iron per year, and generally 90 per cent of it was exported.[12]

These quantities and proportions remained steady until after the end of the eighteenth century. In 1747 ironmasters formed a cartel, the executive body of which was *Jernkontoret*. Its main initial function was to provide interest-free loans to merchants, but funds accumulated from a levy on all exports soon enabled *Jernkontoret* to lend cheaply to ironmasters as well, and to employ a number of *övermästare*. Their experiments, discoveries on *Jernkontoret*-funded tours abroad, and instructions to furnace and forge masters and workers raised the quality of Swedish iron during the second half of the eighteenth century.[13] Output remained almost constant from 1748 to 1803 at just under 50,000 tons a year, and the number of ironworks grew only from 324 in 1695 to 340 in 1803. A small expansion of production was sanctioned in 1803, but by 1808 output had collapsed to 50 per cent of the permitted quantity. Napoleon's Continental System and the expansion of puddled and rolled coke smelted bar iron output in England had severely eroded European markets, and at the opening of our period the industry was in crisis after dominating the European bar iron trade for two centuries.[14]

The first iron deposits to be exploited in Sweden were the ores which are continuously, though slowly and impurely formed in the bogs and lakes on outcrops of basement rocks in the shield of Småland, Närke, Värmland and Dalarne.[15] Rock mining began in the thirteenth century,[16] but the enormous phosphorus-rich deposits at Grängesberg and at Kiruna-Gällivare in Norrland were not exploited until the late nineteenth century. Until then, ores low in phosphorus were required to produce high quality bar iron. They are much less abundant (hence the prohibition of ore, pig and *osmund* iron exports until the middle of the nineteenth century) and occur mainly in small deposits, except at Dannemora, thickly scattered amongst more abundant high-phosphorus reserves in Central Sweden. The southern limb of this crescent-shaped district is much less ore-rich than the northern one.[17] Centred on Finspång, its southern boundary is the fault to the north of Motala Ström. The northern shieldland parishes of our region were on the southern extremity of the metal mining, smelting and forging heartland of Sweden.

Literally translated, a *bergslag* is a 'mountain company', a group of peasants known as *bergsmän* who banded together to mine and smelt iron ore. They had the sole legal right to ore and the charcoal needed to smelt it in the shieldlands where their arable holdings lay, paying a tax of one-tenth of the value of the iron they produced.[18] In areas where these rights existed, the use of timber was restricted by law to conversion into charcoal for the iron industry. There were many companies of *bergsmän* in and immediately outside the metal bearing area, still responsible for about half of pig iron output in the eighteenth century. But by then some of the bigger deposits of ore, such as that at Dannemora, had been taken over by large integrated businesses with the capital necessary to operate Walloon forges.[19] There, mining and smelting were done by specialist workers rather than part-time by peasants, although the latter retained the right to charcoal production in the forests if they had not been bought out by companies.

The water power needs of furnaces and forges limited their operation to spring and summer when stream levels were high.[20] The haulage of charcoal, pig and bar iron was restricted to winter, when frozen land and lake surfaces allowed the use of sledges. Forges consumed 3–4 tons of pig iron and 6 tons of charcoal per ton of bar iron produced. The high cost of haulage and the friability of charcoal limited the radius over which it could be drawn to a maximum of 20–30 km,[21] which restricted the capacity of forges to about 200 tons of bar iron per year. Ironworks were often the first colonizers of the tracts of forest where they sought fuel and power. Scattered at intermittent water power sites, spatially separated by their demands for charcoal, remote from the arable plains and their dense populations, they had to organize and finance the supply of skilled and unskilled labour as well as raw materials and fuel, and to guarantee a steady flow of victuals and other necessities. These widely ramified integrated enterprises were known as *bruk*. Ironmasters incorporated pre-existing farms into their enterprises and set up new leasehold farms on pockets of arable land in the forest over which they held the charcoal rights. As late as 1908 their 'owners and managers concentrated their interests on agriculture and forestry while the production of iron and steel . . . was regarded virtually as a side-line'.[22] Many who were granted ironworking rights in the seventeenth century were already noblemen, and many others were ennobled in the following century. Their tenants paid rents in charcoal burned in the ironworks' forest and worked part-time in haulage and menial jobs about the works. Wages and payments to freehold farmers for charcoal were often made in barley, rye or other foodstuffs such as salted fish, credit being extended by *bruk* truck-shops against future work or charcoal deliveries. Ironmasters often purchased arable farms on the plains to guarantee the food to exchange for charcoal and haulage. Grangärde parish, between Grängesberg and Falun, produced only 30 per cent of the grain consumed there and nearly 60 per cent of all men's working time (amounting on average to 150 days per year) was spent working for ironworks, rather than in farming.[23]

Forge smiths (and workers at *bruk* making brass, paper and glass) were highly skilled and had valuable privileges: they 'knew they were better off than farm workers and were surrounded by an aura of superiority'.[24] Each German forge could be operated by four full-time workers: a master smith (*mästersmed*), a master's helper (*mästersven*) and a couple of charcoal hands. Most ironworks had two hearths, and therefore eight full-time workers. The dual-hearth Walloon process employed ten persons per forge. Under the Regulations for Hammersmiths of 1766 the perquisites of forge workers were stipulated as follows:

a master [smith] was entitled to a cottage with a bedroom, a kitchen and a cattle shed. Masters' assistants and married forge hands not boarded by a master were to have a room, a kitchen and a cattle shed. They were also entitled to free feed for their cattle – a master for two cows, a master's assistant and forge hand for one cow each – either in the form of pasture rights and an allotment of meadow for haymaking, or a fodder allowance [in] cash.[25]

Forest grazing and patches of arable land for turnips, grain or potatoes were also usually allocated to *bruk* employees. Smiths were free from military service and certain forms of taxation,[26] and under the heavily patriarchal *bruk* régime work of some kind was always provided if at all possible; if not, wages were paid on a reduced scale. On retirement or death, a smith or his widow was given housing, meadow and arable land. On the other hand, smiths could not leave without their employer's permission, nor ask for increased wages,[27] and truck-shops were exempt from regulations limiting retail trade to chartered town markets.

The whole workforce at a *bruk*, full and part-time, formed a highly diversified but closely integrated community, all involved in farming, forge production and raw material supply. Although the arable land of the shield was of little value compared with the plains, other resources were the basis of a very distinctive form of livelihood, which was by no means inferior to arable farming as a provider of guaranteed sustenance.[28]

Between the poles of *bergsmän* and *bruk* lay a less common stage of the industry: about 10 per cent of bar iron output was retained for conversion into sheet iron and steel, bolt, hoop and rod iron.[29] Iron rods were forged from bars with hammers (*knipphamrar*) in Sweden, rather than rolled and slit, as in England. These forges required less charcoal and water power than furnaces and bar iron forges. Some *bruk* contained their own rod hammers, and forges for converting the rods into finished articles (*manufaktursmedjor*). However, the exigencies of charcoal supply meant that it was common for rod hammers and manufacturing forges to be run separately, by people of far lower social standing than the ironmasters. This was also true of the finishing stages of the industry. Although cannon and other armaments were made at large specialist works such as Finspång in Östergötland and Åker in Södermanland,[30] nails, hinges, locks, bolts, cutlery, tools and so on were usually made by independent smiths. Putting-out had developed by the late eighteenth century, especially in cutlery making, by *bruk* and guildsmen in towns where the goods were sold to their consumers. The peasant-smiths of Dalarne in northern Bergslagen were by far the most important source of manufactured ironwares in the eighteenth century.[31] The town of Eskilstuna was founded by the government near the southern shore of Lake Mälaren in Södermanland in 1771. Incorporating two pre-existing *bruk* and the congeries of cutlers and other smiths they supplied, it was set up for free smiths, independent from ironmasters and merchants, who could benefit from the external economies of scale available to a large group of craftsmen in the same trade.[32]

The government also fostered the growth of large-scale units of production in other industries such as tobacco, sugar, glazed pottery and textile manufacture. The last accounted for over 80 per cent of all employment in these establishments. Known generically as 'manufactories' in the late eighteenth century, they 'were treated like pampered pets',[33] with privileges including free land and credit (funded by a tax levied specifically for their support); freedom from customs duties on exports, from normal restrictions on retail sale by producers and from guild membership, and tax concessions. Competing imports were restricted by duties or banned. People without any professional training could set up manufactories; they were not obliged to employ master craftsmen and the laws under which they operated, codified in 1739 as the Manufacturers' Privileges, ordered that all people 'of whatever sorts and conditions' should be allowed to work at manufactories.[34] Being part of the government's policy to foster urban growth, they were mainly set up in Stockholm, Göteborg and Norrköping, but some filtered down into smaller towns. Manufactories were collections of hand-workers under the control of one establishment, which might spread across several buildings and include home-workers, especially women. Generally master craftsmen were employed to direct less well-qualified hand-workers. The entrepreneur was responsible for capitalization, raw material supply and product sales. Workers were by law under the same strict control as those on a noble estate, and they were similarly free from military service: 'the prerogatives enjoyed by the nobility were taken as a prototype ... [and] the entrepreneurs became a class of aristocrats in their own right'.[35]

The size and complexity of Swedish manufacturing industry is, therefore, grossly under-represented in aggregate occupational statistics, and large-scale industry was continuously encouraged by the government throughout the seventeenth and eighteenth centuries. The urban manufactories, like the smithies of Eskilstuna and other towns, were divorced from farming and forest work. However, the iron industry was vastly predominant in manufacturing, and most of the people involved in it, whether as freeholding peasants, *bruk* tenants or smiths, were completely integral with the agrarian economy. In the forested regions where they operated, agricultural survival was dependent on supplementary income from industry and the supply of its raw materials, often, as in the example of Grangärde, to an overwhelming extent.

Handicrafts, trade and towns

The products of the *bruk* were made largely for export, and the manufactories, particularly those making textiles, 'were much too ... "genteel" in their production, and accordingly catered almost exclusively to the rich, or in any case the more prosperous noblemen, gentry and burghers'.[36] Peasant households were capable of making a very high proportion of the made-up goods they consumed, or of acquiring such things as shoes, cloth and baskets contractually from crofters. The provision of housing, subsistence and clothing to hands and *statare*, and of land for cultivation to soldiers and crofters, meant that the consumption needs of subservient livelihood positions in agrarian society were also largely satisfied without recourse to market exchange.

The movement of goods in any but small quantities was made very difficult by rocky terrain, dense forest, widespread bogs (especially during and after the spring thaw), and the severity of winter freezing. Most heavy haulage had to be done on sledges in winter. According to Marshall, 'there are few kingdoms in Europe where [the roads] are so bad',[37] and in the first half of the nineteenth century a bad harvest brought severe dearth to inland areas because it was impossible to haul large quantities of grain from other areas or the ports.[38] Even more extremely than in the classic model of peasant societies, a large proportion of the small amount of necessary formal exchange in Sweden involved small quantities and was conducted at fairs. Although it criss-crossed the whole of the Scandinavian peninsula, making the specialist products of the whole array of ecotypes available to each them, this system of exchange gave rise to very few permanent livelihood positions independent of agrarian households. By law only craft goods made in by-employments in farming households could be sold beyond the boundaries of the maker's rural parish, and it was only in 1766 that even a peasant was 'able to dispose of his goods and agricultural products at any place suitable to him', rather than in his nearest town market.[39]

Retail and wholesale dealing were the prerogative of town merchants, and full-time craft production for sale in town markets, to merchants or at fairs was restricted to members of urban craft guilds. The towns were as cosseted in Sweden as anywhere in Europe; indeed, probably more so, even in the seventeenth century, and certainly for longer, until well into the nineteenth, as part of the mercantilist policy of encouraging specialization of function, the substitution of imports by home manufactures, and raw material exports by finished products.[40] On setting up the national machinery of government and administration in the sixteenth century, King Gustaf Vasa determined that all craft work, commerce and shipping should be confined to towns.[41] Certain coastal towns were designated as 'staples'. All foreign trade must pass through their customs houses, where duties and tariffs were collected. More numerous mainly inland towns were known as *uppstäder* and confined to internal trade. Of the staples, Stockholm had exclusive privileges over trade in the Gulf of Bothnia.[42] Göteborg developed as the staple for North Sea trade and soon grew to be Sweden's second city on the basis of Sweden's increased trading links with England and Scotland in the eighteenth century. The third staple to grow beyond the rest in size was Norrköping, in Östergötland. It had been the base of the trade, shipbuilding and manufacturing operations of Louis de Geer, who came from Amsterdam to run the Swedish armaments industry in 1627 and built up an empire of industrial enterprises stretching from Bergslagen through Södermanland to Östergötland before his death in 1654.[43] In 1614 and 1617 King Gustaf II Adolf set out a definitive list of staples and *uppstäder*. Town tolls were introduced at both in 1622. The number of towns was almost doubled to sixty-four at about the same time. The list of towns and their privileges remained unchanged until 1765. Internal tolls were abolished in 1810.

From the sixteenth century the craft work over which the towns had a monopoly was put in the hands of guilds, which spread to the smaller *uppstäder* by the early eighteenth century.[44] To protect guildsmen (as well as manufactories and rural domestic textile production) domestic weaving for sale was forbidden in the towns.[45] After widespread complaints about the prices charged by urban craftsmen, guilds lost control over the pricing of their goods in 1762,[46] but they remained influential, their privileges still otherwise intact, throughout the eighteenth century. Guild masters had exclusive rights to the business of their trade in a town and its hinterland: no outsider was allowed to set up a workshop on his own account. The only means of access to guild membership, to serve as an apprentice and journeyman with a guild master, then submit a master-piece to the guild for approval, required considerable outlay.[47]

By the early eighteenth century the trade of merchants in *uppstäder* had often been usurped through direct dealings between rural works in their hinterlands and merchants in Stockholm, Göteborg and Norrköping.[48] The markets of retail merchants and craftsmen were eroded by peasant self-provisioning, the supply of the genteel demands of the rich by manufactories, by craftsmen at rural estates,[49] and by the distribution of homecrafts by peasant producers and *knallar*. Trade between the towns and their rural hinterlands was in general not very active, although 'a tendency to a greater division of labour not only in the towns but also between the towns' had emerged by the early eighteenth century, to produce an emergent urban *system* with reciprocal specializations,[50] and estate owners spent some of their surplus wealth in the towns. The more elaborate craft products they required 'were always turned out by urban artisans; luxuries were strikingly prominent in urban handicraft ... In most Swedish towns there were, for example, gold- and silversmiths; a hatter was seldom absent and the wigmaker was common'.[51] Both craftsmen and peasants must sell their grain surpluses, on two or more occasions each year, only at markets held in designated towns where the price scales according to which taxes and rents expressed in kind, but paid in money, were fixed separately for each area in the country.[52] Some of the income derived at the town market by the peasants was spent there: they took their

surplus to town during the long threshing season from September up to and including March. In exchange, they got salt, herring, tar, malleable iron, smithwork, alum for home tanning, lime to whitewash the houses, building stones for the chimneys, whetstones and grindstones. Of groceries and small wares they bought only small quantities and few kinds. Tobacco and tobacco pipes, some spices, several dye-stuffs, silk ribbons, laces, and small braids ordinarily belonged to this category. A little loaf sugar and some syrup seems to have been bought in some cases. Much more important were vessels of earthenware, iron and copper, the last mostly for conspicuous consumption and for the distilling of spirits.[53]

Thus, despite leakage through trading products made in peasant by- employment through direct sale, peddling and fairs, some of the exchange required to satisfy household needs which could not be self-provisioned was forced by legislative fiat into the towns. All administration above the parish level was also done there. This injected a significant wealthy element into small town society by the early eighteenth century. By that time, urban craftsmen were less wealthy and powerful than the administrators and the small merchant clique which organized links through the whole urban system.[54]

The development of non-agrarian livelihoods, 1810–60

A 50 per cent increase in the output of agriculture, which fed Sweden's rising population and after the 1820s provided surpluses for export,[55] was the mainstay of economic growth during the first half of the nineteenth century. But this was not enough to prevent the country's GNP per head falling, according to the conventional means of measurement, relative to the average for European countries[56] because of the slow growth of plant-based industry. In 1870, 72 per cent of the Swedish population was still employed in agriculture and only 22 per cent in manufacturing, mining, handicrafts, commerce and transport combined. The shift from the first to the second category had been less than 1 per cent since 1751.[57] The proportion of the population living in towns increased by barely 1 per cent, to just over 11 per cent, between 1810 and 1860.[58] We have already seen that aggregate statistics of this kind are misleading about the significance of industry, but the stability of these figures does imply that there was no significant structural change until after our period ended. The customary conclusion has been that 'the economic expansion of the 1870s can be characterized as an industrial revolution';[59] 'in the 1870s we can identify the watershed between the old Sweden and the new'.[60] However, there could not have been a sudden unheralded release from stifling archaism into comprehensive growth and change. In fact, the freeing of trade and other restrictions began as early as 1815, when the free movement of foodstuffs internally was allowed.[61] And notwithstanding the apparently unequivocal message of economic statistics of the kind outlined above, there were some significant developments before 1860, which accelerated as that year approached. Like the changes in agriculture, many of them were of a kind which would lead us to expect them to occur in our region.

Iron and other manufacturing industries
Napoleon's Continental System and rapid innovation in Britain caused the overseas markets for Baltic iron to collapse. *Jernkontoret* searched frantically for technical innovations. The Walloon process was developed into the Franche-Comté technique, which economized on charcoal consumption, allowing larger scale and hence cheaper unit production. Of more significance was the eventual outcome of experiments with rolling and other aspects of forge work begun in 1815. They resulted before 1830 in the introduction into Sweden of the 'Lancashire process'. Government restrictions on the outputs of individual forges were relaxed in 1829 to accommodate the use of these new techniques. The Lancashire process was perfected in 1845 by Gustaf Ekman, who applied gas converters to the tunnel furnaces of Lancashire forges. This allowed gases from wood or charcoal produced alongside the forging hearth to be forced into the iron, bringing temperatures as high as those in coal-fired forges. It yielded a product which was soft enough for rolling without having absorbed impurities from the fuel.[62]

Improved Lancashire forges and rolling mills ran most efficiently if they were kept continuously supplied, and thus brought a further increase in forge capacity. Whereas 200 tons a year had been a large output in the eighteenth century, twice that was produced by the largest Franche-Comté forges and pre-gas converter Lancashire forges of 1844. By 1860, the largest Lancashire works were producing 1,000 tons a year, with steam-powered hammers and rolling mills.[63] Until the railway network was built after the 1860s, such outputs were only possible at works with water transport to assemble charcoal from a wide area.[64] But these developments caused Swedish iron exports to grow strongly again, to 90,943 tonnes in 1846 and 136,835 tonnes in 1860.[65] The government freed ore mining from restrictions in 1855, the export of pig iron in 1857, and all remaining legislative constraints were lifted from the iron industry in the following year.[66]

By the 1820s 20 per cent of Swedish bar iron was worked up at home and the proportion continued to increase, especially after 1840.[67] Some of this manufacturing remained in the hands of independent smiths making cutlery,

Plate 3. Motala Engineering Works in the 1890s.

locks, clocks, bolts, hoops and so on, especially in Dalarne and Eskilstuna.[68] Many old-established *bruk* producing armaments diversified their outputs as the Franche-Comté and Lancashire processes allowed them greatly to increase their manufacture of bar iron. The origins of many of the great Swedish engineering concerns of the late nineteenth and twentieth centuries lay in engineering works[69] established in the 1840s. Outside Stockholm, Göteborg and Malmö, they were often associated with pre-existing *bruk* which were moving into the production of rolling mills and agricultural machinery as well as rifles; others diversified into the production of nails, horseshoes, spades, axes and so on, through stoves and other domestic ironware to farm machinery such as harrows, threshing machines and ploughs, on to pipes, steam engines, dredges and steamboats.[70]

The engineering works of the 1840s had a few precursors of great importance, which introduced British ideas, managers and workmen to Sweden and trained many of the entrepreneurs, administrators, engineers and craftsmen who set up the works of the 1840s. Two of the early works pre-eminent in this 'nursery' function belonged to the Motala Company in Motala and Nyköping. The first, which was initiated in 1822 and eventually became Sweden's premier engineering works, was situated in our northern subregion between the Göta Canal and Motala Ström. It will be dealt with later. The works at

Nyköping, east-north-east of Norrköping, were developed as a subsidiary of Motala in the early 1830s, primarily for the construction of steamships. They were closed before 1860, but the net effect of the steam engine technology developed there and at the Owen and Bergsund Works in Stockholm is testified by the coal import figures for Sweden, which rose from 13,000 tons a year in the early 1830s to 591,000 tons a year in the early 1870s.[71]

Whilst ironworking was ramifying from smithy hearths to engineering works, textile production was moving from cottage firesides into spinning and weaving mills. Spinning machines were first installed at Lerum, near Göteborg, in 1797, by workmen tempted from Danish mills which had copied English machinery. The development of cotton manufacturing was stimulated by low import duties on raw cotton, high tariffs on cotton cloth and bans or high tariffs on finished goods, introduced in 1816. By 1820, 40 per cent of Göteborg's industrial production comprised textiles, and there were seventeen spinning mills, as well as thousands of handworkers, in and around the city by the 1850s. Swedish raw cotton imports increased from 4,800 tons a year in 1851–55 to 6,200 by 1856–60.[72] One of the putters-out of Sjuhäradsbygden built the first Swedish weaving factory at Rydboholm, near Borås, in 1834. The original looms and other machinery came from Belgium and Germany and the skills of powerloom weaving were initially taught by workers brought from England. By the 1850s, when there were many other weaving factories in Sjuhäradsbygden and spinning mills had been set up to supply them, Borås had developed into a textile town reminiscent of those in northern England, supplied from Göteborg by a 'long train of little one-horse carts, each with his bale, or bale and a-half, of cotton, driving laboriously to the spinners and factories ... [of this] ... out-of-the-way spot'.[73]

Large-scale merchant-organized woollen manufacturing was based in Stockholm in the early nineteenth century, but mechanization drew it to Norrköping, where Motala Ström generates massive water power as it tumbles into the Baltic, and where the cost of living was cheap and there was a tradition of work in manufactories going back to de Geer's time. The first spinning jennies were brought from England to Norrköping in 1799. Woollen production then grew rapidly and was joined by cotton manufacturing. Andrew Malcolm, who came from Scotland to Motala Works, moved to Norrköping in 1842 to make self-acting mules and water turbines, which convert stream power more efficiently than wheels. In the year Malcolm arrived, 504 factory horsepower were already installed at Norrköping mills on Motala Ström. This increased to 1,799 by 1868. The town's population grew from 9,000 in 1800 to 17,000 in 1850,[74] when there were 122 cloth manufactories in the town. Some used jacquard looms, and some power looms of the English type, although many were still dependent on handlooms because weaving did not become heavily mechanized until the 1860s. Mechanized textile production was also important in and around Halmstad, on the coast between Göteborg and Malmö, and in Gävle, on the Gulf of Bothnia. The number of workers employed in textile factories grew at 3.5 per cent per year between 1830 and 1860, and the value of output at 10 per cent per year. Half of the capital invested in joint stock companies between 1848 and 1860 went into textile factories.[75]

The third industry to grow substantially during our period was sawmilling. Mechanical pulp was not produced until 1857, chemical pulp until the 1870s, but the export of timber, largely as planks, boards, spars and props, increased fivefold between 1820 and 1860 to become Sweden's major export earner, ahead of grain and iron, by 1840.[76] Norrland, which was effectively colonized by the sawmilling industry at that time, was responsible for much of this output; but timber production increased from all the forested shieldland in the country. Careful government regulation ensured that it did not encroach on the iron industry's charcoal supplies. Water-powered fine-bladed frame saws were used to cut deals, planks and boards from the late eighteenth century; steam saws were not introduced until the 1840s, and hand-hewing continued to produce squared timber such as balks, spars and sleepers. The industry stayed tied to water power and coastal and inland waterways until the introduction of steam sawing and railways.

A number of other industries which became prominent in the late nineteenth century also began to grow earlier. The manufacture of phosphate matches began in 1844 at Jönköping, on the southern tip of Lake Vättern, where the safety match was invented and came into mass production in 1852.[77] Large-scale lager brewing was introduced from Germany in the 1840s and, as we have seen, ten years later restrictions on home distilling pushed the production of spirits into highly capitalized commercial enterprises.[78] The first paper making machines in Sweden, made by Bryan Donkin of London, were installed in Skåne in 1832, together with the usual complement of English workers. Other hand paper making works were forced to follow suit, and Donkin machines were soon at work in Bergslagen, Norrköping and Småland (the traditional centre of hand paper making). The English Pontifex and Wood company was the source of the first sugar refineries using vacuum cooking methods and steam power to be installed in Sweden, at Göteborg by the brewer David Carnegie in 1839. Again, the first innovation was quickly followed, in this case by D. O. Franke at Landskrona in 1840.[79]

The distinctive form of labour organization represented by the *bruk* continued through the nineteenth century in iron making and forging: the 'small Bergslagen world, with its sense of *bruk* loyalty and pride of place and status, was transformed during the breakthrough of industrialization' after the end of our period.[80] The early engineering works had similar labour organization to the *bruk*: workers had a high degree of freedom and security, and employers were paternalistic. Contracts of employment were usually for three years, and 'production continued to rely on the physical strength and skill of the individual worker . . . [and] . . . craftsmanship . . . was indispensable until far into the twentieth century'.[81] Even in precocious Eskilstuna, 'craftwork's social and cultural legacy was strongly established . . . until the end of the 1880s'. Work was task-oriented, organized and carried out by master craftsmen and their helpers, with no rigid time-keeping or production-line work. Housing, too, was usually provided for at least some of the workers, plus injury, sickness and other benefits, schools and places of worship and recreation. However, workforces of several hundreds could not be provided with the agricultural resources provided by *bruk* proprietors.

Like the manufactories, textile mills employed a considerable amount of female labour, and often combined put-out with in-house work. Children were also employed, and in many ways the organization of work within early Swedish factories was similar to that in textile mills elsewhere in Europe,[82] although perhaps due to the pervasiveness of the traditions of the manufactories and *bruk*, it was very common for factory proprietors to be involved in community philanthropy.

It would be wrong to exaggerate the effects of these industrial developments on national economy, culture and society in Sweden. Although they propelled Swedish manufacturing output to an almost fourfold increase between the early 1830s and the early 1860s,[83] the statistics given earlier demonstrate that their impact on aggregate national measures was imperceptible, and many of the towns these industries spawned were very small by contemporary English and later Swedish standards.[84] It is also true that 'wide sectors of the population suffered a final bitter taste of Old Sweden's poverty and limited horizons' in the 1860s, when crop failure at home and the disruption of foreign markets by the American Civil War brought widespread famine, unemployment and hardship.[85] Nonetheless, it is worth bearing in mind that it is now argued that British economic growth had no significant effect on national aggregate figures or economic structures until the 1830s.[86] Perhaps it is reasonable to compare what happened in Sweden in the first half of the nineteenth century with what happened in Britain between 1780 and 1830: economic changes affecting aggregate statistics and national structures had to await the development of dense cheap haulage railway networks; earlier changes were limited to places with access to a far sparser network of cheap water transport.[87] Most of the places where precocious development occurred outside Bergslagen and Sjuhäradsbygden were situated on the coast, lake shores or canal banks. The Göta Canal was of considerable significance in this respect: 'as yet the richer agricultural districts lay . . . mainly grouped along the axis formed by the Göta Canal',[88] and its effects must have spread beyond agriculture.

Urban handicrafts and trade, 1810–60
The industrial expansion and trade of Sweden in the early nineteenth century were not of a kind which augured well for urban development. Increased

homecraft output and the trade it spawned was heavily concentrated in the countryside, and the 'merchants and dealers of the Swedish towns in the 1820s did not carry on an amount of business that was much greater than forty years previously'.[89] The craft products of the towns retained their reputation for high quality, high prices and exclusiveness, and the numbers of apprentices and journeymen in craft workshops increased only from 0.5 and 0.7 to 0.6 and 0.8 per master between the middle of the eighteenth and the early nineteenth centuries.[90] The government's 'efforts devoted to restricting the crafts to the towns had been ... determined and unflinching',[91] but in the nineteenth century attention shifted to the encouragement of homecrafts. The removal of guild control over full-time craft work culminated in the throwing open of craft occupations to rural people and the abolition of the guilds in 1846. The right to operate workshops training and employing non-family members was still reserved to town craft masters, but anyone, whether resident in country-side or town, could now practise any trade from their own domestic premises. The last big bastion of townsmen's privileges was their legal monopoly of full-time retail trade in general merchandise. This, too, was relaxed in 1846, when it became possible to get permission to set up retail stores beyond 30 km from a town.[92] This distance limitation was lifted in 1864, when 'country stores sprang up like mushrooms'.[93]

All cannot have been gloomy for Swedish towns in the early nineteenth century. Town fairs and markets continued to be important places of sale for craft goods and agricultural surpluses. The large-scale dealing, specialist retailing and professional activities which were confined to towns throughout the period must have benefited from development in their hinterlands. In Northern Sweden, for example, merchant Abraham Stenholm of Luleå had, at the time of his death, booths in the coastal markets of Luleå, Kalix and Råneå and the inland markets of Jokkmokk and Gällivare. Debts were owed to him by 532 individuals, of whom 380 lived in villages on the coastal plain and river valleys, or isolated settlements in the forest.[94] The more that people specialized in working in a particular line for sale, whether in agriculture, industry, forestry or crafts, the more they must buy what they had previously produced for themselves, from weavers' reeds and thread to food, clothing and furniture.

There is a lot of evidence that formal market exchange was penetrating further into the provisioning of rural and urban households, and that this was leading to greater subdivision of labour and an increase of specifically urban full-time livelihood positions in, for example, bakeries, restaurants and inns.[95] Some of the customers must have been peasants spending more of their time and money on 'Z goods' in towns: 'an extra source of income has nearly always been immediately reflected in the living standards of the peasantry. They have dressed more expensively and improved their houses'.[96] New farm implements, such as Scottish ploughs, and more elaborate waggons and carriages bought by the wealthier peasants must have been made by specialist urban craftsmen. By 1850 'the use of cash rather than kind became realized among the peasant community',[97] especially south of Lake Mälaren, where in 1857 they were 'in the process of transition from *bauer* to gentleman, and they felt very self-important ... the wealthy freehold Bonders [sic] ... claim they have a right to their silks and champagne as much as the gentry'.[98] This was fully evident by the 1850s among the peasants who were expanding their farms in our region. The inventory of Anders Larsson of Broby in Strå itemized mahogany furniture and gold and silver items worth 1,644 Rdr – more than the value of a category III farm. Estate owners spent some of their increased incomes with an equal concern for the finer things in life. The inventory of War Commissioner Lallerstedt of Kungs-Starby had gold and silver goods worth 1,300 Rdr, less than Larsson's holding of precious metal, but Lallerstedt's droshky waggon, worth 400 compared with Larsson's 166 Rdr was more splendid, and he had a launch, also worth 400 Rdr, on Lake Vättern at the end of his garden.

Agricultural development brought a much higher propensity to consume the kinds of goods that could only be got from full-time specialist craftsmen and merchants in towns. After the lifting of restrictions on craft production in 1846, town dwellers as well as rural people were more fully able to respond to changes in demand, and the emergence of banking and the telegraph must also have given a boost to urban economies even before the development of the railway network. In regions which were developing in other respects, this would be especially so. Swedish towns grew by 53 per cent in population between 1800 and 1850, slightly faster than the national average; the fifteen towns on the Göta Canal and the shores of Lakes Vänern and Vättern, which it linked to the Baltic and Atlantic, increased in size by 91 per cent between 1800 and 1850 (82 per cent if Göteborg is excluded).[99]

There is, therefore, good *prima facie* evidence that the towns and industries of our region grew during our period. Its agriculture was being rapidly commercialized and there were abundant forests and water power on the shield to the north of the canal. How did agrarian changes affect the ancient *uppstad* of Vadstena on the shore of Vättern on the fertile plain, and what was the nature of the new industrial settlement at Motala, on the canal fronting the shield?

Non-agrarian livelihoods in western Östergötland in the early nineteenth century

The situation described above leads us to expect few specialist non-agricultural livelihood positions in our region at the start of our period; that those which did exist would be concentrated in the town of Vadstena and in the forested shieldland of northern Aska, and that the latter would have strong repercussions on the amount and nature of agricultural settlement. How far are these expectations fulfilled by evidence from the taxation registers?

Rural crafts and services

Parish craftsmen were very rare in 1810. They were completely absent from nearly half the parishes of the region, and a glassblower (at an estate) and a forester were the only supplements to seven tailors and five shoemakers shared between all the others. There were only a shoemaker and a tailor in the whole of northern Aska; on the plain only Västra Stenby, Hagebyhöga, Allhegona and Strå had two parish craftsmen each. The multiplier effects out of peasant agriculture into full-time rural non-agrarian livelihood positions were negligible. Neither did any of the estates in our region contain little craft colonies of the kind depicted by Hanssen in Skåne.[100] Kungs-Starby estate supported six non-agrarian livelihood positions in 1820, but only one of them, a tailor, was a craftsman.[101] Full-time service and administrative officers were also absent from the countryside. The ecclesiastical and legal systems were run by paid ministers and judges domiciled in the countryside, but as we have seen, both were partly remunerated with Crown land and usually involved in farming. The vergers, organists and sextons of the parish churches earned negligible amounts of money and, though they rarely owned farmland, we saw in chapter 4 that they were completely integral to the rural economy.

At the start of our period the peasant and estate economies supported hardly any full-time livelihood positions in the countryside outside agriculture – eloquent testimony to the prevalence of self-provisioning and the extent to which government policy was reflected in reality.

Ownership and production structures of rural industries

Resources outside agriculture did support some full-time non-agrarian livelihoods, especially in the shieldland where water power and timber were abundant. Non-farming properties were itemized but not valued in the taxation schedules of 1810. In 1820, when thirty of the forty-seven non-farming properties listed were assessed, they accounted for 3.4 per cent of the total value of property (see table 5.1). Even if the unassessed properties were worth the average for their type, the aggregate would not have amounted to over 4 per cent of the total because most of the omitted values relate to sawmills and watermills, which were not generally worth very much (see table 5.2).

All of these industrial properties were based on water power and forest resources, so that the vast majority were in northern Aska, although the meagre water power resources of Dahl turned four mills (see figure 5.1). That in Väversunda in the extreme south was valued at only 195 Rdr, but the average value of the three mills at Broby in Strå parish on the Mjölnaån, which carries the overflow of Lake Tåkern into Lake Vättern, was 3,121 Rdr and Mjölna mill itself was the second most valuable in the whole region. There were also two windmills in Dahl, in Väversunda and in Nässja parish adjacent

Table 5.1. *Values of different types of property*
in Aska and Dahl in 1820

	Rdr	Per cent
Agricultural land	2,230,225	89.2
Industrial plant	84,945	3.4
Urban real estate	130,706	5.2
Other[a]	54,649	2.2
Total	2,500,525	100.0

Note:
[a] Includes a health spa with a taxation value of
22,000 Rdr and various much less valuable inns, eel
fisheries, turbaries and so on.

Table 5.2. *The values of rural non-agricultural properties in Aska and Dahl in*
1820

	N	N valued	Total listed value in Rdr	Average value in Rdr
Watermills	28	16	31,625	1,977
Windmills	2	2	899	450
Saws	6	4	1,357	339
Ironworks	3	3	45,298	15,097
Brickworks	3	3	5,333	1,778
Quarries	2	0		
Limekiln	1	0		
Chamois leather mill	1	1	100	100
Spa	1	1	22,000	
Totals	47	30	106,612	2,752

to the town of Vadstena. The larger of them was worth only one-twelfth of
Mjölna watermill. The only other non-agricultural properties on the plains in
1810 and 1820 were a stone quarry at Borghamn, on the lakeside in Rogslösa,
and a brickworks at Tycklingen, adjacent to Vadstena in St Per.[102] There was
not a single industrial property in the whole of southern Aska between
Vadstena and Motala Ström. The plain contained remarkably few sources of
livelihood outside farming. Industrial properties supported only thirteen
adults, including hands and their wives in 1810. Even in Strå, where the milling

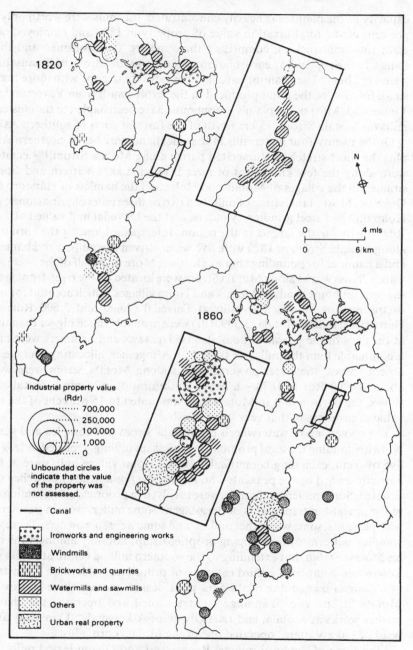

Figure 5.1. Non-agrarian properties in Aska and Dahl in 1820 and 1860.

capacity of the plains was heavily concentrated, the mills were worth only 10 per cent of the total taxation value of property in 1820 and employed only eight full-time workers, comprising three millers, two workmen and three hands. In their splendid near-isolation on the plains, the Broby mills must have been very busy. They cannot have had the capacity to cope with more than a small fraction of the grain produced in the fertile swath from Väversunda to Vinnerstad. Most of the plain's output must have been hauled to the cluster of mills on Motala Ström, 15 km north of the furthest parts of southern Aska.

Of the twenty-four watermills in the shieldland parishes of northern Aska plus the short stretch of Vinnerstad parish along Motala Ström,[103] eighteen were along the few kilometres of river between Lakes Vättern and Boren, situated in the villages of Motala and Holm and the hamlets of Hårstorp and Duvedal. Most of this string of mills had two or three pairs of grindstones; one Holm mill had steel grinding cylinders, and the Duvedal mill valued at 7,600 Rdr was by far the largest in the region. Interspersed among the corn mills along Motala Ström in 1820 were five water-driven sawmills, a rod hammer, and a hammer for pounding chamois leather. More than half of the twenty-six water-driven works along Motala Ström were located on the river-frontages of the peasant homesteads of Motala and Holm villages, which account for most of the missing values. The mills at Duvedal (7,600 and 3,600 Rdr) and Herrekvarn mill in Motala (3,309 Rdr) were probably much bigger than those at the ends of the peasants' homesteads. The saws and hammers were much less valuable than the mills (see table 5.2). Altogether, allocating averages for absent values, the water-powered works along Motala Ström were worth about 44,000 Rdr in 1820 – a heavy concentration of capital. As table 5.3 shows, the portion of it in Motala itself amounted to 15 per cent of the total value of property in that very large parish.

The ownership of waterwheels on Motala Ström was complex and varied. Notwithstanding the legal prohibition of peasant milling, repealed in 1828,[104] five of them, excluding communally owned Holm village mill, were at least partially owned by the peasants who farmed the homesteads where they were situated. Some millers and smiths were involved in proprietorship, reminiscent of the peasant-entrepreneurs of Bergslagen. Some millers were in partnership with peasants, some with other millers, and some were sole owners. There were complex patterns of overlapping proprietorship, not only amongst the mills on Motala Ström, but extending to the southern milling centre at Mjölna.[105] Townsmen comprised a third category of proprietor. Alderman Regnstrand of Vadstena owned one of the Strå mills, leased another there and two on Motala Ström; two Skänninge tanners owned and operated the chamois leather works at Motala, and Lars Olsson of Askersund, 50 km away at the head of Lake Vättern, operated one of the Motala corn mills.

Like some of the local millers, Regnstrand and Olsson leased mills from estates, to which by far the largest proportion of Motala's mills belonged in the

Table 5.3. *Numbers and values of non-industrial properties in the parishes of northern Aska in 1820*

	Västra Ny		Kristberg		Motala[a]		Total	
	N	Rdr	N	Rdr	N	Rdr	N	Rdr
Watermills	4	1,660	1	1,380	17	31,490	22	34,530
Saws	1	300	1	345	4	1,390	6	2,035
Ironworks	1	8,850	1	36,048	1	400	3	45,298
Other ind.	2	1,333	0		2	2,100	4	3,433
Spa	1	22,000	0		0		1	22,000
Totals	9	34,143	3	37,773	24	35,380	36	107,296
Share of all property	21%		43%		15%		18%	

Note:
[a] Averages from table 5.2 substituted for unvalued properties.

early nineteenth century. The noble estates of Baron Fleetwood of Carlshult in Motala, Count Gyldenstolpe of Ulfåsa in Ekebyborna and Baron Lagerfelt, who owned Charlottenborg just outside Motala but lived at Lagerlunda in Kärna parish (outside our region), owned many of the mills on Motala Ström, including the largest at Duvedal. All of these were leased out to millers, the two at Duvedal plus adjacent saws to miller Falk in 1810 and miller Molander in 1820. The estates and lesser holdings of non-noble persons of standing in Motala parish worked as well as owned their mills. Judge Duberg of Lindenäs estate owned and operated two mills on Motala Ström. Unlike noblemen, persons of standing used their mills as parts of integrated agricultural enterprises, turning their estates' own grain into flour.

Further north, deep in the shieldland of Västra Ny and Kristberg parishes where arable land was very sparse, industrial plant accounted for a higher proportion of total asssessed property values than in Motala. In Kristberg they were almost equal in value to agricultural land (see figure 5.1 and table 5.3). There were seven mills and saws there in 1820, clustered in the settlements of Medevi in the extreme north-west of the region, and at Bona and Karlström amongst the tangle of lakes and streams in the densely forested centre of northern Aska. None of these installations was valued at more than a few hundred *Riksdaler* in 1820, but all three settlements also contained a much larger investment in the resources of the shield. At Medevi there was a health spa valued at 22,000 Rdr, whilst the ironworks worth 8,850 Rdr at Bona and 36,048 Rdr at Karlström were the largest industrial plant of the region, making northern Aska a microcosm of forested Sweden as as whole.

As elsewhere, the iron industry was going through hard times in the first two decades of the nineteenth century. Bona *bruk* in Västra Ny, which received its privileges in the boom conditions of 1782, comprised one bar iron hearth in 1810 and 1820. A complete record of its output is not available, but given state policy it must have followed the same trend as the larger Karlström's *bruk* in Kristberg parish, which had existed since the late seventeenth century when it contained a smelting furnace as well as a forge.[106] In line with national trends, Karlström's *bruk* had prospered in the late eighteenth century, but its average licensed output of 96 tons a year in the early 1790s was reduced to 71 tons a year between 1806 and 1810, then halved to 36 tons a year between 1811 and 1815. It must have been running well below the capacity of its two taxed forging hearths at the start of our period. Nonetheless, the clusters of mills, saws, forges and hammers in the interior of the shield were impressive concentrations of industrial activity: those at Karlström's *bruk* were worth almost as much as the twenty-six mills and other works along Motala Ström. The drop in bar iron production was reflected in the fortunes of works which processed it into ironware. Karlström's *bruk*'s two manufacturing hammers were still operating in 1820, as was the Motala rod hammer of smiths Falk and Swartling. But these were the only ones left. Bona apparently did not work up its bar iron at that time; Blommedahl manufacturing forge was listed as closed in the tax register of 1810 and omitted from that of 1820, and Motala village's two rod hammers of 1810 had been reduced to one by 1820. A seventeenth-century tool-making hammer at Kärsby in Motala parish had its production privileges renewed in 1784 and was taxed in 1810, but had closed by 1820.

All the non-agrarian resources in the interior of the shield were developed directly by the estates on which they lay: where resources other than farmland existed on estates, they were exploited. Indeed, only four of the eleven estates of over 25,000 Rdr value which existed in 1810 or 1820 did not contain some non-farming activities. On most, these comprised windmills, watermills, brickworks and rod hammers worth no more than a few thousand Rdr: altogether, of negligible value compared with the agricultural property of the estates. However, the spa, mill, brickworks and limekiln on Admiral Virgin's Medevi estate were worth 24,000 Rdr in 1820, compared with the 35,960 Rdr of the land belonging to it. Bona and Karlström's ironworks were typical *bruk*. The first belonged to a member of the important ironworks-owning family of Grill,[107] whose major Östergötland operations were based at Godegård, 15 km north of Bona outside our region. In 1810, 29,250 Rdr's worth of land was owned by the Bona estate, including a 13,000 Rdr holding of prime plainland arable in Örberga parish, far to the south in Dahl, and sheds on the lakeshore in Vadstena. The larger Karlström's *bruk*, together with a mill and in 1820 a sawmill, belonged to ironmaster af Burén in 1810. In 1820 it had passed to widow Busck, though af Burén retained ownership of various installations at Vadstena used to ship ironworks goods throughout the period.[108] Although

Table 5.4. *Non-agricultural employment in northern Aska in 1810 and 1820*

		Total employed population		Non-agricultural employment	
		1810	1820	1810	1820
Västra Ny	N	361	392	23	27
	%			*6.37*	*6.89*
Kristberg	N	121	120	25	17
	%			*20.66*	*14.17*
Motala	N	429	559	19	32
	%			*4.43*	*5.72*
Totals	N	911	1,071	67	76
	%			*7.35*	*7.10*

four times as large an industrial enterprise, Karlström's *bruk* owned only about one-third as much agricultural land as Bona in 1820.[109]

Thus, in all but the three cases when they were in the hands of urban entrepreneurs from Vadstena, Askersund and Skänninge, the industries of the region were closely integrated into its agrarian economy, either through forward linkage into grain milling or backward linkage into charcoal and food supplies. Most were part of the compound economic enterprises of estates, the others of peasant holdings. Although one or two specialist millers and smiths leased plant from both kinds of proprietor along Motala Ström, the vast majority of industries were still fused with agriculture into single units of capitalization and operation.

This intimate interrelationship carried through into labour supplies. Valuable though some of them were, the industrial works of the region provided few full-time livelihood positions. The reason for the complete absence of people listed under about half the works in the taxation records of 1810 and 1820 is that they were operated by part-time labour during lulls in the agricultural work of peasants and crofters, primarily in winter. This is also why only four charcoal burners and no sawyers, tar burners, lumberjacks or hauliers were listed. Although industrial plant accounted for 3–4 per cent of the taxation value of property, they provided livelihood positions for only 2 per cent of the region's occupied population. In shieldland northern Aska in 1810 non-agricultural employment accounted for 7.35 per cent of all livelihood positions, falling slightly to 7.10 per cent in 1820 (see table 5.4). In Kristberg the proportion was much higher, although the difficulties of the iron industry brought a slight fall in the total population of the parish and a 30 per cent reduction in the proportion dependent on industry. In Motala, on the

Table 5.5. *The industrial labour force of Aska and Dahl and its dependants in 1810 and 1820*

	Workers		Dependants		Unproductive			Total		
	Men	Maids	Wives	Children	Males	Females	Children	M	F	Ch
1810	97	24	33	57	12	10	8	109	67	65
1820	119	43	48	91	3	6	3	122	97	94

Table 5.6. *Ratios of children to wives by occupational group in 1810 and 1820*

	1810	1820
Farmers	2.00	1.89
Crofters	1.63	1.71
Industrial workers	1.73	1.90

other hand, where most of the industrial plant were mills to process rising agricultural output, both the number and proportion of the population dependent on non-agricultural activities increased between 1810 and 1820. Of the ninety-seven men occupied at industrial works in 1810 (see table 5.5), eleven held executive positions as owners, tenants and managers, thirty-three were skilled foremen, master craftsmen, journeymen or apprentices, and fifty-one were labourers or hands. In addition, there were twenty-four maids and twelve apparently unproductive men (presumably retired workers) living at the various works. In 1820, there were nineteen executives, thirty-three skilled and sixty-six unskilled workers, forty-three maids and three unproductive men. At both dates the hands were obviously mostly young, few being married and even fewer having children. All the apprentices were single, and so were most journeymen. Thus, a nucleus of less than fifty adult married men (indicated by the number of wives on table 5.5) was supplemented by a much larger complement of juvenile, and therefore probably transient, labour. Judging by the ratio of children to wives, the adult labour force seems to have had a similar age profile to that of the region's farmers and crofters (see table 5.6). The largest rural industrial labour forces, and the most skilled, were at the ironworks. The Karlström's *bruk* workforce of seventeen men in 1810 and fourteen in 1820 were master hammersmiths and their skilled helpers, plus five charcoal hands and stokers. Bona *bruk* had eight male workers in 1810 but only three in 1820. In principle, Karlströms *bruk*'s three hearths could have produced and processed 400 tons of iron per year, employing six hammersmiths and skilled assistants, each with a hand, and about twenty charcoal burners working full-time through the winter season, plus sled-hauliers for pig and bar iron, charcoal and timber.[110] But it produced far less and employed fewer permanent workers than that at this time. The part-time employment it generated for the agricultural population of the shieldland must have been commensurately smaller.

Despite the depression in ironmaking, the effects of the *bruk* in generating part-time employment for the surrounding farming population are still clearly evident in the farming statistics of northern Aska: the vast majority of holdings were in category III, and crofters and cottagers were much more numerous than on the plains. We have already shown that this was partly due to the sparsity and low quality of arable land on the shield and the importance of grazing and other forest resources, and the domestic industry they supported. However, another part of the reason was the typical effects of *bruk* in generating work for farmers and crofters in their vicinity through their demands for charcoal and the haulage of timber, charcoal and iron.

The vast majority of work generated by the industries of northern Aska was hidden in this way, even in Motala, on the very edge of the shield. Few of its mills had more than one or two workers (the largest had one miller and five hands), the brickworks and quarries no more than double that. Even this large concentration of water-powered plant contained few workers compared with the farms, crofts, and cottages of the agricultural holdings on which they were located. The eleven settlements bordering Motala Ström contained seventeen farmsteads, only one of which was larger than a category III holding, and twenty-six industrial works. The farms also contained forty-nine crofts and thirty-two cottages, housing altogether 139 adult males and a total population of 479. The works, which must have been worth, in taxation value, not far short of double the farmland, housed only thirty-six men and ninety-six persons in total. Even in the most industrialized place in the region, the population was apparently much more dependent upon agriculture than industry. But this, again, is an illusory picture. The high ratio of crofts and cottages to farms and the smallness of the farms implies that many of their occupants derived some of their livelihood from the machines turned by the river. For example, Motala Aspegård farmstead was valued at only 750 Rdr: barely enough to support a peasant family at subsistence level and too small to require hired labour. Owned by Mr Bromander of Bromma estate (probably for the mill on the river at the end of the homestead) it was leased by Carl Holmström, who occupied it with his wife and two children. This was Holmström's only holding, but even so he employed a maid and a hand, and there were two crofts and two cottages on Aspegården, housing three crofters with wives and five children, a day-labourer and a widow. The bulk of this labour force must have spent most of its working time on tasks outside agriculture, in and about the mills along the river. The hammersmiths of

Motala village, like those of *bruk*, worked agricultural land, although they had to lease or buy their farmsteads in the village. Again, we come up against the inseparability of industry from agriculture in the early nineteenth century. The only places where this was not true in Sweden at the time were the *uppstäder* thinly scattered across the plains. How many and what kinds of livelihood existed in Vadstena, and what were the resources on which they were based?

Vadstena

Figure 5.2 shows that Vadstena was dominated by two massive royal buildings. The convent, monastery and abbey on royal land to the north-east owed their foundation to St Birgitta, who was canonized in 1391.[111] The town was granted its charter in 1400, and the magnificent abbey church, which thenceforth housed the reliquary of St Birgitta, was consecrated in 1430. The castle to the south-west of the town was begun in 1545 as a royal stronghold next to rebellious Småland and to guard the passage from Danish Skåne along Vättern into the heart of the newly independent kingdom. The castle and its precincts were made part of St Per parish, most which belonged to the royal estate of Kungs-Starby. Vadstena expanded eastwards towards the abbey to compensate the loss of its western end to the castle, hence the eccentricity of the town hall and its square. The convent and monastery passed into Crown hands together with their land. The castle remained royal property, although it was not used as a royal residence after the seventeenth century.

Despite the eclipse of its great establishments, the town retained borough status and therefore the rights to hold markets and fairs, to govern itself, and to organize guilds. It was the only *uppstad* in our region (although Skänninge lay alongside Allhelgona parish) and therefore the only place within it where full-time craft occupations, retail and wholesale trade were permitted, and its landing stages linked the fertile surrounding plain with the long perimeter of Lake Vättern. The presence of large royal buildings with land holdings attached to them ensured that its tertiary functions served areas well beyond its immediate hinterland. By the early nineteenth century most of the old monastery and convent buildings were used as a royal hospital, a sanatorium for people with venereal diseases, an old soldiers' home and, after 1819, a workhouse[112] and prison.

According to the census of 1801, the population of the town was 1,313, third largest in Östergötland, half the size of Linköping, the provincial capital, and one-sixth the size of Norrköping.[113] According to our calculations, based on the reconstitution of household data from the tax registers, Vadstena's population was 1,224 in 1810 and 1,438 in 1820, 11.1 per cent and 10.5 per cent of the Aska and Dahl totals.

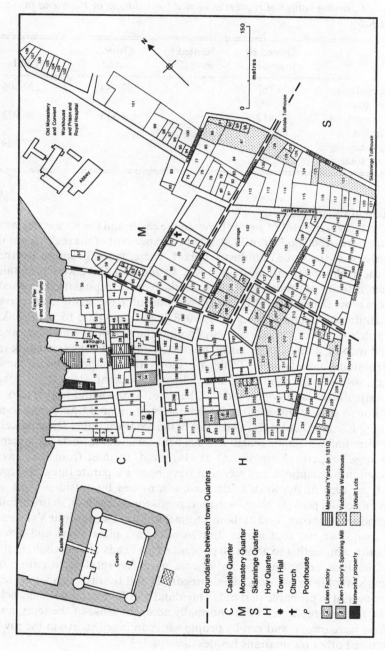

Figure 5.2. Vadstena in 1820.

Boundaries between town Quarters

C Castle Quarter
M Monastery Quarter
S Skänninge Quarter
H Hov Quarter
✳ Town Hall
✝ Church
P Poorhouse

▨ Linen Factory
▨ Linen Factory's Spinning Mill
▦ Ironworks property
▥ Merchants' Yards (in 1810)
▨ Vadstena Warehouse
░ Unbuilt Lots

Old Monastery and Convent
Workhouse and Prison and Royal Hospital
Abbey

Town Pier and Water Pump
Lake Tollhouse
Market Square
Vicarage

Castle
Castle Tollhouse

Skänninge Tollhouse
Motala Tollhouse
Hov Tollhouse

0   150
metres

N

Table 5.7. *Taxation values of resources owned by residents of Vadstena in 1820 (in Rdr)*

	Owned by townsfolk	Rented by townsfolk	Owned from outside	Total
Vadstena real estate	106,762		23,944	130,706
Vadstena salaries	86,900			
Vadstena farmland	32,812		4,111	36,923
Other farmland	20,824	16,376		
Vadstena total				167,629
[Townsfolk total =	247,298]			

The nature of urban resources

The real estate on the grid of streets between the castle and the monastery was valued for tax at 130,706 Rdr in 1820,[114] about 5 per cent of the regional total and nearly twice the value of industrial plant. These buildings supported the bulk of the livelihood positions in the town. In addition, there were hospitals and other institutions funded by the Crown (the Royal Hospital itself received rents from more than 20,000 Rdr's worth of land within the region); a damask linen manufactory; state salaries of 8,690 Rdr per year paid to townsfolk; 36,923 Rdr's worth of agricultural land within the parish of Vadstena, and rents and income derived from farmland owned and tended by townsfolk in the surrounding countryside. Table 5.7 shows the composition of these resources, with salaries capitalized at a yield of 10 per cent.

The agricultural land comprised 272 hectares of townland arable and a hundred or so grass plots, separated from the built-up area by a continuous fence along Rännevallen, marking the line of the old town walls. These parcels of land were unfenced and located in open fields, without farmsteads, except for the large property distinguished as Hagalund, furthest from the town, which contained buildings and seems to have been a separate integral farm. Only 11 per cent of the town's farmland was owned by outsiders. Just as countryfolk could provide many non-food goods for themselves, so townsfolk could produce their own foodstuffs and animal fodder. Twenty-four Vadstena people owned arable land alone, four owned grass plots alone and seven owned both, amounting to about 10 per cent of the heads of household in the town. Classified according to the scheme used for rural farms, four of Vadstena's were in category I, six in category II and twenty-one in category III. However, it is probable that only Hagalund and four rural holdings held from the town were 'farms' in the normally accepted sense of the term, and even these were owned and run by people who continued to live in the town and followed other occupations besides farming.[115]

Owners of townland lived scattered across the whole town. Some of them had barns (see figure 5.4). Only one was valued separately from other buildings on the same plot, but at 520 Rdr it was worth more than most houses. The unbuilt lots holding the barns were probably also used for grazing; cattle, pigs and horses must have moved continually between urban lots and the town-fields, and harvests were brought into the barns.[116] The hands and maids who worked this land also lived in the town, and there must have been enough slack in the urban labour market to provide for the seasonal peaks in the farm labour schedule. Vadstena had a decidedly rural aspect: a similar amount of land supported about 250 people in the rural villages, about 17 per cent of the town's population. However, this land was part of diversified portfolios of resources rather than the sole source of livelihood of its owners. In this respect, the urban farm holdings were like small person-of-standing estates rather than peasant farms.

Urban real estate was worth almost four times as much as the town's farmland in 1820. It was split up into 274 plots, plus half a dozen or so unnumbered properties at the toll houses and in their vicinity 'outside the fence' along Rännevallen (see figure 5.2). The thirty plots not yet built on were on the north-eastern fringe of the town and in a discontinuous blunt wedge from the northern periphery towards the market square. The locations of the castle and monastery were echoed by the public buildings of the town, giving a western orientation to civic power and an eastern one to spiritual authority and wider social services. The market square lay at the centre, where the main streets and the boundaries of the quarters which followed them met. The landing stages on the lake were to the west, along the shoreline of Castle Quarter, near the lake customs house, a merchant's yard, and property belonging to ironworks owners. Just inland were other merchants' yards (only distinguished in 1810) and the damask linen manufactory. The town looked to Lake Vättern and the massive ancient structures which overshadowed its shore, rather than towards its fields and the plain beyond them. This orientation was reflected in the pattern of property values (see figure 5.3), which ranged from 25 Rdr for cottages to 5,000 Rdr for the building on plot 181 near the market square. Generally, values were highest in Castle Quarter, especially along the east–west spine of Storgatan, which marked the boundary between Castle and Hov Quarters before running through the market square and on to separate Monastery from Skänninge Quarters. Some plots contained buildings owned by several people. On the other hand, groups of plots, such as 149–153, were given a single value ascribed to one owner.

The ownership of Vadstena's real estate was more widely distributed than its farmland, amongst 164 people living in the town. Fourteen of them were wives, children, journeymen and apprentices living in households headed by others, so that 150 (43 per cent) of the 347 households in the town owned property within it. The rest rented housing or lodged with other families.[117]

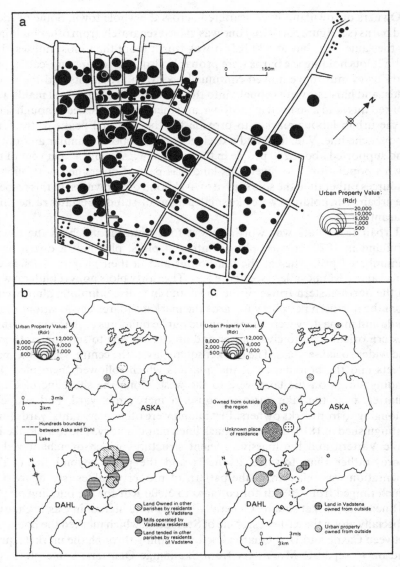

Figure 5.3. (a) Plot values in Vadstena in 1820. (b) Rural property owned and tended by residents of Vadstena in 1820. (c) Places of residence of outsiders who owned property in Vadstena in 1820.

Forty-six townsfolk owned more than one plot, mostly adjacent ones, which might initially have comprised a single plot (see, for example, 1–5 and 189–91 on figure 5.2). Others, such as 35 and 36, might always have been separate. Twelve people owned non-adjacent properties, eight of them three or more, but only two multiple holdings were so valuable (over 6,000 Rdr) and scattered as to suggest that they were real estate investments, made deliberately to yield rental income.

A much larger proportion of the town's real estate than its farmland was owned by non-residents (see table 5.7): 18 per cent was in the hands of twenty-five outsiders (see figure 5.3(c))[118] whose holdings ranged from the cottage of Grenadier Ek of Fivelstad to the three town plots and six grass plots of Widow Jaensson of Arneberga in Örberga parish in Dahl. The reason for most of these acquisitions by outsiders can only be guessed. The ironworks owners' sheds and waterside properties were obviously for the transit of goods to and from their *bruk*, 20 km to the north-east, and Major Vult von Steijern of Kungs-Starby's town properties were used in connection with his linen manufactory. The two properties in the hands of non-resident heirs might have been bequests from erstwhile residents of Vadstena, whilst those of rural estate owners like Lieutenant Colonel Rosenborg of Naddö in Nässja, District Judge Duberg of Lindenäs in Motala, Commissioner Myrtin of Kvissberg and wealthy Widow Jaensson of Arneberga might have been acquired to facilitate business and/or social life in the town. On the other hand, were these properties bought as investments to yield rents; or were the owners from Vadstena families before marriage, outmigration or inheritance (Widow Jaensson lived in Vadstena before moving to Arneberga); or had they bought them with the intention of retiring to the town from the countryside? Only longitudinal study of the people concerned can answer such questions about why people scattered from Gränna in the south to Linköping in the east and Godegård in the north-east owned so much of Vadstena's urban property.

Urban social structure and property ownership
Occupational or status labels attached to names in taxation registers make it possible to classify people in these terms. Of course, this has a number of limitations. A dozen or so of the townsfolk were described in the *mantals-längder* as following a dual occupation, such as apothecary and winetrader, or ascribed one occupation there and another in the *taxeringslängder*: for example, innkeeper in one and brewer in the other. We assigned such people singly to that category in which they paid most tax, or, if that is unknown, the commoner of the two. Many townsfolk were assigned no gainful occupation, but described as widows, spinsters or formerly of a particular activity, though some of them paid tax on income from either the occupation of their deceased husband or that which they 'formerly' followed. They were removed from the

'unproductive' category and classified with the occupation for which they paid tax.

We ended up with household heads classified according to single occupations, with a residual 'unproductive' category of widows, spinsters and retired persons. The occupations were then classified into the five groups listed in table 5.8,[119] which are reasonably distinctive in principle, if not always in practice. First, salary earners in the state and civic administrative service, plus those following legal, church and medical professions; secondly, merchants and others such as apothecaries and innkeepers who lived from buying and selling rather than making wares; thirdly, craftsmen; fourthly, people in menial occupations, and fifthly, those who were apparently unproductive. The first three groups comprised the bourgeois mainstay of a pre-industrial urban economy. People in each of them paid taxes on salaries or incomes from businesses as well as property; they manned the town's courts and councils, and the second two groups would have contained the members of its guilds. These 'basic' bourgeois urban economic activities seem to have had powerful local multiplier effects, because bourgeois households accounted in total for less than one-third of those in the town, and there were more in both the menial and the unproductive groups.

Although a high proportion of all groups owned property, a hierarchy is suggested by the data in tables 5.8–5.10. Certainly, the occupied groups are ranged quite clearly in table 5.8. The dealers had more property owners and higher valued urban holdings than the other groups, although the craftsmen owned more land in the townfields. The dealers' incomes on which taxes were paid, although not the taxes themselves (which were not completely listed), were higher than those for all other groups (see table 5.9). On the other hand, table 5.10 shows that the administrative and professional group, besides paying higher average taxes, also owned and tended considerable areas of land outside the town. The net effect was that the dealers, administrators and professional people of Vadstena, as in Swedish towns generally, were wealthier than the rest, with larger town properties and substantially higher incomes than the craftsmen, whilst the menial workers were ranged much further below the craftsmen.[120]

This very clear arrangement was complicated by the existence of substantial holdings of property amongst the large apparently unproductive group. On average, its urban property holdings were almost as large as those of the craftsmen; its members owned a majority of the farmland of the town, and they owned and tended more rural farmland than any other group. This suggests that quite a few farmers *manqué* were hidden beneath the status and retired designations by which members of this group were labelled, and/or that there was a large *rentier* class in the town, living on incomes from capital invested in urban property and farmland.

Table 5.8. *Occupations of heads of households in Vadstena in 1820 and the urban property holdings of each group*

	N		% total		Urban real property				Arable and grass		
Group	1810	1820	1810	1820	% owning property	Total value	Average value	N	Total value	Average value
Administrative and professional	38	28	11	8	43	13,397	1,116	8	4,447	501
Dealing	26	16	7	5	63	19,953	1,995	3	2,620	873
Crafts	49	50	14	14	50	18,936	757	5	6,928	1,386
Menial	89	111	25	32	35	5,954	153	1	58	58
Unproductive	154	142	43	41	46	48,522	753	13	18,759	1,443
Totals	356	347			44	106,762	702	30	32,812	1,093

Table 5.9. *Incomes from and taxes paid on incomes from salaries and businesses in Vadstena in 1820*

| | | Salary or business income | | | Taxes paid | | |
	N	Number	Sum paid (Rdr)	Average Rdr	Number of payers	Sum paid (Rdr)	Average Rdr
Administrative and professional	28	24	8,690	362	24	418	17
Dealing	16	6	2,683	447	12	195	16
Crafts	50	4	900	225	43	274	6
Menial	111	9	512	56	5	30	6
Totals	205	43	12,785	297	84	917	11

Table 5.10. *Farmland owned and tended outside Vadstena by the occupational groups within the town in 1820*

| | Owned | | Tended | |
	Number of owners	Value in Rdr	Number of tenants	Value in Rdr
Administrative and professional	4	10,331	2	7,232
Dealing				
Crafts			1	1,680
Menial	3	10,493		
Unproductive	7	20,824	1	7,466
Totals	14	41,648	4	16,378

What was the composition of these groups, and how were income, property and other possible sources of livelihood distributed within them?

Administrators and professonals   Twenty-four of the twenty-eight persons in this category received taxable salaries. Altogether, 8,690 Rdr flowed into the urban economy through this channel, of which the vicar, Dean Kinnander, received nearly one quarter. He owned 163 Rdr's worth of townland arable, 2,332 Rdr's worth of rural arable land, and four urban properties worth 900 Rdr in addition to the large vicarage and adjoining plots, which he occupied as a grace and favour residence. The next highest salaries were those of the prison commander (700 Rdr) and three accountants (600, 500 and 500 Rdr). Curates

Tingwall and Harlingsson, two schoolmasters and a hospital doctor each got 300 Rdr per year and grace-and-favour houses. Tingwall also owned 774 Rdr's worth of property in the town, 32 Rdr's worth of townland arable and 4,666 Rdr's worth of rural farmland. The postmaster and army accountant, a judge, a canal book-keeper and an alderman[121] earned between 100 and 300 Rdr a year, and the remaining salaries ranged from 43 to 125 Rdr.

The smaller urban property holdings of this group in general, compared with the dealers and craftsmen, were due to the omission of their crown land grace-and-favour residences from the tax valuations and, perhaps, to the fact that goverment officials moved around the country during their careers, which may have acted against acquiring property in any particular place before retirement. However, their rural property does suggest that some of them invested to provide extra income, as do their multiple urban properties. Even so, the largest of their urban holdings (that of Judge Gezelius) was, at 2,565 Rdr, worth much less than half of the largest amongst the townsfolk. It is not possible to discover from tax records whether these people were engaged in other economic activities, but the inventories of two churchmen's widows show that they were farming, and Alderman Regnstrand was heavily involved in both farming and milling.

Dealers The taxation records list twelve *handlande*, which can be translated as dealer, trader, merchant or shopkeeper. Four of them were ascribed an income from trade and taxes due on it, five taxes only, and three neither. The occupation cannot have been free of risk because three, all owning property and paying tax, were in the hands of bankruptcy trustees. Perhaps this and the considerable decline in the number of dealers between 1810 and 1820 (see table 5.8) indicate that Vadstena dealers, like ironmasters on the shield, were going through difficult times. Some were prospering: *handlande* Wallberg earned 1,000 Rdr a year from his business, on which he paid 50 Rdr tax, and two other dealers earned over 300 Rdr per year, as did the innkeeper and apothecary/ winetrader who have been classified with them. The remainder either earned, or paid taxes on earnings equivalent to, less than 100 Rdr per year.

The decline in the number of merchants and the difficulties of some of them might also have been related to the beginning of a shift of port activities from the lakeshore in Castle Quarter to the castle moat, accessible to shipping from the lake, and the location of the Vadstena Warehouse registered in St Per (Dahl). In 1810 it was called the Royal Warehouse and housed only the occupier and his family. In 1820 Alderman Regnstrand and his family, a stores foreman plus an assistant, a housekeeper, two hands and four maids lived there. It must have been used in connection with Regnstrand's farming and milling business, and might well have handled other goods.

The small merchant-cum-shopkeeping community, like the administrative and professional group, was highly differentiated. At its apex, half a dozen people had larger incomes than all the administrators and professionals except

Dean Kinnander. They owned nearly all of the urban real estate, townland and rural farmland belonging to the dealers as a whole. Judged from their property holdings, two of the three bankrupts had belonged to this sub-group. However, two-thirds of the dealing community were much poorer; three owned no property and apparently earned too little to be liable to income tax.

The businesses of the apothecary/wineseller, innkeeper and lacetrader are self-evident. The first paid far more tax on his pharmacy than his wine trading. According to his inventory of 1830, his pharmacy licence was worth 13,000 Rdr and included a practice at Medevi health spa as well as Vadstena. Many inventories include small debts owed to him, whilst he owed money to a bank and to apothecaries in Göteborg, Stockholm, Jönköping and Skänninge, as well as to a trading company in Stockholm and to four people in Norrköping. The location of nearly half of the dealers in Storgatan and the market square (see figure 5.4) implies that shopkeeping and retail trade with the rural hinterland on market and fair days was important for some of them. The merchants' yards on the waterfront (see figure 5.2) and another cluster of their dwellings there (see figure 5.4) suggests that some of them were involved in long-distance trade. *Handlande* Ringström certainly was. His inventory lists 392 types of goods worth 2,357 Rdr in his rented shop, ranging from thread, buttons and cloth, to snuff, vitriol and treacle, but mainly haberdashery. Debts owing to Ringström, carefully listed as secured or unsecured, totalled 560 Rdr, with the latter kind accounting for 440 Rdr. Mostly of a few *Skilling* or *Riksdaler*, they were owed by 137 people: some of them were weavers and weavers' journeymen, but most were maids, wives, widows and spinsters, and the largest debt by far, of 48 Rdr, was owed by a maid. The places of residence of forty-one debtors are listed. All lay outside Vadstena, implying that those who were not given a place of residence lived in the town. Only nine of Ringström's customers lived outside our region, four of them in parishes adjacent to its boundary. Ringström's debts were quite different. There were only twenty-nine, but they amounted to 6,637 Rdr. Nine were owed to people in Vadstena, including rent to the owners of his room and his shop, and those to three merchants and an apothecary, to whom he owed 525 Rdr, far more than to all his other local creditors combined. He owed similar or greater sums to manufacturers, merchants or companies in Stockholm, Karlshamn, Göteborg, Borås and Örebro, but his largest debts were to Norrköping, where he owed 1,172 Rdr, and Askersund, where he owed five merchants and/or aldermen 2,098 Rdr, some of it for commission goods.[122] Perhaps some of this trade was related to the Göta Canal. Although it was not completely opened from Lake Vättern to Göteborg until 1822 and to the Baltic until 1832, some impact on Vadstena from its construction is suggested by the existence of a canal book-keeper and a canal provision master in the town in 1820, and the itemization of canal shares in some inventories of the 1820s.

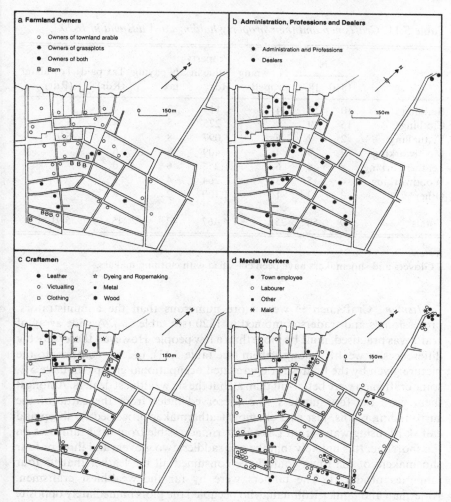

Figure 5.4. The distribution of owners of farmland and various occupational groups in Vadstena in 1820.

The dealers' property holdings suggest that they were less involved as *rentiers* than the administrators and professionals, presumably because they invested accumulations of capital in stock-in-trade. The high value of most of their urban properties was due to the size of their dwellings and business premises. Only one owned more than one property not on adjacent plots, and although he owned seven they were worth only 755 Rdr in total. Nearly half of the townland owned by dealers belonged to the innkeeper.

Table 5.11. *Craftsmen and their property holdings in Vadstena in 1820*

	N		N owning property	Property value in Rdr	N paying tax	Tax paid (Rdr)	Townland (Rdr)
	1810	1820					
Building	10	11	4	114	7	4	
Clothing[a]	15	9	1	225	8	4	
Victualling	2	7	6	1,097	8	8	5,910
Metalworking	7	9	4	409	8	6	
Leatherworking	8	6	6	1,258	6	10	
Woodworking	4	5	3	264	4	5	
Other	3	3	1	1,100	2	8	
Totals	49	50	25	4,467	43	45	5,910

Note:
[a] Glovers and shoemakers have been classified with clothing makers.

Craftsmen Craftsmen[123] were more numerous than the administrators, professionals and traders combined in 1820 (see table 5.8). A wide array of crafts was practised, none by more than a few people. However, there were big differences in wealth amongst them (see table 5.11); indeed, contrary to the picture given by the figures for aggregated occupational groups on table 5.8, some craftsmen were better-off than all but the six wealthiest dealers. Amongst them were the four tanners who preserved hides from the cattle of the surrounding plains, just as a chamois leathermaker processed its sheep and calf skins, using water from Lake Vättern, accessible from their tanneries on the shoreline. It is unlikely that the one saddler, two glovers and three or four shoemakers of the town could have consumed all the leather that its four tanneries produced. The tanners were by far the wealthiest craftsmen. Alderman and tanner Röhman, who occupied the plots immediately opposite the castle, was taxed at 25 Rdr on a craft income of 500 Rdr, which was half of the tanners' total taxes and an income greater than all but one of the dealers. The tanners' properties were also more valuable than those of other craftsmen, because of the value of their dwellings-cum-tanneries in Castle Quarter rather than accumulations of more than one property.

The 'others' (two dyers and a roper) were the second weathiest group, followed by the five brewers and two bakers who comprised the victuallers.[124] The number of brewers suggests that like the tanners they provided part of the export base of the town. Two of their premises were located hard by the shore, and they were more than twice as wealthy as the bakers. Although the wealthiest of them had an income only 10 per cent the size of that of tanner

Röhman, all the townland arable belonging to craftsmen was theirs – presumably because their craft needed grain.

The dyers and roper were as wealthy as the brewers. Like the tanners and brewers, they were engaged in processing produce from the peasants of the plain, in this case the raw cloth made in their farmsteads and their crops of flax and hemp, using the water supplies of the lake. The inventory of Alderman and fine dyer Hellström, taken in 1829, gives an idea of his business and other activities. His three plots by the lake and a building he owned on Rännevallen were worth 1,600 Rdr. They housed his farm implements and animals as well as a dyehouse, press house and colour house, together worth 200 Rdr. The last contained small quantitites of ten different colouring and bleaching agents. The 155 small debts owed to him at his death suggest that he processed small quantities of thread and cloth for people from the town and a surrounding area stretching from Motala in the north to Västra Tollstad, just to the south of our region. These debts totalled 3,562 Rdr, more than twice the value of his property. He owed 2,744 Rdr to twenty-six people, much of it in small quantities to manufacturers and merchants, but 324 Rdr to *handlande* Wallberg of Vadstena and to Lönngren and Eskilsson of Norrköping, probably for goods supplied, and 975 Rdr, presumably a loan, to Judge Duberg of Lindenäs.

The remaining craftsmen must also have performed basic central place activities for customers from the surrounding region and the town itself. Amongst the metalworkers were two goldsmiths, a watchmaker, two copper-smiths and a tinsmith's widow as well as blacksmiths. The goldsmiths paid less tax than the brewers, but more than the other metalworkers, and generally this group contained fewer people with property than the craft groups discussed so far. The inventories of a watchmaker and a master smith (taken in 1828 and 1829) were a far cry from those of the *handlande* and fine dyer. Small debts owed to smiths, journeymen and apprentices suggest that the blacksmith was still active, but his total inventory amounted to only a few hundred Rdr, with far fewer debts and credits (the largest, a debt of 22 Rdr, owed to a merchant) and built property worth only 52 Rdr.

The clothing makers were equally varied, including glovers, shoemakers, a clogmaker, hatters, and tailors, who all presumably depended, like the metalworkers, on selling what they made retail in the town. A tailor's and a hatter's inventories were of the same magnitude as those of the master smith and watchmaker. The joiners and carriagemakers among the woodworkers must also have depended on retailing their products in the town, although the turner probably supplied at least some of his output to other craftsmen. The carpenters and coppersmiths, like the bricklayers, glaziers, painters and tile stove maker, might have been mainly dependent upon the demands of the townsfolk themselves.

The varied craft occupations of the town were clearly graded in terms of property, income, tax payments and inventory wealth, reflecting their differing market orientations. At the apex were tanners, brewers and dyers who processed raw materials from the plain, probably for sale by the town's merchants as well as locally. There were no more than ten of them, all with premises near the lake shore in Castle Quarter. Another group of about the same size, including goldsmiths, hatters, glovers and a watchmaker, produced luxury goods for direct sale to the wealthier people of the region. More than half the craftsmen were concerned with satisfying much more mundane demands for bread, clothes, shoes and housing. They were much poorer in terms of property and tax liabilities. The bulk of the craftsmen, and all the wealthiest, lived between the merchant-dominated main street of Storgatan, the market square and the lake shore, though a minority was scattered amongst the plots well to the east of the town's main thoroughfare, presumably amongst their urban customers. Castle Quarter and the small parts of the other quarters near the market square comprised a distinct business and industrial heart of the town, hence the high property values there shown on figure 5.3.

The damask linen manufactory was by far the biggest employer of industrial labour in the region. Typical of the Swedish manufactories of the time, it produced Vadstena's most widely celebrated product, table linen with complex woven patterns and inscriptions which was regularly bought by the royal household.[125] It was set up with Flemish craftsmen in the wings of the castle in 1753 as an assembly of hand weavers under one roof. Like most of the region's industrial plant, it was still part of a large estate in the early nineteenth century, owned and run by the nobleman Vult von Steijern, who was custodian of the castle and occupied the large royal estate of Kungs-Starby to the south of the town. The flax dressing and bleaching grounds of the works lay on the estate (as did the castle itself) but the factory was moved into premises on Townhall Square in 1816. It is likely that spinning, done only by women, was carried out domestically in Vadstena and the surrounding countryside, as for other weaving manufactories.[126] Weavers trained at the works set up factories at Norrköping, Nyköping and Örebro as well as additional works in Vadstena. In 1820 three master weavers, seven journeymen and eight apprentices lived on the premises with their five wives and fifteen children, and a journeyman lived in a small property he owned in the town. None the less, Vult von Steijern's three properties were valued in total at only 1,770 Rdr, despite their favourable locations, and only 8 Rdr tax was paid for the works, no more than the town's run-of-the-mill craftsmen and less than one-third of what tanner Röhman paid. The brickworks just north of the town at Tycklingen in St Per (Aska) was valued at 2,337 Rdr in 1820 and employed only four hands (who with their three wives and two children, lived on the premises).

So far, we have mainly discussed occupations in the export base of the urban economy. It is a remarkable testimony to the self-provisioning capacity of

Table 5.12. *Property holdings of menial workers in Vadstena in 1820*

	N		N with property	% with property	Total value in Rdr	Average value in Rdr
	1810	1820				
Town exployees	18	18	4	22	419	105
Labourers	48	73	28	38	4,548	162
Maids	10	9	4	44	246	62
Others	11	11	3	27	741	247
Totals	87	111	39	35	5,954	153

rural households that the 13,000 or so people of the plains and shield provided markets and materials enough to support negligible numbers of specialist craftsmen in the rural areas themselves and only sixteen dealers and fifty craftsmen in their central place. The ratio of specialist urban livelihood positions to farms was only about 1:9, and of central place positions to farms about 1:2. But if the multiplier effect of the rural economy into urban activities was very small, that of the specialist urban activities themselves into non-basic elements of the town economy was apparently massive because, as table 5.8 shows, there were considerably more menial positions in the town than the combined total of the occupations discussed so far.

Menial workers   Tables 5.8 and 5.9 show that menial workers were much poorer than craftsmen. A workman's inventory of 1828 was worth in total only 90 Rdr, of which his house comprised 50 Rdr. However, it is perhaps more remarkable that over one-third of the menial workers owned the houses in which they lived. Indeed, two labourers owned more than one property, another a small portion of townfield arable, and the town sexton owned more than many craftsmen and dealers. Neither in terms of property ownership nor places of residence was there much difference between the various menial groups which can be distinguished. Table 5.12 shows their property holdings, and figure 5.4 their residential distribution. Their property comprised cottages and other small dwellings, very different from the dealers and wealthier craftsmen in Castle Quarter. Those which were owner-occupied were spread across the back streets and lanes of Hov and Skänninge Quarters. Many menial workers also lived in rented cottages or parts of subdivided buildings put up next to their own much more valuable dwellings by owners of plots in Castle Quarter and near Market Square.

The town employees comprised people required to ensure the order and smooth running of the municipality and its townland. They included four town guards, four night watchmen, a customs guard and the town sexton.

Only the last owned property (worth 71 Rdr) and he was the only one whose salary of 114 Rdr was large enough to be liable to tax. The other town employees, except for the town fisherman, supervised agricultural activities, emphasizing either their importance to the town or the difficulties of combining agricultural and urban activities in one place. Two pig bailiffs, three field guards and a grass plot guard were necessary to prevent animals straying off their owners' holdings and to supervise everyday coming and going between the townfields and barns and dwellings in the town. The 'other' menial workers included a gardener, two paviours, a chimney sweep and three hands who, like the nine maids, did not live in their employers' houses.

Labourers and workmen[127] were far more numerous than any other occupational group and the only one to increase significantly between 1810 and 1820. The proportion who owned property and the value of their holdings in 1820 (see table 5.12) suggest that they were of the same material standing as the other menial workers with whom we have grouped them, but the kind of work they did is not known. They must have gained their livelihood from things such as field work on the town arable; fetching and carrying on fair days and market days, haulage work generally,[128] and helping with the menial aspects of craft work (perhaps even doing it on their own account, but surreptitiously because they lacked the legally necessary guild membership).

Not only males were capable of earning or owning resources. Three or four of the labourers' houses were owned by their wives, who might also have spun for the linen manufactory,[129] or made lace. In 1755 the government gave a grant to encourage lacemaking in the town, and in 1781 a visitor from Stockholm watched 'old men, old women, boys and girls occupied in making lace which will be sold all over the country' at the Old Soldiers' Home.[130] It was difficult to get spinning done for the linen manufactory because 1,000 women were involved in lacemaking, at which they could earn more.[131] Of thirteen inventories of men and three of wives, six and two, respectively, list spinning wheels, with no differences according to wealth. Indeed, the inventory where spinning for the weaving factory is suggested by a debt of 133 Rdr to Vult von Steijern, and involvement in lace putting-out by a debt of 8 Rdr for lace, belonged to a comfortably-off factory master's wife. She did not possess a lace cushion, but one of the wives and one of the men did. Three of the men's inventories and all the wives' also contained looms, one of them two. A large textiles and lace manufacture was hidden from the taxation records in the households of menial workers and the better-off members of Vadstena society.

'Unproductive' households The same must have been true of the huge proportion of the 40 per cent of the town's households headed by people to whom the taxation records attribute no occupation, as listed on table 5.13. In all cases, the proportions holding property and the sizes of the holdings of the retired were smaller than those of people who were still actively engaged in the

Table 5.13. *The unproductive households of Vadstena and their property holdings in 1820*

			Urban real estate		Townland arable		Grass plots	
			Total value	Average value		Value		Value
	N	n	%	(Rdr)	(Rdr)	N	(Rdr)	N	(Rdr)
Former agriculture	10	4	40	217	54	0			
Former administrative and professional	8	2	25	935	468	0		1	66
Former dealing	3	0				1	673	1	600
Former craft	16	4	22	2,070	518	2	327		
Former menial	8	2	25	90	45	0			
Spinsters	25	14	56	4,009	286	0		1	12
Widows	72	40	56	1,204	30	10	16,713	5	368
Totals	142	66	46	8,525	129	13	17,713	8	1,046

occupations concerned. Perhaps they were living off capital during their retirement, although some might have continued to practise their occupation in a small way. The former surmise is supported by the disproportionately small number of ex-menial workers, who would be least able to stop working and live on accumulated capital in their old age. None of the retired males owned non-contiguous multiple properties, so it seems unlikely that any of them had significant rental income besides ex-apothecary Holmberg, who had 1,273 Rdr's worth of agricultural land.

The most noticeable features of table 5.13 are the large numbers of single women in the town and their enormous holdings of property of all kinds.[132] The high relative standing of the 'unproductive' group as a whole on table 5.6 transpires to be almost entirely due to the property holdings of spinsters, and especially widows. Single women were even more relatively numerous in Vadstena than in pre-industrial towns generally.[133] It is difficult to discover why without knowledge of their earlier lives, and because of the opaqueness of the terms widow and spinster: in only sixteen cases do the registers give an ex-husband's occupation, and only longitudinal tracer studies can reveal how many of them lived in the town before widowhood or, in the case of spinsters, before moving out of or inheriting their parents' home.

They were a very heterogenous group in terms of wealth and background. None of the spinsters owned more than 1,000 Rdr's worth of property or owned more than one non-contiguous property, or any town arable or rural land. Just over half owned the houses in which they lived, but nothing else. They must all, therefore, have been dependent for their livelihoods upon savings, spinning, lacemaking, or taking in lodgers. So must the eight widows who owned less than 100 Rdr's worth of property on a single site and the thirty-two who owned none. Indeed, spinning, weaving, and especially lacemaking seem, on the evidence of the few inventories we examined, to have been even more prevalent in the households of widows and spinsters than in those of married women (which perhaps accounts for the large haberdashery component of the stock-in-trade of *handlande* Ringström). Eight of the thirteen inventories of single women contained spinning wheels; a wealthy minister's widow, who employed a number of maids, had four. Three of these households also had looms, one of them two, and seven contained lace-cushions. Of these, three had two cushions, one three and one four, plus three 'lace chairs' (*knyppelstolar*). Perhaps most of the spinning wheels and looms were used mainly for household self-provisioning; but some spinning wheels must have been used to supply the linen manufactory's yarn, and lacemaking was obviously an important source of livelihood for the single women of Vadstena.

Only one lace dealer was enumerated in 1820. He was relatively poor and practised another occupation as well. None of the dealers' inventories consulted contained lace. The lace trade was in the hands of women such as Mrs Hartwick and Mrs then Widow Gylling, to whom many of the women with lace-cushions owed small sums of money.[134] Widow Christina Elfström was another wealthy lace dealer. In her 500 Rdr house behind the main street, she employed a maid and had three lace-cushions, 13 Rdr's worth of unfinished lace, 135 Rdr's worth in store, and a further 536 Rdr's worth unsold, but out for sale (presumably by commission agents and travelling lace-sellers). She was owed 3,431 Rdr at her death. Of this 48 Rdr was due from lace dealer Anders Molander, and over 400 Rdr from twenty-eight people to whom she had put out lace. All were listed without places of residence, and thus presumably lived in Vadstena. Four, including 'the Russian Gregorius', were men, continuing the tradition of the by-now closed Old Soldier's Home. The rest were spinsters, widows and wives across the full social spectrum of the town. The largest debt Elfström owed was 65 Rdr to merchants Roos and Wallberg of Vadstena, presumably for thread and other lacemaking supplies.

A group of sixteen widows owned large amounts of property (see figure 5.5(a)), of various kinds. The widows of Professor Achaius, War Commissioner Kock and Assistant Judge von Steijern, and Widow Tisell owned single large properties or a group of properties on contiguous plots in the centre of town which could have represented their erstwhile family houses. However, others of the sixteen seem to have invested in property to yield rental income. Accountant Dahlström's widow, for example, owned a large block of properties numbered 94–100 running along the eastern edge of the town inland from the monastery, occupied by five families other than her own. They were worth 6,252 Rdr in total, and widow Sondell's three properties were worth 2,614

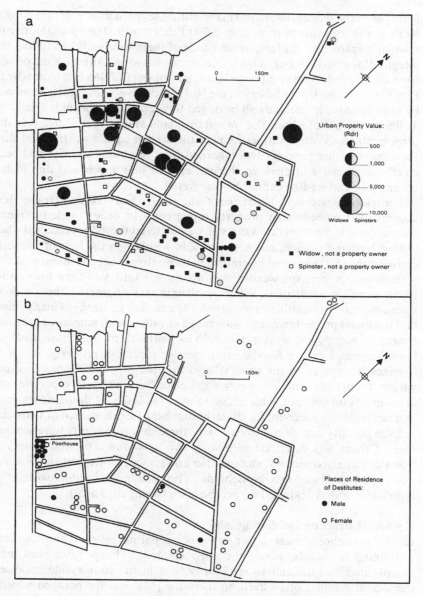

Figure 5.5. (a) Property of widows and spinsters in Vadstena in 1820. (b) Places of residence of destitutes in Vadstena in 1820.

Rdr. Ten of the sixteen owned townland arable, seven of them more than 1,000 Rdr's worth and two of them over 3,000 Rdr's worth. The vast majority of townland arable was, in fact, in the hands of these widows,[135] including the integral Hagalund holding of trader Lindström's widow. Although the person from the unproductive group who leased a farm worth 7,466 Rdr in nearby St Per (Aska) was ex-trader Edström, the 10,493 Rdr's worth of rural land owned by unproductive households all belonged to widows Gyllensköld, Tisell and Tollbom, the last of whom also owned townland arable worth 3,692 Rdr and three contiguous urban properties near the market square worth 2,670 Rdr. How much of her property was inherited from her husband (who had been chief provisioner and clerk to the castle), and how much was due to her lacemaking and trading cannot be discovered.

Thus, sixteen widows (at least one of whom was a lacemaker and trader) had accumulated considerable property, apparently in order to derive rental income from it. Perhaps this was a safe way to invest the money received when a dead husband's urban business was sold – although the lack of property amongst the ex-traders and ex-craftsmen themselves is not consistent with this hypothesis. Or perhaps wealthy widows moved into Vadstena from other towns and from rural areas. Wherever their money originated, these widows were amongst the wealthiest people in Vadstena. The existence of many urban and townland properties over a wide range of values, with numerous potential tenants to occupy and workmen to help look after them, must have made the town a prime location for the investment of capital in property, whilst its liveliness compared to the rural villages and the preponderance of women within it must have made it a congenial place from which to supervise such investments on the spot. This added to the availability of domestic spinning and lacemaking to account for the large number of single women in Vadstena.

Even so, fifty-one of the town's sixty-three destitute adults were women. One of them was described as a wife, two as widows, the rest simply as 'females'. The two widows, an ex-soldier and his wife, and a destitute Russian maid lived in their own households. The remainder lived-in with other families[136] widely scattered across the town (see figure 5.5(b)).

Household structure and demography
Urban households were a far cry from separate cells of consanguinity containing husbands, wives and their children. Things were even more complicated and difficult to see clearly than in the countryside because a number of families often lived on the same plot, and the point at which a number of distinct households living in separate parts of a dwelling became a complex multiple family household sharing a dwelling is impossible to fix. This problem is eased by the designation of many individuals as journeyman, apprentice, hand, maid and living-in:[137] that is, people who were unequivocally members of the households of others. However, it is a moot point

whether two or three widows and spinsters living in one property formed one or a number of households; indeed, they might well have shared some aspects of domestic life and kept others separate. The basic units of social organization in Vadstena are therefore difficult to reconstruct, partly because of the nature of the records, but partly too, we suspect, because they were indistinct in many cases. In any event, the nuclear family household was by no means common.

Table 5.14 shows that salaried, dealing and craft households contained many more people; hence, perhaps, in part, the greater value and therefore size of their properties. The clear distinction in terms of household size and composition between people with specialist urban livelihood positions and those in menial and unproductive positions is even clearer in the averages given on table 5.15. The mean family sizes of the craftsmen and menial groups suggest that they were nearest to an average middle-aged population. The number of children per family of craftsmen was insignificantly different from those of farmers and industrial workers given on table 5.6, although other urban groups had smaller families than those of economically active rural households. The fact that craftsmen had more older children at home than the menial workers could imply either that they were further into their life cycles on average or, more probably, that older children were more likely to stay in the parental homes of craftsmen, whilst those of menial workers moved out to earn their own living as hands or maids in the households of others. The family structures of the administrative and professional group were also quite distinctive, with relatively few young and an average number of older children, which is consistent with their being on the whole older than the craftsmen or menial workers. Some of the dealers, on the other hand, were childless single males, hence their small average family size. Tables 5.14 and 5.15 show that 'unproductive' families contained quite a few children in both age groups. They nearly all belonged to widows: the imperative to seek the kinds of livelihood that Vadstena could provide for single women with or without capital was presumably greater for those with children to support, whilst the wealth of some of them would allow them to keep older sons and daughters at home as companions and helpers.

The differences in family sizes was compounded by the presence of more non-family members in the households of the bourgeoisie. They had many more hands and maids per household, and the craftsmen and dealers housed apprentices or shop assistants and journeymen at the rate of nearly one per household.[138] The non-productive workers also housed a few hands and many more maids. However, the fewness of hands, in particular of those living with the unproductive households who owned most of the townland, implies that much of the work traditionally done by living-in hands was already being done for wages by labourers living in their own homes, explaining their numerousness in the town. In contrast to the hands and maids, the vast majority, and all the males, who lived-in were in non-productive households,

Table 5.14. Household composition in Vadstena in 1820

Occupational group	N	Heads of households				Journeymen and apprentices[a]				Hands and maids[b]				Living-in				Totals				
		M	F	Ch1[c]	Ch2[d]	M	F	Ch1	Ch2	M	F	Ch1	Ch2	M	F	Ch1	Ch2	M	F	Ch1	Ch2	T
Administrative and professional[a]	28	28	18	8	20	1				4	30				5			33	53	8	20	114
Dealing	16	15	7	2	18	11				9	21				5			35	33	2	18	88
Crafts[e]	50	47	41	27	64	63	7	2	10	7	33		1		6			117	87	29	75	308
Menial	111	102	99	13	130					1	6		3		13			103	118	13	133	367
Unproductive[f]	142	43	128	44	107					11	56			55	48		2	109	232	44	109	494
Institutions[g]	2	2	3	4	3					11	26	15	6	44	7			57	36	19	9	121
Total	347	237	296	98	342	75	7	2	10	43	172	15	10	99	84		2	454	559	115	364	1,492

Notes:

[a] Excludes those living in institutions.
[b] Includes nurses and hospital hands.
[c] Children of eighteen years and over.
[d] Children under the age of eighteen.
[e] Includes linen factory.
[f] Includes Hagalund, owned by a widow.
[g] Inmates = 'living-in'.

Table 5.15. *Average household and family size of occupational groups in Vadstena in 1820*

	Average number of people per household	Average family size	Number of children		
			> 17 yrs.	< 17yrs.	Total
Administrative and professional	4.07	2.64	0.29	0.71	1.00
Dealing	5.50	2.19	0.13	1.13	1.25
Crafts	6.16	3.58	0.57	1.28	1.82
Menial	3.31	3.10	0.12	1.17	1.21
Unproductive	3.47	2.17	0.31	0.75	1.06
Averages	3.95		0.28	0.99	1.27

largely those of widows and spinsters. Their roles can only be guessed at. The males probably did heavy work around the house, hardly different from hands except in terms of their lack of wages, stage in the life-cycle and, perhaps, an expectation that they would do what they could to supplement the household's budget from pensions[139] or odd jobbing. The females probably acted as companions, shared chores, and were as heavily involved in domestic craft work as the women with whom they lived.

Households were not headed by providers and rendered complex by festoons of dependants and hangers-on. All members except small children did something to support the group they lived with. Households in every social group provided supplementary livelihood positions of this kind. Generally, the wealthier the household, in terms of salary or craft income and property value, the more supplementary members it contained.[140] The wealthiest craftsman, Alderman tanner Röhman, had a very large household: besides himself, his wife, and four children, there were two journeymen, three apprentices, three hands and two maids on his complex of plots opposite the castle. Dean Kinnander had one hand and four maids to minister to the needs of himself, his wife and four children. The wealthiest widows, Dahlström and Tollbom, also had four maids and a hand each, and the former shared her house with Hospital Director Kullbeck, his wife, three children, hand and two maids, plus a spinster and two destitute females who lived-in with her.

Summary: the regional non-agrarian economy in the early nineteenth century

We can get a clear picture of Vadstena's economy and society at the beginning of the period, although that of non-agrarian activities in the countryside is rather less distinct. In the town, the tightly packed, fenced-in grid of streets and lanes contained a craft, business and commercial sector around Market Square and across Castle Quarter to the Town Hall and the shore. Here the most valuable properties contained large complex households with journeymen, apprentices, hands and maids helping the families who occupied them. Inland, Hov and Skänninge Quarters had the much less valuable properties of menial workers and most of the widows and spinsters. The last, too, had complex households, most with single men and/or women living-in and the wealthiest with hands and maids.

The economy of the town was widely based. Its medical and penal institutions served the whole of Östergötland and beyond, and the remits of its religious and legal officers extended far outside its boundaries. Its port and burghal trading privileges linked Vadstena with places as far afield as Göteborg, Karlshamn, Örebro and Stockholm, but most intensively with Norrköping and Askersund, the nearest ports on the Baltic and Lake Vättern. These formally sanctioned contacts brought in goods for distribution to consumers, whilst informal 'hidden' links took out its lace and linen. The tanners and brewers probably sold in distant markets, though through what channels we cannot say, whilst the dealers and other craftsmen sold to and finished goods for people throughout our region, on the shield as well as the plain, even though their numbers do not suggest an intense stimulus from the farming economy into specialist urban livelihood positions. The linen factory employed about one-third of the town's apprentices and journeymen, and the debts and credits listed in inventories imply that much more business was done among the townsfolk than between them and the denizens of the countryside. Ironworks owners from Västra Ny and Kristberg in the extreme north of the region, and from Godegård beyond, used Vadstena for shipments, and District Judge Duberg of Lindenäs, almost as far away, owned property and was owed large debts in the town. The basic economic activities of administration, dealing and craft production multiplied into the proletarian livelihoods of a majority of the town's gainfully employed households.

However, as in the countryside, there was a large and varied hidden economy which is not revealed in counts of individuals in occupational and status groups. Spinning and lacemaking (as well, undoubtedly, as sewing curtains, cushion-covers, clothes and so on) provided income for wives and single women, whilst rents from urban, townland and rural properties helped support the wealthiest widows of all. Just as many of the non-farm goods that would later be acquired in towns were as yet produced in the farm households of the countryside, so many of the townfolk were engaged in agriculture: the widows and a few of the bourgeoisie in the ownership of arable land, the menial population in farming it as hirelings or tenants. The ownership of farmland in the countryside by townsfolk, and of urban real estate by countryfolk, further blurred the distinction between town and country which seems so clear in Swedish Law at the time and in the occupational ascriptions of the tax records.

Similarly, the non-agrarian activities of the countryside were not as starkly separated from agriculture as their specification in the taxation registers and the listing of specialist workers under them imply. Most workers in those industries were employed part-time, and many of the ancillary tasks such as timber felling and haulage were done by peasants, crofters and the menial populations of town and countryside. The almost complete lack of craftsmen in the countryside testifies to the amount of domestic craftwork done in farms, crofts and cottages. Most of the industrial plant, including the Vadstena linen factory, were owned and run as part of large mainly agrarian estates, and ironworks had large farm holdings to ensure timber and food supplies.

If specialization of labour and business enterprise are the hallmarks of a capitalist economy, then our region was indubitably pre-capitalist in the early nineteenth century, even in its non-agrarian aspects. Townsfolk as well as rural people occupied complicated livelihood positions supported by the resources of a variety of stations; most stations, whether workshops, professional or *rentiers*' households, or town arable land, supported livelihoods indirectly through wage payments and directly through the consumption of their owners and family and non-family members. Generalized and varied livelihood positions such as these and large-scale access to resources other than through wage-earning are the hallmarks of an economy which is weakly penetrated by the market. Although it was differentiated in wealth, the urban bourgeoisie did not contain fortunes as large as the biggest peasant holdings of the plain, or half the minimum value we set for a rural estate.

However, there were some clear indications of change. If guild restrictions and privileges were meant to ensure that each practitioner of each craft achieved a sufficient competence to fit them for similar stations in life, they were no longer having that effect. Although there was a reasonably distinct hierarchy of urban wealth and household composition from dealers and professional people through the craftsmen to the menial workers, it was blurred by wide differences in each group. The dean and one or two other members of the first group, half a dozen dealers and nearly twice as many craftsmen in tanning, brewing and dyeing were much wealthier than everyone else in the town, except for sixteen widows who, with them, owned most urban property. The large and apparently growing number of workmen, the small number of living-in hands, and the absence of shop assistants, apprentices and journeymen from many of the dealers' and craftsmens' households signify that proletarianization was occurring amongst the labour force, or that tasks which were by law restricted to guild master craftsmen, their journeymen and apprentices were surreptitiously being done by others.

258   Peasantry to capitalism

Table 5.16. *Values of different types of real property^a in western Östergötland, 1820–60 (in millions of Rdr taxation value; 1860 × $\frac{4}{3}$)*

	1820	1830	1840	1850	1860
Total value	2.5	2.3	3.0	2.7	3.9
% in industrial plant	3.4	4.2	7.3	8.6	27.5
% in Vadstena real estate	5.3	7.6	6.0	7.8	8.6
% in Motala *köping* real estate	0.0	0.1	16.9	2.2	6.9
% in agricultural land	91.2	88.1	69.8	81.4	57.0

Note:
^{a} Excluding spa, inns, eel fishery and turbary.

Table 5.17. *Proportions of the population of western Östergötland employed by each economic sector, 1810–60^a*

	1810	1820	1830	1840	1850	1860
Total	11,976	13,775	16,633	17,846	18,965	21,840
% industrial	2.0	2.2	4.0	4.5	6.8	12.5
% Vadstena	11.1	10.5	10.6	9.8	10.3	9.8
% Motala *köping*	0.0	0.0	0.4	1.5	2.3	4.8
% Rural non-agricultural	1.0	0.8	1.8	0.8	0.9	2.1
% Agrarian	85.9	86.5	83.2	83.4	79.8	70.8

Note:
^{a} The second row includes industrial plant located in Vadstena and alongside Motala Ström and the Göta Canal. The farmland and small farming population of Vadstena were classified as agrarian.

Changes in non-agricultural livelihoods in Aska and Dahl, 1810–60

General overview

In 1810 our region was similar to Sweden as a whole in that the vast majority of productive resources were in agriculture, which sustained an even larger proportion of its population. However, tables 5.16 and 5.17 show that these proportions diminished quite sharply in western Östergötland, especially during the last decade of the period. As agriculture was transformed, its share of both resources and employed population slumped. By 1860 industrial plant and urban real estate comprised over 40 per cent of the region's taxed property (compared with 20 per cent over the whole of Sweden in 1862).[141] Table 5.17

shows that Aska and Dahl were in line with the population structure of Sweden as a whole in 1810, with about 14 per cent the total supported by urban and industrial employment. Although this proportion changed very little in the national totals by 1860, in our region it doubled. The manufacturing export base of the region was supporting nearly 12 per cent of its population, employing nearly 14 per cent of its male workers, in 1860. In 1810 the equivalent regional proportions were 2 per cent and 3 per cent. However, table 5.17 also shows that despite this eye-catching surge, which accelerated rapidly in the 1850s (see figure 5.8 on page 269), full-time employment in manufacturing was smaller in 1860 than in urban and rural handicrafts, trade and services, which supported 17 per cent of the region's population, compared with 12 per cent in 1810, the vast majority of them in Vadstena and Motala *köping*.

To examine these developments in detail requires some reclassification of the subregions. Much of the industrial and tertiary development was along Motala Ström, where the engineering works was opened in 1822, followed by the setting up of a market settlement (*köping*), and then the opening of the eastern length of the Göta Canal in 1832. Although it was not officially designated as a town until 1888, the straggle of industry and settlement along the river and canal was undoubtedly urban in function long before that. We have pulled together entries for the various components of this settlement,[142] which we will call Motala Town, from the rest of Motala and Vinnerstad parishes, which remained rural.

Changes in Vadstena, 1810–60

The town's share of the region's population declined from 11 per cent to 9.5 per cent between 1810 and 1860, but its population increased by almost 50 per cent[143] to 1,760 and although its changes did not greatly exercise aggregate regional statistics, they were quite profound.

Urban resources
The upper map on figure 5.6 shows considerable change in Vadstena by 1860. To the north-east, the Royal Hospital had been completely rebuilt just beyond the old monastery site in the 1820s to allow for expansion and better medical care. The monastery became the County Hospital of Östergötland, and in 1826 the nunnery became Sweden's first mental hospital.[144] A year earlier, the newly restored abbey was consecrated as the town's parish church and a theatre was opened by Alderman Regnstrand against a corner of the abbey wall. The old parish church of St Per was demolished in 1829, except for its tower, against which a new school was built in 1831. To the south-west, the castle fell into further disrepair and was used as part of the warehouse complex which had developed alongside the moat. This harbour was improved between 1849 and 1855 by providing long moles (made from the castle ramparts) to

ensure a deep channel out into the lake. The restaurant pavilion and the two steam ships which contained restaurants must have added a jaunty aspect to the warehouse-lined castle precincts. Across Slottsgatan from the castle, a large Bavarian lager brewery had opened on what had been Widow Röhman's tannery, and the shoreline of Castle Quarter sported some small private boathouses. The medieval mesh of streets within this topographical efflorescence remained superficially unchanged, but by 1860 it contained a telegraph office and the Vadstena Enskilda (private) Bank at its heart[145] and plots must have been heavily developed to produce the massive absolute and relative increase registered in their taxation value, especially in the last decade of the period (see figure 5.8(b)).

Figure 5.8(c) shows that whereas the citizens' agricultural holdings had made up 36 per cent of their taxable assets in 1820, they accounted for less than 9 per cent in 1860. More and more of Vadstena's townland arable was farmed by outsiders after 1830, growing to nearly 50 per cent by 1860. The first inroads were made in the 1830s by the Lallerstedts' Kungs-Starby estate, which almost surrounded Vadstena's townland. In 1860 about half as much townland was worked from Kungs-Starby as in 1840. Smaller farms in St Per (Aska) and Strå had gained several lesser holdings in Vadstena, but much of the Lallerstedt estate was taken over by Royal Forester Abolin. It included detached Kvissberg, where he lived surrounded by Vadstena's townfields. He owned townland worth 12,407 Rdr in 1860 and Vadstena mill, next to Kvissberg land.

There were still four barns in the town in 1860, and sixteen citizens farmed townland, compared with twenty-four in 1820. One other farmed a category II, and two others category III holdings in adjacent St Per and Strå and nearby Örberga. This was a large shrinkage from the amount of outside land owned and/or farmed by townsfolk in 1850 (see figure 5.8(c)). Nearly half of the farmland remaining in urban hands was in the peripheral integral holdings of Hagalund and Strömslund, where their owner, an ex-apothecary, lived with his workforce quite separately from the town. The smaller Östralund, Snibben and Fridhem holdings had also been carved out of the town fields as integral farms by 1860. Half of what remained of the townland was owned by one townsman, so that the fourteen other holdings worked from the town were all very small. On the other hand, there was a very large increase in the salaries of officials. Although there were, at thirty, only six more than in 1820, their aggregate annual incomes rose five times to nearly 40,000 Rdr in 1860 – more than double the taxation value of the townland still owned by citizens. The value of urban real property grew by a factor of three and increased its share of the total taxable assets of townsfolk from 55 to 70 per cent. If the value of the ships, warehouses, brewery and other non-housing properties are included, it comprised 77 per cent. By the end of our period the interests of the citizens of Vadstena had coalesced around purely urban resources and withdrawn from the townland and farming further afield.

Houses, workshops and shops still occupied the 274 numbered plots lying between the lake shore, Slottsgatan and Rännevallen, with a few cottages beyond the toll gates. The tripling in value of the town's houses was achieved by developing the pre-existing built-up area rather than extending it. Figure 5.6 shows that the old core of the town increased proportionately more in value than peripheral areas in Hov and Skänninge Quarters, despite much higher initial values. Some of the buildings in the peripheral parts of these quarters actually declined in value, though the parts along Storgatan shared in the strong increases. The highest rates of increase of all were in a discontinuous ring around the old core, where plots with low value buildings in 1810 were redeveloped as they were sucked into the core. The purely urban resource of centrality became much more important between 1820 and 1860.

This was reflected in the greater subdivision of plots between owners of different buildings. In 1820 ownership of the 274 town plots was shared between 164 heads of household in the town, 14 dependants and 25 outsiders. In 1860 it was spread across 214 heads of household, 8 dependants and 17 outsiders. In 1810 46 people living in the town owned more than one plot; in 1860 53 people did so. However, the proportion of the town's households owning property decreased by 10 per cent to 33 per cent; none of the multiple-plot holdings was as large as Widow Tollbom's in 1810, and they mainly represented adjacent plots, cottages or unbuilt land. The proportion of the town's real estate owned from outside fell from 18 per cent in 1810 to 12 per cent in 1860. Of this 40 per cent belonged to the nobleman Chamberlain Morath of Klosterorlunda estate in Hov and Lieutenant Lallerstedt of Kungs-Starby. They were entered in the tax records as occupiers as well as owners of their large properties on Storgatan, which seem, therefore, to have been kept as urban residences. Most of the other externally owned houses belonged to people currently living and working as merchants, apothecaries and so on in other towns. Only eight ordinary people from the surrounding rural parishes owned urban property in 1860, none more than a few hundred *Riksdaler*'s worth.

Urban social structure and property ownership
The occupations and status descriptions given in the *mantalslängder* of each decade from 1830 to 1860 were classified according to the scheme used for 1810 and 1820, with the addition of workmen for whom a craft was specified, who were categorized as 'independent craftsmen' (that is, independent of guild training and control, legally allowed to practise after 1846).[146] Multiple occupations and occupations practised by widows and 'retired' males were dealt with in the same ways as in the earlier period. Another problem of classification emerged by 1860: taxes were paid for warehouses, restaurants, the bank, ships and the offices of trading partnerships by named owners who were also taxed at their home addresses, where they were also ascribed an

occupation. To avoid multiple counting, only those occupations ascribed to their places of residence were included in our tabulations. The separation of workplace and home also affects the accuracy with which the occupational structure of heads of households reflects the town's economy, and household structure that of its labour force. Productive single males and females employed at specialist places of work lived in a household without forming part of its workforce. With the general difficulty of defining households in tax records, this means that figure 5.8(a) cannot be taken completely at face value.

Nonetheless, some of the trends on it merit comment. It is tempting to interpret the decline between 1830 and 1840 as a response to loss of trade and custom from what had been the northern part of Vadstena's hinterland to Motala Town. However, the overall decline in households occurred solely in the menial and 'unproductive' groups; the administrative and professional, dealing and craft groups all increased in that decade. Even so, the town's total population (excluding inmates of institutions) declined by about 13 per cent between 1830 and 1840: newly developing Motala Town might have drawn away people from the lower echelons of Vadstena's economic structure, or the flow of in-migrants to it. The emergence of the independent craftsmen, especially during the 1840s and 1850s, is also noticeable on figure 5.8(a). Their numbers waxed as those of men described simply as workers waned. We suggested earlier that many 'workers' might have surreptitiously been doing craft work at the beginning of the period and the increase of independent craftsmen and the decline of plain 'workers' might simply have reflected honest reporting after the monopoly of guilds was abolished in 1846. The two together increased from 89 in 1810 and 113 in 1820 to 130 in 1860, as their share of urban households stayed at just under one-third. Thus, the most noticeable features of figure 5.8(a) might well be illusory, and the occupational composition of the household heads of the town probably hardly changed between 1810 and 1860.

Their relative wealth changed markedly. Figure 5.8(b) shows that the merchants and craftsmen caught up with the 'unproductive' group in the ownership of urban property as the latter's share fell from 45 to 31 per cent between 1820 and 1860.[147] The independent craftsmen did not own much property even in 1860 when they were relatively numerous, and the absolute value of the menial group's was halved after 1840. As the topography of the town changed in ways which suggest that central place functions were growing, those who performed them greatly strengthened their hold on its real estate.

Figure 5.8(d) shows that the aggregate income[148] of the professionals surged ahead in the 1850s, although this was not reflected in their urban property holdings, which fell from 18 to 12 per cent of the total over the decade. This is because many more of the salary earners of 1860 had grace-and-favour accommodation; or they were single lodgers rather than heads of

Table 5.18. *Proportions of each occupational group owning property in Vadstena in 1860 and the average values of their properties (in Rdr of taxation value × $\frac{2}{3}$)*

	N	Urban housing			Other urban property		
		% owning property	Total value	Average value	% owning property	Total value	Average value
Administrative and professional	30	17		39,027	9,757		
Dealing	30	45		95,066	6,790	?	15,910	?
Craft	58	46		84,977	3,147	2	1,532	766
Independent craft	69	9		3,056	509		
Menial	60	17		3,660	366		
Unproductive	230	36		99,966	2,777		
Totals	477			325,752			

households, suggesting that the group's surge in income depended to some extent on an influx of young people in new professions such as banking and telegraphy, which came to the town in the 1850s. The aggregate incomes of the dealers, craftsmen and independent craftsmen also increased in the 1850s. The former two groups each earned about 20 per cent of the total for the town in 1860, the last under 4 per cent. The share of the menial workers fell from 3 to 1 per cent.

Figure 5.8(c)(i) shows that nearly all the townland arable still cultivated from the town in 1860 was in the hands of the 'unproductive' group. Craftsmen and dealers pulled out of agriculture as they invested in urban real estate, and the professionals as they fell in age in the 1850s and, perhaps, as the bank offered a less irksome home for savings. The 'unproductive' group also doubled the absolute value of its urban holdings between 1840 and 1860, though its proportional share fell slightly.

The averages for 1860 in tables 5.18–5.20 show some interesting contrasts with the data for 1820. The average property holdings and incomes of people in the administrative and dealing groups, already wealthiest in 1820, grew rapidly, moving them further ahead of the craftsmen over the forty years. Although members of the 'unproductive' group maintained their large aggregate holdings of urban and townland property (see table 5.17), their rural farmland fell away and the average value of their urban properties increased less than those of the administrative and professional, dealing and craft groups.

The craftsmen advanced more slowly than the two groups above them, but

Table 5.19. *Numbers of each Vadstena occupational group owning and/or tending farmland in 1860 and the average values of their holdings (in Rdr of taxation value × ⅔)*

	Townland arable and grass			Rural farmland		
	N	Total value	Average value	N	Total value	Average value
Administrative and professional	3	766	255	2	4,150	2,075
Dealing	1	1,107	1,107	0		
Craft	6	2,280	380	2	3,800	1,900
Independent craft	0			0		
Menial	0			1	1,287	1,287
Unproductive	13	23,246	1,788	3	2,139	713

Table 5.20. *Incomes of Vadstena occupational groups in 1860 (in Rdr of taxation value × ⅔)*

	N	Total	Average
Administrative and professional[a]	30	40,506	1,350
Dealing[b]	30	10,666	356
Craft	58	11,467	198
Independent craft	69	2,800	40
Menial	60	3,000	50

Notes:
[a] 1 partnership = 2 individuals.
[b] 3 partnerships of 2 = 6 individiduals.

faster than those below. The high proportion owning property implies a more stable older age-structure than the professionals. Although their tabulated average incomes in 1860 were smaller than in 1820 (tables 5.20 and 5.7) this is because income data were recorded only for the highest earners in 1820. If recorded incomes times two-thirds in 1860 are compared with the average tax paid by craftsmen multiplied by twenty (the taxation rate was 5 per cent) in 1820, then craftsmen's average earnings increased from 120 Rdr to 198 Rdr, or by 65 per cent. This was higher than the increase in average dealers' earnings, but the value of the latter's property increased faster than the craftsmen's. The menial and indendent craftsmen groups formed a markedly poorer sector of the working population.[149]

As in the countryside differences between social groups became more marked. In 1820, there had been quite marked gradations within each of the socio-economic levels. Was this still the case in 1860, and why did the fortunes of the groups diverge?

Adminstrators and professionals This was a more varied as well as a much expanded group by 1860. A vicar, curates, schoolmasters, judges, town clerks, a town prosecutor, a town doctor, a veterinary surgeon and two harbour bailiffs ministered to the needs of the town itself. The hospitals housed two managers, a foreman and three doctors; the bank employed a manager and three staff, one of whom doubled as the postmaster; the telegraph office a commissioner and two assistants. The hospitals had 290 patients in 1860 and employed sub-manageresses (who had no recorded salaries) as well as the officials already listed. The poorhouse had thirty-one inmates. The vicar no longer had the highest salary, which was now the 5,604 Rdr a year paid to District Judge Modig.[150] Ten salaries were bigger than the highest earnings recorded for a dealer,[151] seventeen than the highest craftsman's earnings (the 800 Rdr of goldsmith Holm), whilst dealers Björk and Lagerdahl paid almost seven times as much tax on their earnings as harbour bailiffs as they did for the business of their trading company. Of course, the dealer's and craftsmen's taxed earnings represented profit, and generated bigger multipliers into the town economy per 1,000 Rdr than these salaries. Nonetheless, there was a notable surge in the 1850s in the money flowing into the town in return for its serving a wider hinterland with health and financial services, postal and telegraph links, and as it grew the town itself paid a wider range of professional people to serve its own needs.

Many of these officials lived in grace-and-favour residences attached to their posts. The town houses owned by the postmaster and the district judge were impressive affairs near the centre of the town, worth 10,067 and 5,473 Rdr. The inventory of the postmaster's predecessor who died in the 1850s had a positive balance of nearly 20,000 Rdr and included very many debts owing, which suggest extensive investment in loans at interest. It is possible, in fact, to discover who invested in and borrowed from Vadstena Enskilda Bank from surviving ledger books.[152] We analysed that for 1857. Table 5.21 shows that a very large majority of bank shareholders from Vadstena itself did apparently come from the administrative and professional group. Although this impression is somewhat misleading because of the problem of residential ascription, this is itself testimony to how well known the persons of standing of the surrounding countryside were in the town. The impression that the administrative and professional people of Vadstena were more prominent as depositors rather than borrowers must be correct. The most obvious implication of the table, however, shown by the imbalances between deposits and withdrawals in the last four rows, is the importance of the bank and those who

Table 5.21. *Places of residence of shareholders in Vadstena Enskilda Bank in 1857*

	Number of shareholders	% of shares held	Number of borrowers	% of loans held
Vadstena[a]				
Administrative and professional	47	24.1	30	19.0
Dealing	9	5.9	14	8.4
Crafts	10	1.3	5	1.1
Bavarian Brewery	—	—	—	0.9
Menial	3	1.9	—	—
Unproductive	7	0.3	2	2.8
Unknown	4	1.4	7	3.8
Vadstena total	80	35.0	58	35.9
Dahl	32	7.8	28	18.4
S. Aska	35	12.9	42	15.9
N. Aska	17	4.4	31	23.3
Other	113	39.0	16	6.4
Total	277		176	

Note:
[a] Most of the people for whom no place of residence was given were from Vadstena, but some judges, ministers, army officers, noblemen and other eminent people from the surrounding rural area were not given an address. It was not possible to discover exactly who among those without addresses were from outside Vadstena, and they have all been classified as coming from the town.
Source: Register of Shareholders, Vadstena Enskilda Bank, Landsarkivet, Vadstena.

administered it in syphoning funds from outside the region into its rural areas. The pattern of these flows will be examined in detail in the next chapter, but it is appropriate to note here that this new central place function of Vadstena transferred nearly half a million *Riksdaler* into and around its hinterland.

Dealers   The topographical changes discussed earlier, the opening of the Göta Canal in 1832, and the new town harbour of 1855 suggest that the town's commercial economy must also have changed. Although the number of dealers fell from twenty-six to fifteen between 1810 and 1830, it had climbed to thirty by 1860.[153] The dealers' share of urban property rose sharply between 1820 and 1860 (see figure 5.7); they owned nearly all of the taxed non-house property in the town shown on figure 5.6, and they borrowed more from the bank than they deposited in it (see table 5.21). Rather than suffering in

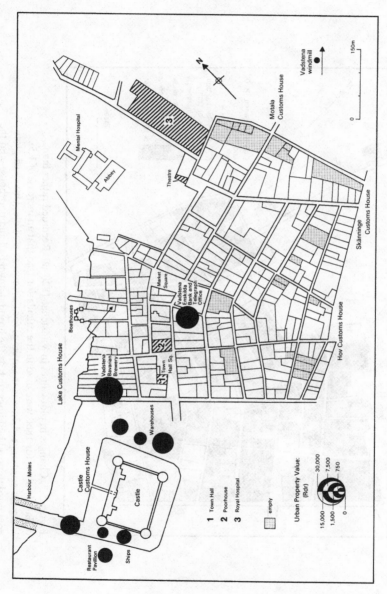

Figure 5.6. Non-residential properties in Vadstena in 1860.

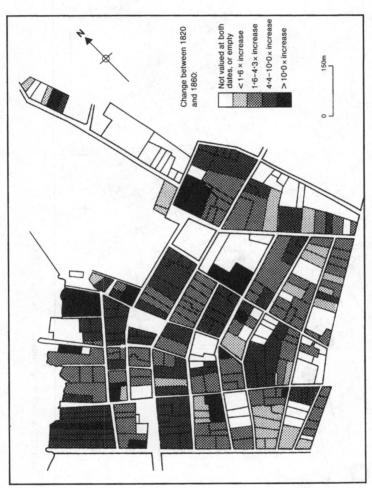

Figure 5.7. Changes in plot values in Vadstena, 1820–60. Recorded rather than deflated values for 1860 were used in the calculations, so that an increase of 1.5x indicates no change. 4.3x was the average for the town as a whole. Unvalued plots contained untaxed buildings such as the hospital, town hall and poorhouse, and the grace-and-favour houses of civil and church officials.

Change between 1820 and 1860:

Not valued at both dates, or empty
<1·6 × increase
1·6–4·3 × increase
4·4–10·0 × increase
>10·0 × increase

0   150m

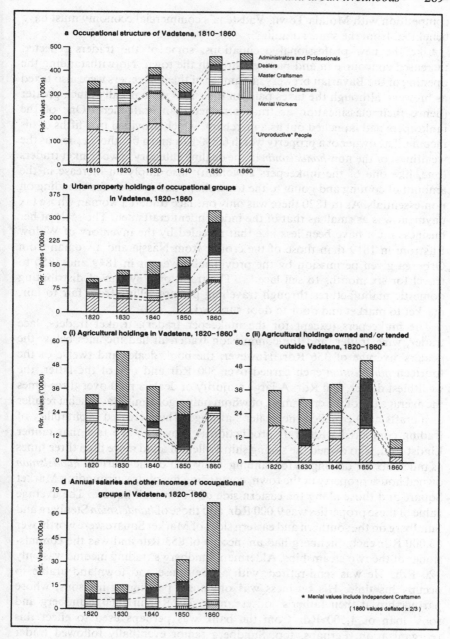

Figure 5.8. Changes in occupational structure, land ownership and income in Vadstena, 1820–60.

competition with Motala Town, Vadstena's commercial economy must have benefitted from the same stimuli.

Like the new professional occupations, some of the traders reflected increased concourse in and passage through the town. Notwithstanding the opening of the Bavarian brewery, a number of the innkeepers were also taxed as brewers, although the taxes paid on their inns were usually much greater (hence their classification as traders rather than craftsmen). One of the innkeepers had launched out as a merchant, which provided two-fifths of his income. The owner of a property worth 6,300 Rdr and a boathouse, he was the wealthiest of the non-*handelsmän* in the dealing category. Two trinket traders (one, like one of the innkeepers a woman), also imply an increase in the amount of coming and going to the town shops and markets for spending on non-essentials. As in 1820 there was only one lace trader, a woman whose tax payment was as small as that of the independent craftsmen. The scale of her business must have been less like that revealed by the inventory of Widow Elfström in 1819 than those of the crofter from Nässja and a woman from Örberga given permission by the provincial governor in 1849 and 1850 to travel for six months to sell lace.[154] The traditional system of distributing domestic manufactures through travelling pedlars selling from fair to fair, market to market and door to door survived.

The innkeepers (except for the innkeeper–trader), trinket traders, lace trader, wine trader, and the commission trader all had incomes below the dealers' average of 356 Rdr. However, the book dealer and twelve of the fourteen *handelsmän* each earned over 500 Rdr and one of the latter, the wealthiest dealer, 850 Rdr. A large majority of dealers paid over three times the average tax of the craftsmen, of whom only a goldsmith (as much a retailer as a craftsman) equalled the dealer's average. The increased profitability of trading compared with craft production is shown by the fact that tanner Lindstrand, who owned the sailing ship Vallentin, paid more than three times as much tax for dealing as for tanning. Only half of the fourteen *handelsmän* owned house property in the town, comprising ten plots surrounding Market Square and those along the eastern side of Town Hall Square. The average value of these properties was 9,000 Rdr, and those of *handelsmän* Stenberg and Sundberg on the southern and eastern sides of Market Square were worth over 13,000 Rdr each. Stenberg had an income of 853 Rdr and was the commissioner of the two steamships. Alderman Sundberg's trading income was only 120 Rdr. He was semi-retired, with a boathouse and townland arable to occupy his time. His business was obviously passing to his sons, whose partnership at their father's address made 533 Rdr in 1860. Sundberg and sons' loan of 4,500 Rdr from the bank was perhaps used to effect this reorganization. Perhaps, too, Sundberg senior eventually followed trader Svensson, one of the wealthiest of Vadstena's dealers in 1850, who had retired to an estate in nearby Strå, where he already farmed 25,000 Rdr's worth of land, retaining only a boathouse and thirty-four bank shares in the town in 1860.

The seven traders with the most central properties paid the highest taxes and owned two of the town's four boathouses. Like the wealthiest *handelsmän* of 1820, they were involved in importing goods for retail sale in the town, such as the cargo sent to Vadstena in 1835 from Norsholm (the nearest point on the Göta Canal to Norrköping) containing nuts, aniseed, snuff, cotton yarn, wine, fine wine, paper of various types and orange peel. Probably the five ploughs plus accessories imported from the canal also passed through their hands. *Handlande* Wallander, who died in 1852, was one of the shopkeeping merchants by the Market Square. His debts of 1,761 Rdr were owed mainly to other Vadstena merchants (the largest, of 979 Rdr, to *Handlande* Sundberg across the square), but debts to three merchants and a goldsmith in Motala suggest that the canal and lake shipping had a role in his business, too. His comfortable inventory balance of nearly 3,000 Rdr and the 500 Rdr's worth of paper money, gold and silver in his house at his death indicate an opulent lifestyle, though he was recorded as a 'bankrupt former trader' and ascribed no income in the *taxeringslängder* of 1850. At least one of the seven central shopkeepers, trader Mauritsson of Townhall Square put out domestic craft work.[155]

Although we cannot be sure that the domestic manufacturing of the plain was being controlled from Vadstena, some of its merchants were certainly growing rich from trading in its agricultural produce and the flour and liquor into which it was made. Two seed traders supplied inputs to the farms of the plain, one of them making 533 Rdr a year from the seed trade alone. Most of the grain shipments along the Göta Canal were bound for Norrköping, and many others for Stockholm. Flour merchant Swartz of Norrköping paid tax on an income of 533 Rdr in Vadstena, and Mr Matérn of Stockholm owned a warehouse alongside the castle moat which, at 12,000 Rdr, was twice as valuable as the two of the Vadstena Trading Company. War Commissioner Lallerstedt's son, who owned extensive property in the region including Kungs-Starby estate and the Royal Warehouse in Vadstena Castle adjacent to it, was taxed in 1860 in Vadstena, but as a merchant *of* Stockholm. Vadstena's role of funnelling out the produce of the plains to the Baltic was, therefore, passing into outsiders' hands by the end of the period.

It was from linking the developing commercial agriculture of the plain with the large markets opened by the Göta Canal that the increasing number and wealth of Vadstena's traders came in the latter part of the period. Undoubtedly, too, it was the spending of what wholesale traders paid to peasants and estate owners for grain, meat and liquor that supported the retail traders by the Market Square, the restaurants, liquor shops, wine and trinket dealers.

Table 5.22. *The property holdings and incomes of Vadstena craftsmen in 1860 (in Rdr of taxation value × ⅔)*

	N	N owning property	Average property value	Average income	Value of other property
Building	7	4	1,623	149	0
Clothing	13	3	2,527	177	613
Victualling	10	3	2,802	284	440
Metalworking	10	7	4,995	189	240
Leathermaking	8	5	2,539	195	1,387
Woodworking	5	3	1,567	176	0
Other	5	2	5,070	192	67
Totals	58	27	3,147	198	2,747

Craftsmen Comparison of tables 5.22 and 5.11 shows that the balance of numbers across the crafts and the total number of craftsmen remained stable between 1820 and 1860. The only craftsmen to shrink in number were builders, whose properties and incomes were at the bottom of the craft group's range in 1860. Craftsmen serving the needs of the surrounding countryside as well as the town itself, such as those in clothing, victualling and metalworking, increased in numbers and/or wealth over the period. Despite some differences in fortune, however, there was a general evening out in the property holdings and taxed incomes of the different craft groups compared with 1810. In fact, this tendency was even more marked than table 5.22 shows because the averages for some of the groups are affected by highly anomalous values. Even with those values craftsmen's incomes were much lower than dealers': *handelsmän* generally made five times as much, some eight times as much.

The incomes of the victuallers ranged between 180 and 240 Rdr except for the 640 Rdr of the miller and the 900 Rdr of the Bavarian Brewery manager.[156] Neither owned property. For these two, large-scale 'industrial' production brought operations of a completely different magnitude from those of the traditional craftsman–retailers of the town. Without the miller and the brewer, the victuallers' average income was 175 Rdr, snugly within the range spanned by the other group averages on table 5.22. The number of brewers declined to two as some of them moved into innkeeping (and one of them on from that into trading) and, presumably, as the Bavarian Brewery pumped its bottom-fermented lager beer into the town. There were five bakers and a confectioner. One of the former earned 240 Rdr and lived in property worth 4,734 Rdr at the hub of Castle Quarter. Generally, the leather workers, particularly the tanners, were still the wealthiest of the craftsmen. There were four tanners earning an average of 230 Rdr. Three of them owned all the non-housing property of the leather workers. This was despite the fact that some tanners, including Alderman and tanner Regnstrand,[157] supplemented their craft work with activities which brought them more income, causing them to be classified in other groups. Tanner Röhman's widow, who had herself worked as a tanner in 1850 before selling much of her property to the Bavarian Brewery Company, was one of the town's wealthiest inhabitants.

As the richest brewers and tanners moved into more lucrative dealing and professional positions, new niches were providing comfortable but less opulent livelihoods for other craft groups. Goldsmith Holm made as much as any but the wealthiest dealers and lived in a large valuable property between the Market Square and the monastery garden. Two coppersmiths and two tinsmiths each paid tax of 180 Rdr and owned property on the outskirts of the town, perhaps because their wares were bound for the stills of the surrounding plains as much as the roofs and water systems of the town. A watchmaker paid tax on only 90 Rdr income, but his 4,000 Rdr property suggests that he had once operated on a larger scale. Amongst the clothing crafts, one of the tailors made 360 Rdr per year, and though the two hatters earned only half as much, one of them lived in a property worth over 5,000 Rdr. The presence of two coachmakers in 1860 reflects both the increased diversity of the craft goods available in the town and the increased mobility which supported it.

Amongst the 'other' craftsmen, two linen weavers were all that survived of the town's old industry after the manufactory and two successor linen works had closed by the 1850s.[158] The wealthiest craftsmen in this group were a dyer, representing the traditional crafts of the town and living in Castle Quarter, and a mirror manufacturer, representing new crafts and living just to the north of the Market Square in Skänninge Quarter. The latter, named Ohlin, was obviously successful: he owned townland arable, his town property was worth over 5,000 Rdr, and soon after 1860 he bought a large farm in Väversunda. His business must have been that set up by glazier and mirrormaker Stenberg, who died in 1852 with 246 mirrors of various kinds, 3 dozen mirror frames, 16.5 dozen cardboard boxes, 'sundry mirrors being worked on' and considerable quantities of glass, all worth 506 Rdr. Besides his workshop, valued at 66 Rdr, Stenberg owned two market stalls. More than twenty people owed him money at his death, including noblemen and craftsmen, presumably for mirrors bought on credit. He owed four debts. One was to a Stockholm manufacturer and another to Råby glassworks, presumably for raw materials. The largest, of 350 Rdr, was owed to Mr Ohlin, who succeeded to the business. At the other end of the spectrum from mirrormaker Stenberg was twinemaker Forsberg, who also died in 1852. The total value of his inventory was less than 100 Rdr. The main credit was his working clothes, valued at 20 Rdr; the main debts were owed to local *handelsmän*, presumably for raw materials bought on credit.

Thus, although apparently stable in numbers and nomenclature, the craft community of Vadstena changed significantly between 1820 and 1860. The most successful of the old-style craftsmen were moving into dealing, just as some dealers were moving into office-holding. The appearance of more numerous and wealthy specialists such as confectioners, hatters, coachmakers, bookbinders, girdlemakers and watchmakers must have been a response to an increase in the volume and disposable income of passing trade. The mirrormaker, fetching raw materials from long distances to his workshop and selling from market stalls, epitomized the new crafts which emerged during our period. Ten craftsmen had bank shares, and although they averaged only 3.3 each compared with the dealers' 16.4 and the administrators' and professionals' 12.9, they owned more of the bank's funds than craftsmen borrowed from it.

Figure 5.9 shows that craftsmen with and without property lived scattered through the town, although there were more in Castle Quarter than in the northern Monastery and Skänninge Quarters combined. Castle Quarter and Storgatan contained their most valuable properties. Even so, there had been some spread of craft workshops into the interior of Hov and Skänninge Quarters, which housed twice as many craftsmen in 1860 as in 1810. The leatherworking and victualling crafts stayed in the old craft centre of Castle Quarter, whilst metalworkers, woodworkers and clothing makers were more common in Hov and Skänninge Quarters, where the properties of a coppersmith and a watchmaker were amongst the most valuable owned by any craftsmen.

Independent craftsmen   Figure 5.9 also shows that a large majority of the independent craftsmen lived in the two inland quarters of the town: only eleven lived (all in rented accommodation) in Castle and Monastery Quarters. Table 5.23 shows that there were more independent craftsmen than guildsmen in three craft groups, especially in building and woodworking. In the first they were heavily concentrated in carpentry, where there were fourteen, and in the latter in joinery, where there were ten. The tax registers do not reveal if they were employed by master craftsmen or worked on their own account. Perhaps we can surmise that the two coachmaker's labourers were employees, but the falling number and relative poverty of building craftsmen suggest that independent building workers were competing with rather than working for them. All independent craftsmen paid 1 Rdr tax on incomes of 40 Rdr a year except for joinery labourer Kalm, who paid an extra *Riksdaler* for a beer and coffee shop: even workmen moved into trading. Very few owned property, only two tailor's labourers more than 1,000 Rdr's worth. Quite clearly, independent craftsmen were much poorer and in similar economic circumstances to menial workers.

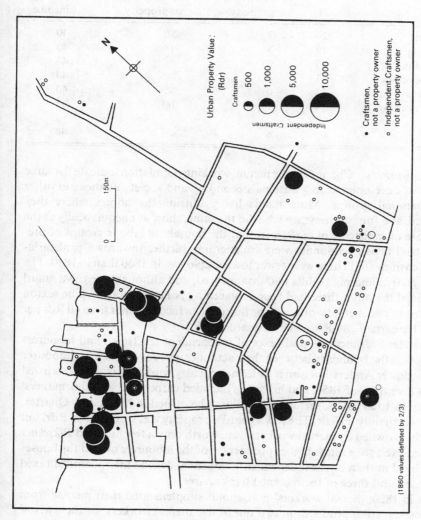

Figure 5.9. Property holdings of independent and master craftsmen in Vadstena in 1860.

Urban Property Value:
(Rdr)

Craftsmen

500
1,000
5,000
10,000

Independent Craftsmen

• Craftsmen,
  not a property owner

○ Independent Craftsmen,
  not a property owner

(1860 values deflated by 2/3)

Table 5.23. *Independent craftsmen and their property in Vadstena in 1860 (in Rdr of taxation value × ⅔)*

	N	N owning property	Average value of property	Average income
Building	26	2	90	40
Clothing	19	3	914	40
Victualling	5	0		40
Metalworking	1	0		40
Leatherworking	4	0		40
Wookworking	14	1	100	43
Totals	69	6	509	40

Menial workers   Changes in the menial working population indicate the same kinds of development in Vadstena's economy and society as those in other occupational groups. More maids living outside the homes where they worked, and four journeymen who did the same, show as unequivocally as the increase of independent craftsmen that the household labour groups of pre-industrial urban economies were fragmenting further towards a proletaria-nized cash-nexus. There were fewer town employees in 1860 than in 1810. The fence guards and pig bailiffs had disappeared, and although one town guard remained the other three had been replaced by police constables. The sexton was one of two menial workers to be taxed on an income other than 40 Rdr per year. He earned 280 Rdr, a bank guard 400 Rdr.

The identical incomes and taxes of independent craftsmen and labourers imply that the boundary between their activities was very indistinct. Labourer and widower Anders Jonsson is called a carpentry workman in the debts listed in his inventory of 1850, and his assets included carpenters' and shoemakers' tools. His total assets, including a building he owned in Monastery Quarter, were worth only 66 Rdr. The town guard's property was worth 1,680 Rdr, but all others owned by menial workers were worth only a few hundred *Riksdaler* or less. Even so, a much higher proportion of the labourers owned the houses they lived in than of the independent craftsmen, despite their identical taxed incomes, and three of them owned bank shares.

As in 1820, menial workers' households supplemented their income from domestic textiles production: two out of five menial workers' or their wives' inventories contained valuations of spinning wheels. None of this (admittedly very small) sample contained looms or lacemaking equipment, whereas similar numbers of inventories of craftsmen, dealers and their wives contained four with looms and one with a 'bobbin chair' as well as two with spinning wheels. The inventories of six out of eleven maids contained textile-making

Table 5.24. *Menial workers' property and incomes in Vadstena in 1860 (in Rdr of taxation value × ⅔)*

	N	N with property	Average property value	Average income
Town employees	12	2	573	90
Labourers	25	8	346	40
Journeymen	4	0	0	40
Maids etc.	19	0	0	40
Totals	60	10	366	50

equipment, including lace cushions and a 'lace chair'. One maid (who had a loom) had a positive inventory balance of 478 Rdr, but the average was only 60 Rdr and two were negative. Inventories confirm the impression of table 5.24 that living-out maids were among the poorest people in Vadstena, and the supplementary income of menial households generally must have been very small: their average value was under 100 Rdr.

'Unproductive' households   Households headed by people to whom no gainful occupation was ascribed increased from 43 per cent and 41 per cent of all households in 1810 and 1820 to 48 per cent in 1860. The group's composition and the distribution of property across it changed. Retired agricultural workers increased in number, but only slightly in percentage terms (see table 5.25). Only four were former farmers, three of whom owned the houses in which they lived; three had been grenadiers, three crofters and six farm foremen. One of the ex-crofters owned his house, another 1,900 Rdr's worth of townland.

Although, like their still-active counterparts,[159] only a few of the former administrative and professional personnel owned property, they were obviously a distinguished bunch. Five were former military officers, one of whom owned a property worth nearly 8,000 Rdr. There were also two barons (one of whom lodged), a retired district judge and a marshal of the court. With the still-active military officers, two local nobles who had houses in Vadstena, the Vult von Steijerns, Widows Bergenstråhle and Tisell (who retired to large town houses as dowagers from the Medhamra estate) and numerous demoiselles, they must have formed a glittering company. All lived near to the centre of the town and they were responsible for most of the bank deposits and loans of the unproductive category. A still-salaried protocol secretary must have ensured that *bon ton* was maintained, and the theatre assuaged their aspirations after high culture.

Table 5.25. *The 'unproductive' households of Vadstena and their property in 1860 (in Rdr of taxation value × ⅔)*

Former occupation or current status	N	With urban property	Urban total	Property average	N with town land and grass plots	Total agricultural value
Agriculture	19	11	12,026	1,093	2	1,960
Administrative and professional	16	4	13,913	3,478	1	1,233
Dealing	6	4	21,473	5,368	2	14,060
Craft	13	3	4,620	1,540	0	
Menial	31	6	567	95	0	
Property owner	28	24	13,147	548	1	227
Spinster	43	11	10,893	990	4	2,840
Widow	74	19	23,327	1,228	3	2,926
Total	230	82	99,966		13	23,246

The increase in the number and wealth of the former dealers reflected the fortunes of their active counterparts, although more than half of the urban property and the vast majority of the townland all belonged to one of them – ex-apothecary Holmberg, who owed 8,000 Rdr to the bank. Perhaps the fall in the number of retired craftsmen between 1820 and 1860, and the much smaller proportion of them who owned property, also reflected the changing relative fortune of that occupational group: maybe they now had to work until much older before having a competence on which to retire. The same reasoning would predict a fall in the number of the former menial workers. However, although few owned property and those who did owned little, their number increased from eight to thirty-one and from 6 per cent to 13 per cent of all 'unproductive' households. Over one half were former bell-ringers or night watchmen. No more than half a dozen had ever provided these services simultaneously between 1820 and 1850, before they were substituted by police constables (of whom there were three retired examples in 1860). Thus, these were honorary posts given for a short while to still able-bodied men who had little other means of support: a night watchman was owed for four days work by property owner Johansson at his death, implying that they were simply day-labourers with honorary titles.

'Property owners' (*gårdsägare*) began to appear in the taxation registers by 1860, and there were even four 'former property owners' by that date. Perhaps the very low proportion of independent craftsmen with property is because 'property owner' was preferred to 'joinery workman' by those who had managed to acquire it. *Gårdsägare* Anders Petter Larsson, who died in 1851 with a total inventory value of 948 Rdr – 772 Rdr in credit – had two acres of townland and a cottage with a barn and smithy (worth 375 Rdr and 500 Rdr), and tools for weaving and joinery. *Gårdsägare* Johan Petter Johansson, who died the previous year, was twice as wealthy. His two urban properties worth 1,363 Rdr were investments: he was owed 96 Rdr in rent by three tenants. He might also have been a moneylender: more than one quarter of his inventory credits were debts owing to him, ranging from 4 Rdr to 100 Rdr. Further insight into the economic role of *gårdsägare* is given by the remarkable inventory of Johan Johansson, who died in 1851. Judged from the location of some of his debts and credits, he moved from Linköping to Vadstena, where between 1848 and 1850 he bought the property on plot 118, one third of that on 157, half of that on 162, 213 and 232, and half of 233. Their taxation value was 1,155 Rdr, though he paid more than twice that sum for them. They were obviously investments. Rent was owed to him for all of them at his death, but they did not yield quickly: Coppersmith Ljungberg owed 200 Rdr – a return of 6 per cent per annum on the purchase price – for five years rent (some carried over from the previous owner), and eleven rents were owed on his six houses, some from as far back as the year of purchase, totalling 230 Rdr. The tenants included workmen, widows, a shoemaker, a girdlemaker, a joiner and three night watchmen, and three of the rents were owed jointly by two people who obviously shared accommodation. In addition to these rents, Johansson was owed debts on certificated loans ranging from 46 Skilling to 82 Rdr by thirty-three people, twenty-four from Vadstena and others from Linköping and rural parishes around the two towns. He was owed eight more debts against pledged pawns, ranging from 1 Rdr 24 Sk against a black dress coat to 10 Sk for two serviettes. Johansson was not exceptionally wealthy. His total inventory credits amounted to 2,660 Rdr, with a positive balance of 2,000 Rdr. He was not the only pawnbroker in the town, either, because hatter Lindells' inventory included 2 Rdr for earrings pawned with Dalecarlian Andersson, who cropped up as a creditor in many inventories.

Of the five property owners' inventories examined, three contained weaving equipment, one a spinning wheel and one a quilting frame. Like the workmen and independent craftsmen, none had lacemaking equipment. Thus, the property owners had much in common with the lower layers of Vadstena society with whom they were linked by their letting and lending. However, their capital allowed them to fill the niche made available by the increasing number of non-bourgeois livelihood positions, which required rented housing and the facility to take out small – sometimes minute – loans. But the length of time that many of these loans were left due implies an absence of the rapacity conjured up by the terms 'capitalist landlord' and 'loan shark'.

We inferred earlier that some of the widows and spinsters of the town had fulfilled these roles in 1820, but widows in particular were much less prominent as property owners in 1860. The spinsters retained their 18 per cent share of the 'unproductive' household headships and 10 per cent of their property, whereas the widows' shares fell from 51 to 31 per cent of households and 85 to 23 per cent of the 'unproductive' group's property. Between 1810 and 1860 their number remained constant, their urban property fell to about a half, and their rural property to a quarter. Even so, they owned more urban property than any other group of 'unproductive' people in 1860, and the average value of their holdings was more than twice as large as that of the *gårdsägare*. Widows and spinsters ranged across the whole social spectrum. Spinsters' houses ranged in value from 27 Rdr to 4,000 Rdr, widows' from 20 Rdr to over 6,000 Rdr. One or two of them were members of the town's colony of noble/military families. Others, such as spinsters Tollbom, Lidbom and Östberg and Widow Hartwick (neé Tollbom), inherited their townland arable and/or grass plots and urban property. Some of them could have derived incomes from letting house-room and farmland, but there was no equivalent in 1860 of the sixteen widows who had headed Vadstena's property holders in 1820.

Perhaps the bank now provided a safer home for capital, and restaurant-keeping a more congenial or profitable form of enterprise than landladyism. However, single women stayed heavily involved in domestic textile production, and they continued to dominate lace production and dealing. We examined twenty-eight inventories of widows and eleven of spinsters who died in the 1850s, of which only eleven and five, respectively, did not contain spinning wheels, looms, ribbon looms, lace-cushions or sewing tables. Most also contained quantities of thread and cloth, although none contained finished lace. We have already seen that cloth was put out to at least one of the town's widows by Trader Mauritsson, and some of the single women had a number of looms of different types. As many had lace cushions as any other form of textile-making equipment: most of them had more than one, and one had four. Some had extra bobbins, and Widow Hoffstrand had three lace-cushions, diverse lace patterns and four smoothing irons, and owed or was owed 131 Rdr for 'lace collected for sale'. Another widow had 300 Rdr's worth of put-out lace (*spetsförlag*) at her death. This does not mean that lacemaking was exploitative and spiralling towards its practitioners' penury. Lace was made in the wealthiest households, such as those of Widow Gylling, Widow Röhman and Mrs then Widow Hartwick, who each had three or four lace-cushions.[160]

Women had comprised a very large majority of the town's out-poor at the start of our period, and the eight male and single female inmates of the poorhouse in 1820 became eight males, forty-one females and thirty-nine children by 1840. The move of the poorhouse into what had been the linen manufactory was accompanied by a fall in the number of in-poor to nine males, twenty-nine females and six children in 1850. In 1860, there were only twenty-one adults and no children. As in 1810 and 1820, other destitute people were scattered around the town, in households of all levels of wealth. Only three destitute couples lived in their own households, one of which, that of a labourer and his wife in the innards of Hov Quarter, contained a further destitute widow and maid. The growing prosperity of the town's economy in the 1850s was accompanied by a decline in the proportion of destitutes in the adult population from 5.9 to 4.3 per cent. A maximum of 7.8 per cent had been reached in 1830, so that the proportion of poor in the population, and the falling representation of young women within it, seems to have been a good barometer of the state of Vadstena's economy.

Changes in household structure and demography in Vadstena, 1810–60
Table 5.26 shows the household structure of Vadstena in 1860 in the same format as table 5.14 in 1820. The growth of the economy of the town over the period was reflected in an expansion of 44 per cent in its total population. The growth of high order services for the town's hinterland accounted for much of this growth: its hospitals housed 18 per cent of Vadstena's total population in 1860 compared with 8 per cent in 1820. The legal freedom for all housholds to practise crafts was associated with a decline of menial households' share of the total from 24 to 9 per cent, although the independent craftsman category did not account for the whole difference. Apart from these changes since 1820, the proportions of the town's population accommodated in each of the occupational categories remained quite stable. Households supported by the backbone of a pre-industrial urban economy – administration, the professions, dealing and crafts – still housed considerably less than half of the population, whilst those of apparently unproductive people accounted for nearly 30 per cent.

Nuclear family households were still rare. The menial and craft groups most closely matched that model, with few non-family members per household. But the roughly equal numbers of adult males and females tabulated under heads of households does not indicate a prevalence of married couples amongst them: eight of the independent craft households were headed by single women (mainly seamstresses and bakers), and twenty-one menial households were headed by maids. The greater number of households than adult males was due in most occupied groups to households headed by gainfully employed women. The appearance of productive households headed by women might have been because the legislation of 1846 meant that they no longer had to be hidden in the unproductive category. However, increasing bustle in the town also created service sector niches which were customarily filled by women at the time.[161]

Even so, as in 1820, the families of the productive males of the town, augmented by their journeymen, apprentices and maids, provided the core of its households and its demographic dynamic. Table 5.26 shows that ninety-eight out of a hundred journeymen and apprentices were employed and housed by dealers and craftsmen. In 1820 there had been 0.69 per dealer, but in 1860 there were 0.75; excluding those in the linen manufactory, there were 0.98

Table 5.26. Household structure in Vadstena in 1860

Occupational group	N	Heads of households				Journeymen and apprentices[a]				Hands and maids[b]				Living-in				Totals				
		M	F	Ch1[c]	Ch2[d]	M	F	Ch1	Ch2	M	F	Ch1	Ch2	M	F	Ch1	Ch2	M	F	Ch1	Ch2	T
Administrative and professional[a]	36	30	15	12	35					13	50		2		6			43	71	12	37	163
Dealing	30	24	18	5	29	18	1		1	12	34		7	1	12			55	65	5	37	162
Crafts	58	55	47	18	112	80	10		8	12	55		11	2	13			149	125	18	131	423
Independent craft	69	60	56	3	78	2					3			1	2			63	61	3	78	205
Menial	60	43	49	1	82						7		7		9		4	43	65	1	93	202
Unproductive[e]	231	111	146	32	159					24	67		22	6	37		16	141	250	32	197	620
Institution[f]		6	8	5	2					10	24		4	150	178			166	210	5	6	387
Total	484	329	339	76	497	100	11	0	9	71	240	0	53	160	257	0	20	660	847	76	579	2,162

Notes:

[a] Excludes those living in institutions.

[b] Includes nurses and hospital hands.

[c] Children of eighteen years and over.

[d] Children under the age of eighteen.

[e] Includes Hagalund, owned by a former apothecary.

[f] Inmates = 'living-in'.

Table 5.27. *Average household and family size of occupational groups in Vadstena in 1860*

	Average number of people per household	Average family size	Children per family		
			> 17 yrs.	< 17 yrs.	Total
Administrative and professional	4.53	2.56	0.33	0.92	1.25
Dealing	5.40	2.53	0.17	0.97	1.14
Craft	7.29	4.00	0.31	1.93	2.24
Independent craft	2.99	2.87	0.04	1.14	1.18
Menial	3.37	2.92	0.02	1.64	1.66
Unproductive	2.68	1.94	0.14	0.69	0.83
Total	3.69	2.54	0.15	1.03	1.18

per craftsman in 1820, and 1.45 in 1860. The craftsmen also employed nearly twice as many hands and maids in 1860 as in 1820, although the number of craft households increased by less than one-fifth. Thus, even though their wealth seems to have declined relative to that of the groups above them, the master craftsmen's workforces grew. On average, too, they had more children living at home than in 1820 and than other occupational groups in 1860 (see table 5.27). The craftsmen's households were much fuller than those of any other occupational group, with an average of 7.29 persons in 1860, compared with 6.16 in 1820.

The next largest households after the craftsmen and dealers, averaging 4.53 members, were those of the administrators and professionals who, despite the number of single men in new occupations such as telegraphy and banking, also had many more children at home in 1860 than in 1820. So did all the productive groups except the dealers (many of whom were also young single men or women), despite a shrinkage in the number of people living-in with other families from 132 to 89. Only the 'unproductives' saw a decrease in the size of households, causing a slight fall in the average size between 1820 and 1860. This could have been because more retired people moved into the town. The number of hands and maids employed by the group, as well as the wealth and social status of some of its members, suggests that Vadstena had become something of a 'leisure town' by 1860.[162] Overall, increases in household size between 1810 and 1860 reflected the greater intensity and variety of economic activity.

Changes in rural crafts and services, 1810–60

The proportion of the region's rural population supported by non-agrarian pursuits doubled, all the increase coming in the last decade of the period (see table 5.17). Sextons, organists and schoolmasters all grew in number, especially the last.[163] Removal of prohibition on full-time craft work outside towns in 1846 saw the number of rural craftsmen increase from twenty-five to fifty-eight between 1840 and 1850, then to eighty-one in 1860. In 1860 there were still no craftsmen at all in Hov (Dahl) and Bjälbo, which were small parts of much larger parishes; Nässja and St Per (Aska), next to Vadstena; Ask, which was remote and still had something approaching a classical peasant economy, and Orlunda, where many tiny farms heavily involved in craft by-employments apparently made full-time craftsmen unnecessary. Most rural craftsmen in 1860 served the needs of neighbours for shoes and clothes, and many were still called *gärningsmän* (parish craftsmen). They had been joined by bricklayers, painters, carpenters and blacksmiths, and their increased number must have meant greater market penetration into the household economies of crofts, cottages and peasants. In 1860 some parishes had two shoemakers and tailors; Örberga in Dahl had three of each, and seems to have been becoming a small rural service centre, with a painter, carpenter and joiner as well. The first parish blacksmiths appeared in 1820; by 1860 there were eight. Consistent with its slower agrarian development, the whole of southern Aska had only one blacksmith. In fact, the 407 farms of this subregion supported only 23 rural craftsmen, whereas Dahl's 163 farms supported 26. Some 32 rural craftsmen were interspersed amongst the 172 farms of northern Aska, including as well as the usual shoemakers, tailors and carpenters some who worked on forest resources: at his death in 1856, a Västra Ny joiner and wood-carver had six stools not yet ready for sale as well as numerous woodworking tools. Being far enough from Vadstena to benefit from the partial relaxation of restrictions on rural dealing in 1846, Västra Ny also had a trader. Based in Stora Ingesby, which was remote even within Västra Ny, *handlande* Ekman's stock-in-trade was mainly cutlery (worth 60 Rdr) and feathers (worth 100 Rdr). His total inventory was valued at only 516 Rdr – a far cry from the wealth of Vadstena's merchants.

Ekman's inventory provides no evidence of craft or farm work, but those of craftsmen and their wives show them to have been fully involved in agriculture, as well as earning whatever was possible on the side. The inventory of a master builder's widow had seven sheepskins, flax, wool, thread and cotton cloth, plus a spinning wheel, loom and various carpentry tools. Country craftsmen's families followed the model set by farmers and crofters. Their small increase by 1860 did little to dilute by-employment and self-provisioning in the region's rural households.

Table 5.28. *Number and value of rural non-agricultural properties in 1860 (in Rdr of taxation value × ⅔)*

	N	N valued	Total listed value
Watermills	26	20	39,176
Windmills	12	11	9,238
Saws	8	4	15,945
Ironworks	8	7	764,111
Copper smithy	1	1	667
Brickworks	6	6	26,333
Quarry	1	1	10,000
Oilworks	1	1	1,333
Breweries	3	3	44,111
Papermill	1	1	66,667
Gunpowder works	1	1	25,000
Spa	1	1	24,000
Total	69	57	1,026,581

Changes in the ownership and production structures of rural industries, 1810–60

The value of industrial plant climbed steadily until 1850, when it accounted for nearly 9 per cent of the region's total taxation, then jumped to 28 per cent in 1860 (see table 5.16). Ironworks provided more than two-thirds of it (see table 5.28), and Motala *mekaniska verkstad*, valued for tax at 720,853 Rdr, was reponsible for 94 per cent of the ironworks total, twice as much as the region's sixty-eight other industrial premises, and more than all the farmland of Dahl. Even so, there was considerable expansion in other industries (see figure 5.1). The three cornmills on the Mjölna were joined by a fulling mill at Broby in Strå parish (Dahl) in the 1840s, serving the domestic *vadmal* weavers of neighbouring Örberga.[164] The Broby cornmills were considerably expanded after coming into peasant ownership between 1820 and 1830.[165] Anders Persson, of Östra Starby in adjacent St Per (Aska), owned Mjölna and Nykvarn mills and the fulling mill, together valued in 1850 at 9,000 Rdr, which he leased out to Anders Hermansson. Broby mill itself was owned by Anders Larsson of Broby, who farmed over 20,000 Rdr's worth of land and operated two very large stills. By 1860, the fulling mill and Broby watermill were closed (having probably been badly hit by the lowering of Lake Tåkern's level in the process of land reclamation). The latter had been replaced by a windmill, as had Mjölna mill, and there was a further windmill in nearby Kedevad village. All but one belonged to Anders Persson and Larsson's heir. Persson owned no land, leasing his 6,000 Rdr farm. He also owned the brickworks at Tycklingen, next to Vadstena. Windmills were not built only in Strå, but all over the plain within a 10 km radius of Vadstena, which contained fourteen in 1860. They were built relatively late in the period. In 1840 there were four windmills on the plain (worth 4,137 Rdr), supplementing the three watermills (worth 10,733 Rdr) at Broby.[166] In 1850, there were seven windmills (worth 10,085 Rdr) and three watermills (worth 8,500 Rdr). The windmills were mainly owned by peasants, including the two largest at Östnässja (3,000 Rdr) and Mjölna (3,150 Rdr). Even better than growth in the number of blacksmiths, the expansion of milling capacity testifies to the development of arable farming on the plains: distilling was not the only way peasants could add value to grain crops. When the opportunity arose, the peasants of the plain invested in means of increasing their incomes with alacrity.

The distribution of mills on the lower map of figure 5.1 suggests that grain must still be taken for milling to Motala Ström from the northern parts of southern Aska in 1860. Although the mills beyond Motala Ström did not increase in number, they grew considerably in capacity. The largest in the region, valued at 6,200 Rdr, was in the heart of the shield on the site of Karlström's *bruk* in Kristberg. Like the mills at Bona, Medevi and Kärsby in the shieldland, it was owned by an estate. The biggest saws were also in the heart of the shield at Karlsby, Tryfall, Ubbelsby and Kavlebäck. The last was converted to steam power in the 1850s. All the saws for which values were given in both 1820 and 1860 became much more valuable: Karlsby's increased from 200 to 5,000 Rdr; Tryfall's from 666 to 1,333, and Kavlebäck's from 500 to 2,000 Rdr.

As in Sweden as a whole, the iron industry recovered after the difficulties of the first two decades of the nineteenth century. By 1860 there were no less than eight ironworks in northern Aska, together accounting for 74 per cent of the taxation value of industrial property there. Bona *bruk* increased in taxation value from 8,500 Rdr in 1820 to 15,000 Rdr in 1830, and Blommedahl manufacturing hammer, valued at 666 Rdr, was brought into operation again in 1830 after being stopped for two decades. Blommedahl works grew to 1,333 Rdr in value in 1840, and stayed there for the rest of the period. Bona *bruk* doubled in value to 30,000 Rdr between 1850 and 1860, when it had forging and manufacturing hammers and employed a locksmith and a clockmaker to convert some of its rods into finished articles. Karlström's *bruk* near Karlsby had been much bigger than Bona in 1810 and 1820. It still was in 1850, when its bar and rod hammers were valued at 19,600 Rdr, but this was a big drop from its heyday. In 1820 it had two forging hammers, three hearths and two manufacturing hammers on two water power sites known as the upper and lower works. Between 1820 and 1825 it produced 145 tons of bar iron per year – more than four times as much as a decade earlier.[167] Its output rose further

Plate 4. Bona ironworks in the mid-nineteenth century.

to 175 tons a year between 1831 and 1835 to reach its peak of 199 tons per annum 1836–40, when it was valued at 28,600 Rdr. The upper works were closed in 1848, and before 1850 it had passed into the hands of *brukspatron* Tisell (who also owned Kvarn's *bruk* in the part of Kristberg parish outside our region and Hättorp's *bruk* near Tjällmo, a few kilometres to the north-east of our region). The value of Karlström's *bruk* slumped to 19,666 Rdr in 1850, and it had closed by 1860.[168]

Thus, although there was a resurgence after the depression of the early nineteenth century, ironmaking faded again in the heart of the shieldland before the end of our period, where it had traditionally been important. This region was early to feel what later became a general tendency for the Swedish iron industry to shift from isolated *bruk* sites to become concentrated in places well located for transportation. Where they survived on the shield, the mode of organization of *bruk* changed little from what it had been two hundred years earlier. They continued to be closely integrated with agriculture, with very small full-time labour forces. The large mill and saw in Karlsby housed only a hand and a maid, and the brickworks at Medevi, valued at 6,000 Rdr, no workers at all. The eleven industrial properties collectively valued at nearly 54,000 Rdr in the parishes of Västra Ny and Kristberg contained only thirty-eight full-time employees. The closure of Karlström left Bona as the largest industrial employer on the shield. Besides *brukspatron* Grill and his family there were an engineer, a hammersmith and his helper, three workmen, seven hands and six maids there in 1860, with three hands and their families at an adjacent mill and a locksmith living alongside the *bruk*, which also housed six aged or destitute men and women (presumably retired workers and their widows). Altogether, there were thirty-six adults and nineteen children at Bona in 1850, a sizeable concentration of industrial population deep in the forest. A large proportion of the labour on the sixteen farms of the *bruk* estate, which supported seventeen crofts, four cottages and eleven hands as well as the peasant families, must also have spent much of its working time in hauling and charcoal burning for the ironworks.

Blommedahl also became a significant forest settlement by 1860. Its mill and works were worth only 2,000 Rdr and employed only five men, three of them hands, full-time. Besides Squire Söderpalm, his family and a household workforce of nineteen, there were eighty-two adults and sixty-one children in the crofts and cottages of Blommedahl in 1850. Countess Mörner's Medevi estate supported an even larger non-agricultural community. Its saw, brick-works and spa, plus a copper smithy at nearby Stora Baggeby, formed the nucleus of a settlement of crofts and cottages which housed ninety-nine adults and sixty-seven children in 1850, including a coppersmith, blacksmith, spa hands and a miller.

The relationship between industry and agriculture moved both ways. The inventories of a master hammersmith and a clockmaker at Bona both contain valuations of crops, farm tools and animals, and that of a works foreman's wife listed two spinning wheels, two looms, 52 ells of cotton cloth and 52 more not yet finished. Ten out of eleven inventories of people with non-agricultural occupations on the shield contain textiles equipment and cloth, three of them numerous axes and saws. The industrial communities of the shield obviously still had traditionally mixed household economies in 1860, which must have been the sources of the cloth and 200 kg of table linen shipped on the Göta Canal from Motala to Göteborg in 1856.

The emergence and early development of Motala Town

Alongside this traditional shieldland industrial region, a settlement of totally different character developed along the banks of Motala Ström and the Göta Canal. In fact, almost the whole of the expansion of industrial plant in the region as a whole occurred in Motala Town along these two waterways and the long thin artificial island separating them.

Economic base

The engineering works   Motala Engineering Works was opened in 1822 by the Göta Canal Company to supersede a number of workshops on the western

Table 5.29. *Capital assets and debts of Motala Engineering Works, 1825–60 (in 000s of Rdr)*

Year	Value of stock in trade	Value of fixed capital	Total asset balance	Total debts[a]	Share capital	Profits
1825	46	113	113	33	112	—
1830	327	419	766	600	166	—
1835	114	652	839	519	320	—
1840	426	423	880	615	228	37
1845	271	1,318	1,854	802	1,105	−21
1850	761	1,789	2,629	940	1,501	188
1855	1,433	1,844	3,488	1,403	1,396	180
1860	2,507	3,803	6,439	4,691	1,491	292

Note:
[a] The escalation of debts after 1850 was due to an increase in customer advances.
Source: Gårdlund, *Svensk*, 94–5.

stretch of the canal, which was completed in that year. The first buildings were erected to service and repair a steam dredge, delivered at Thomas Telford's instigation from Bryan Donkin in London by Daniel Fraser, who also brought some metal-turning lathes. Fraser was a well-qualified and skilled engineer by the time he came to Motala at the age of thirty-five. When it opened, the workshop for the dredge and other canal requirements employed twenty-two men and comprised a machine shop and forge with three hearths. It rivalled Finspång as the largest ironworks in Östergötland.[169] It grew rapidly in the 1820s, although it was not profitable (see table 5.29). By 1830 the site between the river and canal to the east of Motala village (see figure 5.10) contained a machine shop, driven by a waterwheel turned by a sluice between the canal and the river, a coke oven, two forge hearths, a smithy hearth, a puddling foundry, another foundry with two cupola furnaces, a steam boiler-making plant, a sandstone mill, a file works, drawing office and pattern shop, together with an inn, brewery and bakery to service the needs of the workforce.[170] Still under the directorship of Fraser, Motala Works marked time through the 1830s without making any profit. In 1840 its book value was 880,000 Rdr (compared with a taxation valuation of 100,000 Rdr) and its workforce numbered 176, supplemented by 33 seasonal workers in summer. Irked by lack of profit and heavy capital demands at the works, the Göta Canal Company sold it in 1840 to a consortium of eleven people. The new company included the son of von Platen, who had initiated the canal, ironworks owner Tisell of Karlström's *bruk*, two other ironworks owners from outside our region and four noblemen,

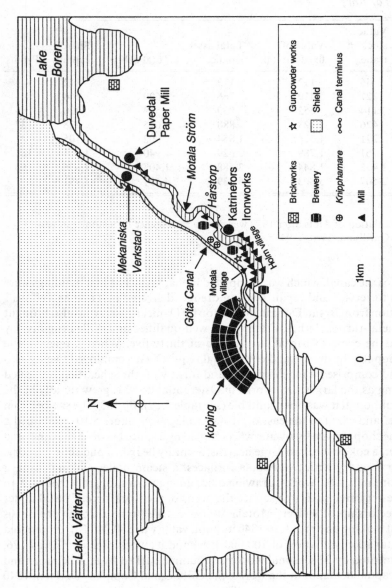

Figure 5.10. Industry and settlements in Motala Town in 1860.

two of whom were local. Fraser retired in 1843, to be replaced by Otto Carlsund, also trained in England, who ran the works until 1870. He was one of the most successful engineers in Sweden during that time, and designed and built the first steam saws in Sweden at Motala.[171] The works was also known for iron bridges, but its main products, apart from canal requirements, were marine steam engines for installation at a shipyard on the canal at Motala, at Motala Wharf shipyard in Norrköping, acquired in 1840, at the company's shipyard at Nyköping, and at the Lindholmen shipyard in Göteborg, which was bought with an adjacent engineering works in 1858. Heavy investment also occurred at Motala itself, especially in the 1850s (see table 5.29). Its taxation value leaped from 120,000 Rdr in 1850 to 721,000 Rdr in 1860, when it was the largest industrial works in Sweden.

The canal was used for gathering raw materials and shipping out marine engines. During April to May and September to October 1835, 119 tonnes of pig iron were shipped to the works' own wharf from Norrköping, Oskarshamn and Sjötorp, and scrap iron from Stockholm. Limestone was brought from Borghamn quarry in Rogslösa (which increased greatly in value after it began to supply the works) and from Gotland in the Baltic. Charcoal came in from Brevik across Lake Vättern, and a wide miscellany of cargoes including spars, roofing tiles, coal tar and smiths' bellows came from numerous points of origin on the canal, the lakes it connected and the Baltic coast. Outwards went 36 tonnes of foundry goods (the canal toll category for engines) to Göteborg and Mem at either end of the canal and to Forsvik, plus 20 tonnes of bar iron and cargoes of nails to Göteborg, Stockholm, Mariestad and Brevik.[172] There was a phenomenal increase in the number of boats entering and leaving the works' wharf between 1835 and 1856. Ten boats arrived and six departed in the four months for which data were abstracted from the canal accounts in 1835; in the same months in 1856, seventy-one boats entered and ninety-nine left.[173] But this is not a true reflection of the increase in manufacturing activity at the works. Of the boats which left, seventy-seven went empty after delivery, and thirteen arrived empty before collecting goods, whilst thirty-five of the fifty-eight boats delivering goods brought building materials required for the expansion of the works indicated by table 5.29.[174] A better index of the growth of output is the weight of raw materials and finished goods shipped in and out: 465 tonnes of pig iron and 33 tonnes of scrap in the months noted in 1856, compared with 119 tonnes in 1835, plus coal, charcoal, limestone and firewood. Most despatches in 1835 must have occurred during the unnoted months June to August; only 292 kg of foundry wares, 25 tons of iron plates and a cargo of nails left along the canal.

Other river- and canal-based industry The engineering works completely dwarfed all other industrial enterprises along the river and canal (see table 5.30). But even without the giant works there was more industry along Motala

Table 5.30. *The number and value of industries in Motala Town in 1860 (in Rdr of taxation value × ⅔)*

	N	Value
Watermills	17	23,867
Saws	5	9,500
Motala Engineering Works		720,853
Other ironworks	6	10,480
Brickworks	3	15,667
Breweries	3	41,667
Papermill	1	66,667
Gunpowder and lampblack works	1	25,000
Tannery	1	14,000
Total	37	927,701

Ström than in the whole of the rest of the region. Corn- and saw-milling continued to grow upstream of the works, and two-thirds of the 50,257 Rdr's worth of milling capacity in northern Aska were along Motala Ström. Some of the Holm mills on the south bank of the river had three pairs of grindstones, plus steel cylinder mills and grain flaking machinery by the end of the period. Like the mill at Broby, one of them was converted to fulling *vadmal* for a decade or so before 1850. In 1860 as in 1820, values were omitted from the taxation registers for many of the peasant mills. The four which were valued were all small, ranging from 300 to 900 Rdr, the latter for Motala Stjärnorpsgård mill, with two pairs of grindstones and a grain flaking mill. Some of the mills downstream of Motala and Holm villages towards the engineering works were much bigger in value (though in these cases the equipment they held was not specified). Duvedal Mill, Motala Herrekvarn and Nykvarn were valued at 3,250, 5,000 and 5,200 Rdr, respectively. The last was larger than the biggest Strå mill.

The ownership of these mills was as mixed as it had been in 1820, but only three peasant owners remained. J. P. Johansson of Holm Västergård owned three mills, including Herrekvarn, the largest on the river, and leased that belonging to Vadstena Hospital. Whether there were any peasant farmers left by 1860 in the old villages of Motala and Holm is doubtful. The only other farmstead in Holm belonging to a farmer was that of M. Zederström, who owned the mill on his own Norrgård holding and leased that on adjacent Mellangården. Like his neighbour Johansson and Anders Persson of Östra Starby, Zederström was more involved in milling than farming. Apart from Motala Stjärnorpsgård, which belonged with its mill to J. Fredriksson, the

Motala village farmsteads also belonged to people engaged in other activities than farming. For example, J. and C. Kindwall occupied Motala Spolegård and Fiskaregård, valued at 3,126 and 3,000 Rdr, owned and worked a mill there together with Motala Långspång mill, and paid tax as merchants in the *köping*. Three other village mills and one of the village farms were owned by estate owner Berndes of Säter, and two others belonged to C. Ljungqvist, who lived in the *köping*. By 1860, Motala and Holm villages had obviously been engulfed by industrial and commercial development.

The mills gave rise to considerable canal traffic. Grain was imported and flour was exported from the wharf at the entrance to the canal, and there was a large trade in stone and grindstones at the Canal Company's wharf to the east of the mills. Imports from Gotland totalled 119 tonnes of sandstone and 5 tonnes of limestone in the noted months of canal accounts for 1835. Some of the former must have been destined for the sandstone grinding mill at the engineering works, but some was also converted locally into the nearly 100 grindstones working on the river and the smaller streams of northern Aska. District Judge Duberg's Lilla Hals quarry was also the source of some of the thirty-one pairs of grindstones despatched from Motala: sixteen were sandstone, bound for Stockholm and Norsholm (the nearest canal station to Norrköping); five pairs of limestone grindstones were also shipped to Norsholm, and stones of unspecified type to Berg, Lidköping and Borghamn (the last probably bound for the mills at Broby, not far away).

Large quantities of timber were also handled on the canal wharf, much of it in the form of planks, spars, joists, boards and laths imported for building in the locality. Cut timber was also exported from Motala, some of it probably from the three saws at work on Motala Ström. Of these, the two at Holm belonged, like the mills, to Johansson of Västergård. The third, at Hårstorp, was bought by the Göta Canal Company for its own use before 1820 and then greatly expanded: in 1820 it was valued for taxation at 200 Rdr, whilst in 1860 the saw, brewery and mill at Hårstorp were worth 5,000 Rdr.

When the canal was opened in 1832, many more industries joined the cornmills, saws and rod forge which had been there for centuries (see table 5.30). Most of them were established at the time of most rapid expansion at the engineering works, in the 1850s. They included the much smaller Katrinefors ironworks and manufactory, opened in 1858 and valued at 9,147 Rdr in 1860, and three new rod forges, which joined the venerable one on Motala Backegård. Two large breweries, together worth 44,000 Rdr, a gunpowder and carbonblack works worth 25,000 Rdr, a papermill worth 66,167 Rdr, and two brickworks worth 12,000 Rdr all joined the string of mills and saws along the river to make Motala Town an almost complete microcosm of developing Swedish industry during the last decade of our period.

Katrinefors Ironworks were begun in 1858[175] by the Berndes family of Säter estate in rural Motala, who also ran an oilworks at Säter itself, mills at

nearby Kärsby and at Holm Mellangård and Motala Backegård, and a rod
hammer on Motala Ström. Few of the properties apart from the ironworks
were valued in the taxation registers, but judging from the values of similar
works they cannot have amounted to more than 7,000 Rdr or so. Katrinefors
Ironworks was valued at 9,147 in 1860, when it housed three master smiths and
thirteen smiths' helpers and hands. Although much smaller than the engineering works, Katrinefors was much bigger than the traditional rod forges, which
were valued at between 1,000 and 2,000 Rdr and housed only four or five
workers. A brewery and smithy in Motala village, valued at 25,00 Rdr and
housing thirteen workmen in 1860, were also built by a local estate owner, A.
Rosendahl of Skrukarp in Kristberg parish. The paper mill at Duvedal in
Vinnerstad parish, across the river from the engineering works, was owned by
a limited liability company and valued at 66,667 Rdr in 1860. It followed the
pattern set by the Klippan mill in Skåne, where Bryan Donkin machinery was
first installed in Sweden in 1829, rather than the traditional paper *bruk* of
Småland.

Like the *verkstad*, corn mills and saws, these new industries used the canal.
Loads of wallpaper and bottled ale left Motala Works wharf in 1856; bricks
and various types of iron arrived at and left the canal company's wharf
upstream. 8,262 *kannor* of nails were shipped out to Karlshult, bar iron to
Göteborg and places on Lake Vättern, and foundry goods to Hammar. Pig
iron was imported from Råå on the northern shore of Lake Vättern (probably
en route from Bergslagen) and Nyköping on the Baltic. Small quantities of
finished hoop iron and foundry wares were also shipped inwards, despite the
enormous production in Motala Town itself (perhaps the former was destined
for the barrels in which liquor and nails were shipped out). Gunpowder was
despatched in 1835, apparently before the gunpowder works was opened, but
none left during the noted months of 1856.

Motala köping   The efflorescence of industrial activity along the river and
canal was accompanied by a *köping* (market settlement), laid out to the west of
Motala village in a fan shape from where it abutted on the inlet of the river and
canal (see figure 5.10). There were two gracefully curved streets along the east–
west axis of the settlement, with eleven cross streets and a town square at the
riverward end. The taxation records of 1830 show that just before the canal
opened completely in 1832 it had not developed very far: few of the square
plots within the grid of streets had been sold; many of those which had were
owned by people who lived elsewhere, and there were only twenty-two
gainfully employed inhabitants (see table 5.31). Three *handelsmän* settled in
the town from the start, and merchants Wallberg of Vadstena, Holmgren of
Åtvid, Elig of Skänninge, and Larsson of Borensberg had bought warehouses
near the canal terminus and other property. Many local persons of standing
had also acquired plots in the *köping*, including those with the largest estates

Table 5.31. *The development of Motala köping, 1820–60[a]*

Year	Value of property (in Rdr)[b]	Heads of household					Totals	
		Administrative and professional	Dealing	Craft	Menial	Unproductive	Heads of household	Population
1830	3,388	2	3	13	4	0	22	72
1840	505,892	2	8	25	10	12	57	271
1850	57,753	10	8	39	18	23	98	460
1860	260,314	17	21	82	52	60	232	1,041

Notes:
[a] Including Göta Canal terminus.
[b] × ⅔ in 1860.

such as Baron Fleetwood of Carlshult, Berndes of Säter, *brukspatron* Grill of Bona and Kartorp and District Judge Duberg of Lindenäs. People from as far away as Dahl, who must previously have used Vadstena, also acquired plots, including Baron Vult von Steijern of Kungs-Starby and District Clerk Lallerstedt of Bårstad (soon to become War Commissioner and take over Kungs-Starby). An ironworks estate in Ljung, on the canal towards Linköping, also owned property in the settlement, as did Messrs Wallander and Dahlberg, who worked at the Göta Canal Company terminus. When the canal opened in 1832 hopes were sky-high (not least amongst the tax assessors, who grossly over-estimated the value of property in the town in 1840: see tables 5.16 and 5.31). In fact, it grew slowly until 1850 and then, like everything else along the canal and river, took off in the following decade.

Even so, the canal wharf at the southern end of the *köping* was already busy by 1835. Twenty-one boats unloaded a great variety of wares, from bricks, tiles and window glass for the building of the town, through wines and spirits to sugar, salt, spices, medicines and dyestuffs. Outward trade was dominated by the industrial goods already mentioned and agricultural produce. Flour, unprocessed rye, oats, barley, peas and potatoes moved eastward from Motala, mainly to wharves further along the canal, but as at Vadstena liquor bound for Stockholm was the major agriculturally derived export. Traffic through the wharf next to the *köping* did not grow as quickly as that through the engineering works' wharf, nor even as the town's population, if the sampled month's accounts for 1856 represent the full year's traffic accurately. On the canal as a whole shipping increased from 847 vessels in 1835 to 3,212 in 1856,[176] but at Motala wharf the number of ships increased only from 21 arrivals and 29 departures in 1835 to 33 arrivals and 57 departures in 1856 – and 7 of the former and 40 of the latter were empty. As in 1835, apart from industrial products made along the river, the main export was liquor, which increased nearly tenfold from 1835 to 363 hectolitres in 1856, leaving for Göteborg, Västerås, Askersund, Kristinehamn and Hammar, rather than Stockholm. As a counterpoint to rapidly increasing food consumption in Motala Town itself, as well as the conversion of grain to flour in the riverside mills, to liquor in farm stills and lager in Motala's breweries, outward shipments of grain only doubled between 1835 and 1856. Most of it was wheat. Grain and flour also moved inwards in 1856; the flour came from Stockholm and the grains were oats (for the horses required in land-haulage to the canal and in and about the riverside works) and mixed corn (for the breweries). But inward shipments were still dominated by building materials, firewood and the sorts of merchant goods registered in 1835.

Despite the slow increase of traffic through its wharf the *köping* grew quickly in the 1850s. Judging from a comparison of population/tax value ratios for Vadstena, the 90 per cent fall in the assessed taxation value of the *köping*'s real estate between 1840 and 1850 and its subsequent increase to 1860 were more realistic, although still rather high, valuations. By that date, the new market settlement contained half as many people as Vadstena and its taxation value was 80 per cent as large. However, in 1860 four times as much of Motala *köping*'s property as Vadstena's was owned by people who lived elsewhere: 106,834 Rdr's worth, compared with 153,482 Rdr's worth owned by residents. Many of these 'outside' owners lived in Motala village, adjacent to the east, the engineering works eastward again, or rural parts of Motala parish. Twelve people from the village and works owned 11 per cent by value of the plots belonging to outsiders. They included the parish vicar and curate from the village, a puddlesmith and a joiner from the works, and an engineer who must also have worked there but lived in Spolegården in Motala village. The 30 per cent of outsiders' land belonging to people from more distant parts of Motala parish was in the hands of estate owners, like the Grills of Bona and Kartorp (6,000 Rdr), Berndes of Säter (9,000 Rdr) and Widow Duberg of Bromma (5,000 Rdr), plus two peasant farmers and a master builder, who had three plots, obviously for development. The most distant recorded places of residence of plot owners were not far away, in Råå and Godegård. The tradesmen from Vadstena, Skänninge and Askersund who had owned property in 1830 had all got rid of it by in 1860, when investment in land in the *köping* was highly localized.

The connection between urban life and agriculture was repeated in the new market settlement, but it was not as intense as it had been in Vadstena in 1820. Of the seven farms run from the *köping*, three each were in categories II and III, belonging to merchants and craftsmen. The largest by far was that of weighmaster Dahlberg of the Göta Canal terminus, who owned 1,300 Rdr's worth of land in Motala and Ask parishes and leased 5,907 Rdr's worth in Varv.

The occupational structure of Motala *köping* was very similar to that of Vadstena, except in two respects. Consistent with its newness and rapid building, nearby industry, and large transportation role, there were many fewer unproductive and many more menial workers in the *köping* (see table 5.32). Only three independent craftsmen were registered. Perhaps the absence of old guild workshops allowed craftsmen generally to adopt the higher status designation; however, if the craftsmen who paid tax of only 1.50 Rdr (the menial rate) are classified as independent craftsmen, the household heads of the two urban settlements had a very similar occupational structure, especially if the 'unproductive' group, which was much larger in Vadstena, is omitted as in the second pair of columns on table 5.32.

The property holdings and incomes[177] of the occupants of the *köping* are given on tables 5.33 and 5.34. Again, they were very similar, absolutely and relatively, to those of Vadstena (cf table 5.19), although slightly smaller. However, the administrative and professional group were not so wealthy in terms of either property holdings or income. It was still embryonic as a centre

Table 5.32. *The occupational structures of heads of households in Vadstena and Motala* köping *in 1860*

Occupation	Raw data		Reclassified data	
	Motala *köping*	Vadstena	Motala *köping*	Vadstena
Administrative and professional	7.3%	6.3%	9.9%	12.1%
Dealing	9.1%	6.3%	12.2%	12.1%
Craft	18.1%	12.2%	24.2%	23.5%
Independent craft	17.2%	14.5%	23.3%	27.9%
Menial	22.1%	10.6%	30.2%	24.3%
Unproductive	25.9%	48.2%	—	—
N	232	477	172	247

Note:
Reclassified data: Craftsmen who paid only 1.50 Rdr in tax are counted as independent. Unproductives are omitted.

Table 5.33. *Property holdings of occupational groups in Motala* köping *in 1860 (in Rdr of taxation value × $\frac{2}{3}$)*

		Value of urban property			Value of arable land	
	N	% owning	total	average	N owning	total
Administrative and professional	17	21.7	4,987	988	1	2,913
Dealing	21	38.1	43,567	5,446	2	4,187
Craft	42	38.5	55,654	3,710	1	2,173
Independent craft	40	17.5	8,987	1,284		
Menial	52	12.5	5,800	829		
Unproductive	60	31.8	36,487	1,586		

for high order service provision. The incomes of the postmaster, postman and telegraph officer were lower than in Vadstena, and there was as yet no bank; in 1857 the engineering works and a number of people from Motala used Vadstena bank. Neither were there any civil administrators in the town, let alone a Court Protocol Secretary and high ranking military officers like those of Vadstena. Amongst the professionals in the *köping* were two engineers, who chose to live there rather than at the engineering works. They were taxed on incomes of 400 Rdr, less than those of the senior police constable and an actor, and only one owned property.

On average, the dealers in Motala *köping* paid less tax and had less valuable

Table 5.34. *Occupational group incomes in Motala* köping *in 1860 (in Rdr of taxation value × ⅔)*

Occupational group	N with taxed income	Total income	Average income
Administrative and professional	14	12,662	603
Dealing	20	9,480	316
Craft	42	7,440	118
Independent craft	35	2,100	40

property than those of Vadstena although, as in the older town, they were wealthier than the master craftsmen. The averages are, too, misleading because of the wide ranges they hide. At the bottom end was a flour dealer who paid tax on an income of 80 Rdr, lived in a property valued at only 400 Rdr, and employed no assistants. Other specialist dealers, such as two ironmongers and a leather trader whose businesses were obviously linked to the industrial activities along the river paid tax on incomes of over 200 Rdr a year, as did the apothecary who, like his Vadstena counterparts, was amongst the wealthiest of the townsmen. His property was valued at 6,000 Rdr, his taxable income at 400 Rdr. As in Vadstena, *handelsmän* were the wealthiest merchants, especially those in the partnerships of three pairs of brothers which dominated the *köping*'s trading community. Lindahls' Trading Company was the smallest: it paid tax on 400 Rdr of income and neither brother owned any property. S. and O. J. Kjellman paid tax on 933 Rdr of income from trade and a further 400 Rdr from the latter's restaurant. Two spinsters paid tax on an income of 1,500 Rdr from a billiard hall also owned by the latter, which seems to have prospered mightily from catering to the throngs of young men at the works downstream. The Kjellmans owned 12,666 Rdr's worth of property and employed five assistants, four hands and five maids. J. P. and G. Thamsten were similarly wealthy. Their properties were worth 14,950 Rdr and their income was 667 Rdr. Their workforce was smaller than that of the Kjellmans, with three assistants, two hands and three maids, but their business contacts were very extensive: '*handlande* Thamsten' of Motala was granted governor's passes to travel on business to Germany and Denmark for six months in 1849 and again, when he was styled '*handlande* and *brukspatron*', in 1850.[178] Like their Vadstena counterparts, the merchants of the *köping* expanded their enterprises with bank loans: J. P. Thamsten had ten shares in Vadstena Enskilda Bank, but they were worth only one-tenth of the 5,000 Rdr loans that each brother had taken out.

Amongst the craftsmen the two tanners were, as in Vadstena, much wealthier than the rest. Between them they owned 58 per cent of the total urban property belonging to craftsmen and all the farmland; accounted for 12 per cent of total craft income, and employed seven of the thirty-seven craft journeymen. Tanner Lagermark's urban and rural properties were worth more than those of *handlande* G. Thamsten. The remaining thirty-seven craft households (three of which were headed by widows) contained a similar mix of shoemakers, tailors, painters, carpenters, joiners and so on as in Vadstena, although there were two combmakers (perhaps, like the billiard hall, dependent on the trade of the numerous single young men in the works on Motala Ström). The craftsmen's properties ranged in taxation value from the 253 Rdr of bookbinder Eriksson to the 3,000 Rdr of tile stove maker Johansson. Two-thirds of them owned no property, and only eleven employed apprentices or journeymen.

Except for six maids, two night watchmen and a paviour, the menial workers were labourers. More than half of the 'unproductive' households were headed by widows and spinsters. Unlike in Vadstena, none of them was wealthy; a quarter owned the property they lived in, but the largest was valued at only 3,000 Rdr. The retired craftsmen and dealers owned no property of any great value, and only one person was described as a 'property owner'. The wealthiest member of the 'unproductive' group, with a house valued at 6,000 Rdr, was perhaps Motala *köping*'s equivalent of the property lessee and moneylender Johan Johansson of Vadstena.

Motala *köping* seems, therefore, to have grown, especially in the last decade of the period, as a central place serving the farms of the northern (as yet less commercialized) part of the plain, and the industrializing part of the region on the shieldland, northward again. The large preponderance of young wage-earning men in the numerous works downstream made the new *köping*'s economy slightly different from that of Vadstena, but all-in-all its functions were very similar to those of the ancient *uppstad*.

Production structures, livelihood positions and demography in 1860

The köping   Some 232 households contained 1,044 people, and a further 85 lived in the canal terminus, the courthouse and the poorhouse. As in Vadstena, craftsmen had the largest and most complex households. Those of the master craftsmen comprised 18 per cent of the town's households and contained 30 per cent of its population, whilst those of all craftsmen represented 27 per cent and 42 per cent, respectively. On average, each master craftsman's household contained his wife, more than two children under seventeen, more than one journeyman or apprentice and a maid, and was only slightly smaller than those of Vadstena. Some were much bigger. Tanner Lagermark, for example, lived with his wife, an unmarried clerk, four married journeymen, three unmarried hands, nine children under seventeen, a spinster and a widow with two children. Besides his house and workshop, there was also a shop on Lagermark's valuable plot on Market Square. It was leased out to two merchants, who lived there with their widowed mother, two single assistants, two single hands and a maid. On the same plot lived the households of a bricklayer, a district clerk and a widow. Altogether these five households on the one plot contained thirty adults (including hands and maids) and sixteen dependent children. As in Vadstena, many households contained non-family members, and many were indistinctly separated from their neighbours (in the records, and perhaps also in reality). The 'unproductive' households were, however, generally heavily abbreviated, with only a single widow or widower, although they more often contained an adult child than households in other groups.

Overall, new though it was, and old though Vadstena was, the compositions of the households of their occupational groups were very similar. The *köping* was a fully fledged little town in all respects by 1860, although generally less wealthy and without the social cachet of its more decorous neighbour to the south. However, it was only the western extremity of what had become a small conurbation strung out along the river and canal, with the industrialized villages and hamlets of Motala, Holm and Hårstorp to the east, and eastward again the engineering works before Duvedal, Strömsnäs and the shore of Lake Boren (see figure 5.10). Away from the river, old hamlets had also been engulfed into the built-up area of Motala Town by 1860.

The engineering works Motala Engineering Works housed a much larger population than the *köping* – indeed, not many fewer than Vadstena in 1860. It grew prodigiously. The 166 full-time employees of 1840 became 348 in 1850 and 669 in 1860.[179] In effect, it was a gigantic *bruk*. It had farmland for agricultural produce and charcoal and its own inn, brewery and bakery. Amongst its employees were a preacher, a schoolteacher, a doctor and nurses. Housing was provided for many of the employees, and the twenty-nine widows with their children, disabled and aged men and their families, and orphans who lived at the works in 1860 bear strong witness to its continuance of *bruk* traditions of welfare provision for its workers and their families after the end of their working lives. Table 5.36 shows that in total 2,075 people lived at the works in 1860. Unfortunately, the occupational descriptions in the *mantalslängder* of 1860 are not so explicit as some earlier ones. Many of those described baldly as master, apart from the master mariner, must have been puddlesmiths, founders, filers, boilermakers, mechanics, forge smiths, patternmakers and so on, as listed in earlier records. Those classified on the table as administrators include the preacher, doctor and schoolteacher as well as clerks, the chief accountant and managing director Carlsund, and the senior worker category comprises master mechanics of three grades, an engine mounter and a draughtsman. The master's helpers were the equivalent of the masters' helpers of traditional *bruk*. As the table shows, journeymen were by far the biggest group. Many of them had wives and families, some with children more than seventeen years old. They were by no means all newly

Table 5.35. Household composition in Motala köping in 1860

Occupational group	N	Heads of households				Journeymen and apprentices				Hands and maids				Living-in				Totals				
		M	F	Ch1[a]	Ch2[b]	M	F	Ch1	Ch2	M	F	Ch1	Ch2	M	F	Ch1	Ch2	M	F	Ch1	Ch2	T
Administrative and professional	17	17	11	4	25					3	8				2			20	21	4	25	70
Dealing	21	20	11	0	30	15				15	32				2			50	45		30	125
Crafts	42	41	37	5	92	51	6		14	9	46			1	2		2	102	91	5	108	306
Independent craft	40	34	30	1	61	1	1		3	1	8				1			36	40	1	64	141
Menial	52	46	50	1	92					4	6		2		2			50	58	1	94	203
Unproductive	60	34	69	14	69					6	11		1					40	80	14	70	204
Total	232	192	208	25	369	67	7		17	38	111		3	1	9		2	298	335	25	391	1,049

Notes:
[a] Children of eighteen years and over.
[b] Children under the age of eighteen.

Table 5.36. *The workforce and population of Motala Engineering Works in 1860*

Occupation	Head household composition				Hands and maids				Living-in	
	M	F	Ch1[a]	Ch2[b]	M	F	Ch1	Ch2	M	F
Administrative	10	5	1	12	1	7				
Senior workers	9	9	11	17		8	1			
Masters	119	108	38	276		8	1			1
Masters' helpers	16	8	1	15						
Journeymen	270	206	11	367		5				
Apprentices	33									
Hands	203	72	3	83						
Innkeeper	1	1	1	5		1				
Pensioners	8	47	11	54	1	8	1			1
Total	669	456	77	829	2	37	3			2

Notes:
[a] Children of eighteen years and over.
[b] Children under the age of eighteen.

qualified people straight out of apprenticeship, but formed a large pool of fully trained labour to help the master craftsmen; three who owned land in other parts of Motala were, in fact, described as engineers in the entry of their property holdings. The works seems amply to have fulfilled its founder's intention to create a reservoir of skilled workers who could take best-practice technology to the rest of Sweden (whether they actually did so will be examined in chapter 7). Most apprentices were called simply that, although five were specified as filers, one as a founder and two as draughtsmen. The hands probably did general labouring jobs around the works, apart from the garden hand, stable hand and store hand. Judging from the relatively few who were married, they tended to be young. It is remarkable that they comprised only 31 per cent of the works' labour force, with the corollary that 69 per cent were highly skilled in iron forging and machine making.

The pensioners who lived at the works were mainly widows, of whom there were thirty-nine, with ten elder and thirty-four younger children. Some of them were destitute. So were two of the three old men, two maids, two journeymen and two apprentices who were presumably injured or too ill to work. There were also eight boys and five girls amongst the pensioners, all destitute, probably orphans of deceased employees.

Although the engineering works treated its employees in the concerned way that was customary at Swedish *bruk*, hours of work were long. In 1855, the blue collar working day was reduced from 13 to 11 hours, and that of white collar workers from 11 to 9 hours.[180] The mean hourly pay rate was 10 *öre* in 1848 and 14.67 öre in 1860, so that the shorter working hours did not reduce take-home pay. The annual average wage must have been about 440 Rdr in 1860,[181] although there must have been a large range around it. The four mechanics who lived in the *köping* in 1860 were taxed on incomes of 600 Rdr,[182] and the master mechanics were obviously quite wealthy. Following the custom of *bruk* master smiths, two of them owned and tended farmland, as did a chief accountant's widow who lived as a pensioner at the works and a number of journeymen. A puddlesmith and a joiner who continued to live at the engineering works owned plots in the *köping* (worth 2,267 Rdr in total), whilst a mechanic owned an urban plot and Spolegården farm in Motala village, where he lived. Journeymen and pensioned widows living at the works also owned plots in the *köping* worth 3,400 Rdr in total.

Further evidence that this was a relatively affluent labour force retaining many of the customary characteristics of workers at *bruk* is that fifty-two of the employees were sons of fathers who were also working or pensioned there, or of pensioner widows.[183] Numerous shared surnames imply many other family connections within the works labour force as well as those between parents and children recorded in the *mantalslängder*. The fact that there were 270 journeymen waiting for only 144 masterships also suggests that it was considered to be a good place to work.

The industrial villages   The villages and hamlets straggling for 3.5 km between the *köping* and the engineering works were industrializing rapidly. None of the works there was anything like as big as the engineering works, but four, all opened after 1850, had large resident labour forces compared with the normal size of *bruk* and manufactories at the time.[184] No others housed more than one or two workers: even the gunpowder and lampblack factory had only a master, his helper and their wives. Like some of the engineering works' employees, most of the workers in these enterprises must have lived in the clutter of forty-two crofts and forty-seven cottages built on the homesteads and fields of the nineteen old farms near the river. Besides the eighty-four households headed by peasants, crofters and cottagers in these settlements, a further sixty-nine were headed mainly by labourers and day-labourers, though a few had painters, shoemakers and tailors as heads, supplementing the craft provision of the *köping*. Most of the last were the equivalent of rural parish craftsmen, though one or two had workshops as big as those of the *köping*'s master craftsmen: for example, painter C. Molander of Marstadlyckan croft had a journeyman also named Molander, four other journeymen, an apprentice and two hands.

Table 5.37 shows that the population housed in the straggle of works, villages and hamlets between the engineering works and *köping* was larger

Table 5.37. *Populations of the component settlements of Motala Town in 1860*

	Farmer			Other			Living-in			Hands and maids			Total populations
	M	F	Ch	M	F	Ch	M	F	Ch	M	F	Ch	
Köping[a]				262	214	417	15	38	15	43	121	3	1,128
Engineering works				461	331	758	8	49	65	205	109	89	2,075
Other works[b]				69	46	82	8	19	22				246
Villages[b]	84	98	205	69	53	60	31	60	44	98	89	117	1,008
Total	84	98	205	861	644	1,317	62	166	146	346	319	209	4,457

Notes:
[a] Including Göta Canal station.
[b] Including hamlets and isolated farms in vicinity of river.

Table 5.38. *Demography of occupational groups in Motala Town in 1860*

Occupational group	% married	Children per family		
		> 17 yrs.	< 17 yrs.	Total
Senior workers at Engineering				
Works	100.00	1.22	1.89	3.11
Masters ditto	90.76	0.35	2.56	2.91
Journeymen ditto	76.30	0.05	1.78	1.83
Hands ditto	35.47	0.04	1.15	1.19
Industrial workers elsewhere	85.45	0.23	2.38	2.61
Köping craftsmen	81.33	0.10	2.39	2.49

than the total in the latter. In all there were nearly 4,500 inhabitants in the sandwich of new town, industrial villages and factory town along and between Motala Ström and the Göta Canal. This was almost double the population of Vadstena and nine times as many as lived along the river in 1810. Growth had been sustained throughout the period, and did not simply occur after 1850, when there were already 2,939 inhabitants, compared with 2,272 in 1840. However, there was considerable acceleration after 1850: as the *köping* eventually took off, the engineering works expanded, and many new works were established in between them, the population of Motala Town doubled in a decade, a rapid rate of industrial urbanization anywhere in Europe at the time.

The three components of Motala Town had different demographic characteristics. Overall, the ratio of children under seventeen to women was 1.53, compared to 0.64 in Vadstena. It was 1.16 in the *köping*, 1.87 in the engineering works, 1.60 in the other works, and 1.42 in the villages interspersed amongst them. Although it shows that the industrial population of the works, especially the engineering works itself, was most fecund, this statistic is an imprecise demographic measure because its denominator includes unmarried women. Table 5.38 provides more precise information on the composition of families within selected groups. Of course, we must be careful to take any systematic differences in stage of the life cycle into account when interpreting statistics of this kind. Older groups would tend to contain more married members and to have more children at home, especially if affluent enough to keep their older children with them rather than sending them out as hands and maids. Thus, the senior employees at the engineering works were all married and had almost as many older as younger children still at home, and fewer younger children per family suggest that they had passed beyond the peak of their reproductive lives. However, despite these complications, table 5.38 does show that all the adult groups of Motala Town had more children at home than those of

Industrial and urban livelihoods 307

Vadstena, where the largest families were the 2.24 children of the master craftsmen (see table 5.27). People at the engineering works, and perhaps also those at other industrial plant, were especially fecund. High wages, welfare provision and job security for children had the predictable effect of producing a high birth rate.

Also quite predictably, the poor, including those said to be destitute or 'on the parish', were relatively few in Motala Town. The poorhouses in Motala village and at the Göta Canal Company station near the *köping* must have served, respectively, the whole of rural and urban Motala parish, and the employees of the whole eastern length of the canal. However, they contained only thirty adults, twenty of them women, and eight children. Motala Works, a brewery and the papermill housed six destitute men, eighteen women and twenty children, most of the latter orphans at the Works. In total, 3.3 per cent of the male population of Motala Town, 8.6 per cent of its women and 6.3 per cent of its children were destitute. The equivalent percentage for males in Vadstena was slightly lower, but in rural areas it was nearly 6 per cent., testifying to the general youthfulness of the population and the comprehensive welfare provisions of the Works where a high proportion of it lived.

Conclusion

The cutting of the Göta Canal through the region had a dramatic effect on its non-agrarian livelihood positions in two different, but linked, ways. First, the consequent development of estate and peasant farming, especially the escalation of liquor production for export from the region, and the increased output of by-employments, generated greater wealth and social differentiation in the countryside, particularly on the southern part of the plain. Effects multiplied out from this into non-agricultural activities. A few craft positions emerged in the countryside, where there had been hardly any before, and milling, which had been virtually confined to Strå in 1810, spread across the plains, using both wind and water power. Much more significant was the multiplier effect out of agriculture into Vadstena, the old central place of the plain, which increased in population by 50 per cent between 1820 and 1860. By the end of the period Vadstena's economy and society were undergoing profound transformation: high order institutional services pushed the administrative and professional group into greater prominence; shopkeepers and merchants increased in number and wealth from shipping goods into and out of the agricultural hinterland, and the craft sector supplying the plains with everyday consumer goods and services developed rapidly as independent craftsmen became more numerous and master craftsmen's workshops grew. In addition, Vadstena had begun to attract a large and in part wealthy group of retirees, who spent cash in the town which had been accumulated elsewhere. Doubtless, too, the lace produced by the women of Vadstena found a more receptive market as peasant farmers became wealthier and aspired after the consumption patterns of the gentry and bourgeoisie. The development of Vadstena in relative as well as absolute terms was reflected by a rise in the relative value of its real estate from 6.5 to 8.8 per cent of the taxation value of all property in the region, *despite* the spectacular growth of industry, population and a competing central place and port along Motala Ström. Although Vadstena's population shrank in proportion to that of the region as a whole, it increased from 16 to 20 per cent of that in its immediate hinterland on the plains of Dahl and southern Aska.

Such rapid and rich side-effects of the development of estate and peasant farming would seem much more notable had it not been for the way in which the economic geography of the northern Aska shieldland was transformed by the canal. The stretch of canal and river between Lakes Vättern and Boren emerged as a major Swedish industrial centre. Raw materials were brought mainly from the shores of Lakes Vättern and Boren and the stretches of canal most nearly adjacent to them; finished goods were mainly sent to Norrköping, Stockholm and Göteborg, and probably from there to ultimate consumers even further afield in Sweden and Europe. Motala Town's inner hinterland, within which there was intensive shipment and trans-shipment, stretched along the eastern shore of Vättern from Askersund and Hammar in the north to Vadstena and Borghamn in the south, but its outer hinterland, from which medicines, spices and other luxuries were imported as well as to which its manufactured goods flowed, comprised the whole of southern Sweden. Motala Town represented development of an entirely new order, similar to that where canal transport and energy resources coincided in other parts of Western Europe at the time.

The traditional industries of the shield were depressed during the first two decades of the nineteenth century, but recovered after 1820. By the 1850s they were at best marking time, as capital was transferred by the estate owners to whom they belonged to investments in Motala Town: a precocious example of the spatial concentration of production which came to most parts of Sweden only after the railway network was built.

By 1860 non-agrarian activities were responsible for 41 per cent of the taxation value of the region and 30 per cent of its employment. Over 30 per cent of the population lived in Motala Town and Vadstena, 15 per cent in the central places of Motala *köping* and Vadstena rather than the industrial settlements beside the former. What had in 1810 been an urban/industrial sector typical of peasant societies, completely inseparable from local agriculture, had by 1860 become recognizably modern. Secondary and tertiary activities were spawned by peasant and estate agriculture and by industrial development not directly linked to agriculture; the concentration of both in and near towns gave a strong twist to urbanization. In industrial and urban development, our region was way ahead of most of Sweden at the time, and even of the strip along the Göta Canal as a whole. Was this facilitated by complementary flows between the two contrasting ecotypes which were being so comprehensively changed? Or might they just as well have been at opposite ends of Sweden? We will examine flows of capital and labour to help us answer these questions.

6

Investment and money in agriculture

Setting the scene

Introduction

The *sine qua non* of capitalist farming is capital: that is, money spent not simply to yield output to be consumed, but to yield more money. Farming in Aska and Dahl developed many of the secondary traits of capitalist operations during our period: how much capital was needed for this capitalist development, and where did it come from?

Land with a taxation value of 3,500 Rdr was enough to produce a surplus. It cost about 5,000 Rdr, and if, as was often the case, the expansion was connected with enclosure, then could cost 50–100 Rdr per acre on the plains. To stock and sow the land would cost about 500 Rdr, to equip it with best practice technology from plough through cart to still and dairy of appropriate sizes would cost another 300–400 Rdr.[1] Wages for the first season before any income came from the land would add another few hundred *Riksdaler*, bringing the total cost to about 6,000 Rdr plus enclosure costs. The *net* cost (disregarding inflation) of increasing the average value of a category I farm between 1810 and 1860 by the amount calculated in chapter 3 was about 4,500 Rdr per farm, and the *net* cost of expanding all the holdings of category I farms between 1810 and 1860 must have been well over three quarters of a million *Riksdaler*. Of course, these figures will be overestimates due to inflation over the period, but on the other hand we must take into account such things as unprofitable speculations leading to subsequent sale, and the necessity to recombine farms that had been subdivided by inheritance simply to maintain their size. As we showed in chapter 3, too, farmers could often only expand their holdings by initially buying land at some distance away before exchanging it for closer plots. The combined effect of such exchanges would be to make the gross amounts of investment many times larger than the given figures. However, it was not necessary for very large sums to be spent all at once. Plots of land could be acquired piecemeal and first leased before they were eventually bought from accumulated profits. Stock, seed and equipment could be obtained partly from the pre-existing farm and partly from relatives and neighbours on informal credit. Investment by a local peasant who already had a subsistence-size holding could be effected purely from ploughed-back surpluses, whereas a person moving into mainly commercial farming from scratch would need to borrow 6,000 Rdr.

Expansion by investing all possible surpluses would inevitably be slow,[2] even though peasants kept stores of cash and their high propensity for credit dealing is technically the equivalent of borrowing. Increased cash stocks after good years could, too, generate considerable circulation of surplus funds in the exchange of land between peasant families moving through different stages of the life cycle. However, the spending of surpluses from good years on land did not represent *capital investment* in a peasant society; it simply fuelled inflation in the land market. Only when land was bought to produce surpluses for sale through money deliberately earned for that purpose did such spending represent the ploughing back of profits as investment. Furthermore, as farming became more fully integrated into a market economy, land presented an investment opportunity to outside capital: as it became integrated into a wider system of commodity exchange, farming also became integrated into a wider system of capital circulation.

What kinds of flows of money within the farming system would represent capital investment? Those which were a normal part of peasant economy certainly would not. For example, fathers might pass on money to acquire a holding, rather than a holding itself, to their sons, especially in a system of partible inheritance like that of Sweden in the nineteenth century. This could happen for two reasons. First, the availability of profits on well-situated holdings would allow not only for those holdings to be split, but also for younger sons to be set up on purchased subsistence farms elsewhere. Secondly, if a father's holding was not large enough to accommodate more than one son, the eldest could be bequeathed money with which to purchase the younger sons' shares. Money would also be needed if a son were set up in his own household on family land before his parents' death. In all these cases, the father might supplement his own savings with money borrowed from relatives, neighbours and even outsiders or institutions. In all cases, these transfers were intergenerational, necessary when children first set up their own households. In all cases, too, they must have been funded by getting many small quantities of cash from a variety of sources: by definition, the regular surpluses available to peasants were too small to provide either large savings or high creditworthiness, so that they must depend heavily on the trust of relatives, neighbours and friends. It was also normal for peasant farms which were barely large enough for subsistence in average years to borrow heavily after bad harvests. Indeed, this is usually thought of as the first stage of the dispossession of peasant farmers before their land is amalgamated into large holdings.³ In this case, money would have to be brought into the peasant economy from outside because when times were hard for one peasant they would be hard for all of them.

If all profits made in commercial farming were invested in the owners' own farms, loans to farmers must come from outside the farm economy, from towns, industry or overseas. It has been argued that this was the situation in Sweden at the time: that the large exports of oats were funded by an influx of capital into farming.⁴ The same argument could be applied to the funding of the increased market production of grain and liquor in our region. If this was so, where did any loan capital come from? Did it come from other, less profitable, sectors of the regional economy like ironworking and other manufacturing or craft production; from neighbouring regions where farming was less profitable, or more distantly, both sectorally and spatially? Through what channels did these funds flow: how important, for example were townsfolk and urban institutions? Of course, all funds need not have come from outside: there would always be older farmers (or even more markedly, widows) for whom it would be preferable to lend surpluses rather than invest them in their own enterprise. Secondly, given that there were scale restrictions not only in farming itself, but also in distilling and other processing activities, it would pay those farmers operating profitably at a smaller scale to lend their surpluses out at interest until enough capital had been accumulated to fund the jump to the next scale level. Given, also, that diseconomies of scale existed, it would be most profitable for those farmers whose holdings had reached that point to lend their surpluses to others operating at a smaller scale. The difference between these capitalist loans between farmers and those within peasant society would not only be the intentions lying behind them, but also, like loans from outsiders and institutions, their size and the fact that they would be taken out by large and small farmers alike. Furthermore, as members of the petite-bourgeoisie capitalist farmers might find it more profitable to lend their surpluses to non-agricultural enterprises: capital might flow out of as well as into farming.

All this assumes that agrarian society was composed only of freehold farmers. However, about one-third of the land in our region was farmed by estates and persons of standing in 1860 and tenancy was particularly common on their holdings. We might expect the funding of expansion to have been very different for them. Usually the economy of noble estates extended over wide areas and into enterprises such as forestry, ironworking and urban property holding. Intersectoral and interregional flows of capital could occur within these single accounting units. It would be much easier for estate owners to fund the expansion of commercial farming without recourse to loans, but if this were needed they had much greater security for formal borrowing. The farming enterprises of persons of standing were normally supported by salary income as well as farm profits, which might mean that they could fund expansion more easily and much further than freehold farmers. Both noble estate owners and persons of standing were integrated into much more affluent and geographically widespread networks through which money for investment could be drawn than ordinary freehold farmers.

Tenant farmers on estates, on the other hand, would find it more difficult to fund improvement than freeholders. Although it might have been easier for them to expand their holdings simply by acquiring a larger tenancy, unlike freeholders they had no land to offer as security against loans. In addition, because part of their surplus product went in rent they had a smaller income per unit of land from which to repay loans. Tenants of taxed land would be in a rather different position. We saw in chapter 3 that few of them were wholly tenant farmers throughout their careers. Rather, they used tenancy as a step towards a bigger holding of their own through the temporary use of land that was not being farmed by its owner because, for example, it belonged to a childless widow or a group of heirs who had not yet finally settled their inheritance shares. Tenants of taxed freehold land were thus simply parts of the normal peasant or capitalist farming system, unlike estate tenants who might well be without this supportive network of contacts with relatives and neighbours long established in the same village with farm resources and funds of their own.

So, we would expect large differences in the ease with which different sectors of rural society could expand commercial farming projects, and there are ways to distinguish between the use of money in a peasant economy and its use for investment in a freeholding capitalist farming system.

Investment and money in Swedish agriculture in the early nineteenth century

Unfortunately, these differences have not always been drawn in research into the financing of Swedish farming in the nineteenth century, although it does carry some clear implications on these matters. Farmers in two Södermanland parishes were little involved in money transactions outside their own group before the 1860s, and the loan market of the 1820s was still small, with about one-third of the loans coming from retired farmers and the like.[5] Tenant farmers on estates were most indebted. Exemplifying their separateness from the matrix of freehold peasant society, their loans mainly came from other tenants. The sums borrowed increased over the course of the nineteenth century: the average farmer in the 1860s had loans and mortgages for more than 1,000 Rdr and in the 1880s more than 2,000 Rdr. Estate tenants borrowed larger sums at the beginning of the century, on average about 200 Rdr in the 1820s when freeholders only owed 150 Rdr, which might well have reflected the relative ease with which sons of freeholders could stock and equip their own farms. In the 1860s, however, freeholders had loans more than double the size of tenants', probably reflecting their higher capacity to invest in farm expansion for market production. Institutional financiers such as savings banks, cooperatives and merchant banks in the 1860s accounted for only about 1 per cent of the total debt and in the 1880s for 2 per cent. Rural traders, foremen on estates and retired farmers were the most important money-lenders. Agricultural developments in Södermanland were not funded from outside, either sectorally or spatially: surpluses from agriculture, accumulated over individual life-paths were funnelled back to younger farmers and *not* used for investment in other economic sectors. Whether or not this represented the intergenerational transfer of wealth in a peasantry or investment depends upon whether the money was being used to increase farm size and productivity, which we do not know.

The situation was rather different in By parish in Bergslagen where iron mining and iron making had been important since medieval times.[6] There the large farmers, who were also ironmasters, were important sources of credit: in the 1840s to 1860s 90 per cent of the total claims in inventories belonged to them.[7] Many debts in farm inventories were to local traders, extending to towns 70 km away: the strong emphasis on secondary production gave By parish closer links with towns than the estate farming-dominated areas in Södermanland. The loan market would presumably also be better developed in By, with its industry directed towards remote consumers and agriculture of a mainly subsistence character.[8] Debts were quite common and larger among the larger farmers and farmer/ironmasters, who could afford to be indebted, while small farmers and tenants, who had little property on which to take out mortgages, had smaller if any loans. In spite of the developed market contacts of the ironsmelting farmers and the loan-funded agricultural expansion of freeholders, the average debt in By was only 433 Rdr and even the large ironmasters/farmers only owed about 550 Rdr, whilst the Södermanland freeholders in the 1860s owed about 1,100 Rdr.

It is not possible to determine from the work on By or Södermanland how far the debts listed in inventories represented unsettled credit transactions (very common in peasant societies), and how far money loans. Nor is it possible to discover whether any loans that were taken out were to cover consumption, debts in bad years, or investment. However, the way in which the size of loans varied across social groups is very suggestive. Tenant farmers generally had the smallest debts in By in 1840–60, the opposite of Södermanland. The farmers-cum-ironmasters had most debts, but also had the largest fortunes. Freehold specialist farmers were also highly indebted, but without positive balances, because they financed their arable reclamation and enclosure on loans. Tenant farmers were normally compensated by landowners for effort in connection with enclosure and arable reclamation and thus did not need extra capital.

Martinius related inventory debts in nineteenth-century Västergötland to the total value of farm assets, including land,[9] and calculated the indebtedness of farms in different size categories[10] for 1828–36 and 1858–66. In the first period debt was proportionately largest among younger and smaller farmers at 58 per cent of the value of their assets; in the second the largest farms had by far the largest debts and the middle-sized group also had larger debts than small farms. Money was borrowed from different sources by farms in different categories. The largest farms owed about 15 per cent of their total debt to the Bank of Sweden, other central institutions or the local savings bank, smaller farmers only between 4 per cent and 8 per cent. Small and middle-sized farms owed debts worth between 14 per cent and 46 per cent of their total asset value to local farmers. The institutional debts of smaller farms did not grow, even though many more monetary institutions like merchant and mortgage banks were established. Indeed, the level of debt in proportion to total assets decreased in many cases and only the really large farms were more heavily in debt in the 1860s than in the 1830s. At the start of the period, therefore, most indebtedness was of the kind common in peasant societies, where newly-weds require help from family and neighbours in setting up their own holdings. This continued until the 1860s, and was joined by large institutional debts of large farmers, incurred for investment in farm expansion.

Thus, there were strong regional differences in the capitalization of farming in Sweden in the early nineteenth century. On the plains of Västergötland a peasant system of farm finance continued, with capitalist investment beginning before the 1860s, funded by loans from financial institutions. On the shield in By investment rates were lower, and lowest of all amongst tenant farmers. There, farmer/ironmasters provided loan capital. In estate-dominated Södermanland, loans were less common than in By, implying an even weaker development of capitalist attitudes to farming.

Investment and money in agriculture in western Östergötland, 1810–60

The rates of borrowing discussed so far are much lower than our estimates of the costs of investment in the expansion of commercial farms in our region. This does not mean that investment rates must have been higher in our region: the figures for other parts of Sweden include all farmers, not just large commercial ones, and a difference between costs and borrowing could be bridged by ploughed-back profits. We will now examine how far these findings about the significance of loans in the process of farm expansion in other parts of Sweden are appropriate to the various parts of our region, where the plains were probably more extremely agricultural than in the case studies quoted above, and the shieldland was similar to that of By parish, although iron production was in the hands of estate owners rather than farmers.

The 1820s

Mortgages had to be formally registered in local courts by the 1820s, and renewed every ten years through another court procedure. Mainly they were taken out from private individuals, but official institutions and the Bank of Sweden also channelled funds into the mortgage market. Until private clearing and mortgage banks were set up in the 1830s, after which it has been assumed that loans through institutions became more important, the Bank of Sweden acted in these capacities. It borrowed substantial amounts abroad for investment in various sectors of the economy. Some 25 per cent of its total lending in the 1830s represented farm mortgages.[11]

We examined the mortgage registers for Dahl for 1822–28, when seventy mortgages were taken out or renewed, which suggests a total for the 1820s of more than a hundred. There were 357 plots in the hundred at the time, and therefore, because mortgages were taken out on plots rather than farms, under a third of the plots were mortgaged. In fact, because some plots reappear in the registers, only about a quarter were mortgaged: the single plot Tyskeryd farm in Väversunda, for example, appears four times in the registers. It was the most frequently mortgaged plot because its owner succeeded in getting a mortgage from the Bank of Sweden for a higher sum than his previous local mortgage a year after this had been renewed, and then two more from the bank. Of seven mortgages in Herrestad parish in 1831–33, three were financed by the Bank of Sweden, and 15 per cent of the seventy in Dahl as a whole, in 1822–28, were from national official institutions. Only three were from the Bank of Sweden, and instead all sorts of state funds were used, such as the Ecclesiastical Fund for Lappland and the Army Pension Fund. The contracts drawn up for official mortgages were much more complicated than for private ones and included a fixed-term repayment scheme, which private mortgage contracts rarely did. Their rates of interest in Östergötland were 5.25 per cent, compared with a standard rate on private mortgages of 5 or 6 per cent, but they were obviously popular because farmers took them out to redeem private ones. Two small mortgages (of 200 Rdr) were given by Parish Church and Poor Relief funds, but local institutions such as these were not significant in Dahl during the 1820s.

If older relatives and other farmers on the brink of retirement had been a major source of mortgage funds, many relatives of mortgagees would appear in the registers. This was to an extent the case: of the fifty-three local mortgages in Dahl, seven were given by uncles or aunts. Five more given as part of the purchase price to the buyer of a plot were perhaps also part of the local assistance to young farmers through accumulated savings of relatives and neighbours that was common in peasant society. However, some of these twelve mortgagors often appear among the remaining forty-one mortgages to people to whom they were not related and without being part of a land purchase agreement. In fact, thirty-five of fifty-three local private mortgages were given by people who appear at least three times in the register, and about half of all the mortgages in Dahl were obviously given by people who had accumulated enough wealth to permit a few thousand *Riksdaler* to be lent commercially to unrelated people. Twenty of these thirty-five mortgages were given by Alderman Jaensson's widow of Arneberga estate in Örberga and Alderman Röhman of Vadstena, both of whom had made their money in urban tanning. In all, nearly half of the locally funded Dahl farm mortgages, representing more than half of the total sum, were of urban origin. The average urban mortgage was for 2,300 Rdr; peasant-funded ones averaged only 600 Rdr, and those given by the Bank of Sweden and other official institutions averaged just over 2,000 Rdr. These differences might indicate that farmers were more conservative in their lending than banks and townsfolk, who were willing to take a bigger risk in their moneylending. Or it could have been that larger loans from outside farming were investments in farm expansion, and those from farmers an intergenerational transfer within a peasant economy, where part-payments and small loans from relatives and neighbours were being used to set up new households. Whether this was so can only be discovered from the destinations of the mortgages.

Total mortgage lending amounted to about 30 per cent of the total taxation value of farmland in Dahl in the 1820s.[12] It is difficult to say whether this is a large or small proportion because there are no comparable figures for elsewhere, but it is certainly larger than in Aska hundred. Who took them out, and were differences in the size of mortgages related to the size of the mortgagees' farms? Although the exact boundaries of the categories in table 6.1 are difficult to draw, it is clear that most large mortgages given by people and institutions from outside local farming society did, indeed, go to fund the expansion of large farms. Our old friends from chapter 3 who created large compound holdings were involved with urban mortgages. Lars Larsson of Herrestad took out two between 1822 and 1829, and Gustaf Lundberg of Jernevid, Örberga took out mortgages from Widow Jaensson. We showed in chapter 3 that tenancy was important in the early stage of building up large compound farms, as enterprising peasants supplemented their inherited plots, and persons of standing their official ones, with leased land before they could afford to buy it. Many such farmers besides Larsson and Lundberg appeared in the mortgage registers. We can therefore be certain that there were two mortgage markets in operation in Dahl, one embedded in a peasant economy, where farm purchase by young small farmers was assisted by relatives and neighbours, and another in which farmers were expanding their holdings with money borrowed from towns, official institutions, and persons of standing living in the countryside.

We would expect much less borrowing of the latter kind in Aska than in Dahl because plot amalgamation and the build up of large peasant holdings

Table 6.1. *Farm mortgages in Dahl 1822–28*

A: *Sources of total mortgage credit*

	N	Sum (in Rdr)
Vadstena	23	44,878
Total urban	24	56,878
Banks	11	22,435
Local farmers	24	14,477
Local persons of standing	6	10,243
Total farmers	26	15,777
Total persons of standing	7	15,243

B: *Sources of Örberga farm mortgages*

Farm	Mortgage sum (in Rdr)	Mortgage giver
Kasta	533	Erik Eliesson, Mörby, Örberga
Ullevi (Rosenborg)	3,000	Widow Jaensson, Arneberga, Örberga
Ullevi (Rosenborg)	1,800	Ditto
Mörby	1,333	Röhman of Vadstena
Kasta	707	Dean Pettersson, Örberga
Ullnäs	433	G. Andersson, Östnässja, Nässja
Ullnäs	200	Lars Lohm, Krigsberga, Örberga
Mörby	514	Röhman of Vadstena
Örberga Uppgård	225	Erik Larsson, Lundby, Örberga
Örberga Murgård	900	Jonas Larsson, Östnässja, Nässja
Kasta Postgård	666	Ditto
Örberga Uppgård	433	Lars Larsson, Lundby, Örberga
Järnevid	3,333	Widow Jaensson, Arneberga, Örberga
Lundby	500	Jonas Larsson, Östnässja, Nässja
Ullevi (Rosenborg)	5,000	Distr. Judge Engelhart
Mörby Södergård	500	Röhman of Vadstena
Mörby Västergård	200	Ditto
Örberga Murgård	433	Jonas Larsson, Östnässja, Nässja
Lundby	490	Christoffer Ersson, Lundby, Örberga
Ullevi (Rosenborg)	8,000	Widow Jaensson, Arneberga, Örberga
Örberga Murgård	286	Dean Pettersson, Örberga
Ullevi (Rosenborg)	12,000	Blom of Stockholm

Source: Register of mortgages, Dahls häradsrätts arkiv, Landsarkivet, Vadstena.

had not started there by the 1820s, and a high proportion of land was held by estates. The mortgage registers for Aska in the 1820s were searched only for information on farms in the villages we used as examples in chapters 3 and 4. There were many fewer than in the villages of Dahl: none of the farmers in the Götala hamlets in Styra parish had any mortgages but, curiously, there was one from remote Björketorp on the shield in Motala funded by the Bank of Sweden. Clearly, there was already an awareness in Aska that mortgages were available, and of the possibility of obtaining them from institutions, but Dahl was much further advanced than either the shieldland of northern Aska or the plains of southern Aska in these respects.

It is possible to look further into farm finance and capital flows between stations and sectors of the economy through inventories. They reveal, for example, the relative importance of mortgages and other loans for the individual farm, and the proportion of the total value of assets made up by loans of different kinds. The distinction between secured and unsecured loans in inventories gives a clue, albeit an imprecise one, to the relative importance of deliberately borrowed money and unsettled remnants of the mutual favours and debt and credit dealing which were important in peasant society.[13] In line with the mortgage evidence, inventories show considerable variation in the degree of indebtedness of farmers at their death: one with land worth 1,000 Rdr of taxation value and other property worth a few hundred *Riksdaler* might have total debts of only 40 Rdr or so when he died, while another, already in the 1820s, could owe sums exceeding the total value of his assets. However, a few generalizations can be made out within this blurred picture. Farmers who worked only leased land often had *larger* debts in relation to the value of their belongings than freeholders. This was not just because of the lack of land on the credit side of their inventory; even though they could not have mortgages on the debit side, many of them had ordinary loans without landed security, as well as unsettled credit purchases. It is impossible to know whether their loans were taken out to build up their farm stock and equipment, or to cover losses made in bad years. However, in chapter 3 we showed that multiple-plot farms might in their early stages consist almost completely of tended plots. Such farmers needed capital to develop their production in order to earn enough to start buying land, so that a largely tenant farmer with large debts could represent the beginning of what would eventually have become a large freehold farm if he had survived long enough.

There was also a relationship between the size of debts and the size of holdings. In the inventories of category I farmers in Dahl in the 1820s, the average debt was 30 per cent of the inventory total, for category II farmers 22 per cent, and for category III farmers 18 per cent. On average, farmers in Dahl had *smaller* percentages of debt than those in Aska, but in the latter smaller farmers were also less indebted. Larger peasant farmers and persons of standing had larger percentages of debt because they had taken out larger money loans, which accords with our findings from the mortgage registers for Dahl. Thus, the expansion of commercial holdings was by no means funded through ploughing-back accumulated profits. Larger farmers were usually involved firstly in an accumulation of tenancies around a small or non-existent freehold base; then borrowing money to stock and equip them; then gradually converting tenancies to freeholds on the basis of first loans and then, once the acquisition of freehold plots was under way, mortgages. From then onwards, the process was cumulative: more freehold plots meant greater security for mortgages to acquire even more.

We will now give some examples of farmers' debts in the 1820s. Gustaf Lundberg senior of Jernevid village in Örberga parish in Dahl died without issue while still actively farming in 1820, by which time he had accumulated over 8,000 Rdr taxation value of land. His total assets were worth almost 16,000 Rdr, but his total debt was over 8,000 Rdr. His debts were of a kind which show that he was speculating heavily in the expansion of his holding, ploughing-back income into its expansion through interest payments on loans for land-purchase. Much of the debt was a mortgage of 3,300 Rdr from Alderman Jaensson's widow, with another mortgage of 1,870 Rdr from Secretary Kraft of Vadstena. He owed another 900 Rdr to District Attorney Nicolai, 325 Rdr to miller Holmström of Motala, 200 Rdr to his brother, farmer Nils Lundberg of Djurkälla, Väversunda, 64 Rdr to Dean Petersson of Örberga (another notable moneylender, with three mortgages in our sample from the mortgage register), 100 Rdr to Curate Normelli of Vadstena, and 123 Rdr to two neighbours. Lundberg was custodian of 880 Rdr for two children of local farmers, a common practice at that time. Finally, he had borrowed 160 Rdr from the Göta Canal Company. Some of these debts were only a year or two old, while the largest dated back to 1816. The capital that Lundberg had invested in his agricultural business did *not* come from other farms: excluding the sums for which Lundberg was custodian, more than 90 per cent of his large debt was owed to people not completely connected with agriculture. Their wealth might to some extent have come from farming, because professional people and townsfolk often owned farmland, but they also had other sources such as craft, trade or professional employment (or, of course, interest on loans to farmers).

The debts of Johannes Persson, who was also from Örberga and died in 1820, were quite different. He was a category II tenant farmer whose total belongings amounted only to 1,300 Rdr, in relation to which his debt of almost 600 Rdr was substantial. However, very few of his debts were owed to people outside local farming society. He owed 280 Rdr to his brother, 36 Rdr to Dean Petersson, 40 Rdr to a local grenadier, 45 Rdr to Örberga Church Fund, 30 Rdr to Örberga parish warehouse (*sockenmagasin*) and a few other small sums to various people of whom only one lived in Vadstena. Other small farmers were in the same situation: in debt but on a small scale and within a limited local circle.

These differences between the indebtedness patterns of large and small farmers might signify that large sums could not be advanced by local peasants and must be borrowed elsewhere (either because they did not have surpluses, or because they could profitably invest on their own account); or that large, enterprising farmers had quite different business contacts than the small, which could be used to assist investment as well as trade. Without relating this debt material to the ages of farmers, it is not possible to conclude that the small farmers' debts reflected the sort of indebtedness characteristic of subsistence peasant societies. But we can say that the generally positive balances of even small farmers' inventories show that they were not spiralling into insolvency as a prelude to losing their land.

It was not always the case that category II and III farmers of the plains had their debts (if any) safely based in numerous small loans in the local community, whilst large farmers borrowed extensively in large amounts from outside farming. Category I farmer Olof Andersson of Herrestad had debts amounting only to 300 Rdr. Of this, 225 Rdr were owed to his son and another 60 Rdr to a neighbour. At the same time, other people owed Olof 200 Rdr, among them District Attorney Nicolai (another local mortgagor) and two neighbours, one of them his nephew. It might be that however well he represented the expansionary freehold farmers, Olof Andersson followed his small-scale brothers in not seeking urban capital, funding land acquisitions through savings rather than loans. On the other hand, he might originally have built up his large holding through loans which had been paid off before he died.

Only category III farmers are represented in our 1820s inventory material for Aska. Anders Persson, a freeholder from Orlunda village, died in 1824 leaving two small children, so he must have been fairly young. His debts were 930 Rdr, his total assets 1,140 Rdr. This suggests either that he was in the initial stages of building up a larger holding through credit purchases, or that he was a typical young peasant, recently assisted in the acquisition of a holding. The two types of case are impossible to distinguish on this kind of evidence, but the pattern of his indebtedness suggests the latter. His major debts were not to urbanites, frequent money lenders or institutions, but local people such as grenadier Hell of Appuna parish, to whom he owed 244 Rdr, and hand Svensson of Bjälbo, to whom he owed 390 Rdr. Of his eleven other debts, only one was larger than 52 Rdr, and all were owed to local rural people except 11 Rdr owed to a Vadstena trader, which could have represented an unsettled credit purchase. Persson had also borrowed 100 Rdr from his local Orlunda Church Fund. About half the inventories record loans from church funds of between 50 and 200 Rdr, money which obviously came from other parishioners and local agriculture. Security seems to have been of little consequence to these funds, because few of their loans were properly mortgaged, which might imply that cash was generally lent from them to young members of well-established local peasant families.

Only three inventories of shieldland farmers were examined for the 1820s. They show smaller debts than all others. Jöns Larsson at Lilla Ingesby, Västra Ny, who died in 1822 with grown-up children, had only one substantial debt, 222 Rdr owed to Royal Inspector Regnstrand of Vadstena, who owned cornmills on Motala Ström. The remainder of Larsson's total debt, which at 273 Rdr was only 16 per cent of the value of his assets, was in little sums of a few *Riksdaler* to numerous local people. A number of observations can be made on this pattern. First, even small farmers in this remote area made business contacts over more than local distances and with non-farmers. Secondly, although it might be that Larsson, well on in age, had a money-borrowing career paid off and well behind him, his light indebtedness and the fact that his sole major debt was to the owner of nearby cornmills seem most probably to suggest that he was a peasant, with little in common with the increasingly commercialized and indebted farmers on the Dahl plains.

Few of the large creditors in the system were farmers. People like Widow Jaensson of Arneberga estate were the money-lenders, at a normal rate of interest of 5 to 6 per cent. Her late tanner husband had been called the King of Vadstena, where he was the wealthiest resident when he died in the early 1800s. His widow's money was made to work through substantial mortgages to farmers in the vicinity. At her death, however, only about 8,000 Rdr of her total wealth of 126,000 Rdr were lent to local people, whilst her wastrel son-in-law, Major Rosenborg of Naddö (thankfully also dead by this time) had, while acting as her trustee in the late 1820s, burdened her estate with debts of about 95,000 Rdr. Even though the mortgage registers show that she had been far more active in funding farm expansion earlier in the 1820s, when she gave registered mortgages for 27,000 Rdr in only six years, we can recognize a number of people known to us from chapters 3 and 4 as being amongst the most enterprising farmers in the mortgages still in being when she died. They included Anders Larsson of Broby in Strå, two members of the Lohm family, which was widely represented across the Dahl plains, and Per Lundborg of Svälinge in Herrestad. A few people from Vadstena known for their poor economic acumen,[14] including 'factory-owner' Tollstorp and Colonel Boy, also owed the widow about 1,000 Rdr each. Widow Jaensson's own debts totalled only 3,000 Rdr.

Dean Petersson of Örberga vicarage also appeared as a creditor in many farmers' inventories from the Örberga neighbourhood. An inventory was taken when his wife died in 1828. Her husband was still active then as a financier as well as a pastor. He himself had debt of only 1,945 Rdr, compared with total assets of over 36,000 Rdr, of which almost 40 per cent were money loans. The most usual sums were less than 100 Rdr, and thus probably to cover consumption or indebtedness rather than to finance investment. Only 6 of the 127 loans were bigger than 500 Rdr, and 32 were between 100 and 500 Rdr. It might be that Dean Petersson was most involved in loans without security, taken out by less creditworthy people. Unlike most other church ministers, he was one of the persons of standing who built up large holdings, over an unusually large area. He had two plots worth more than 8,000 Rdr in Trehörna parish, 25 km south of our region; one in Fornåsa parish, east of our area, worth 5,450 Rdr and a small plot in Adelöv parish in Jönköping County, about 60 km away. Apart from this, he tended the category I vicarage farm at Örberga. Quite a few of the people who borrowed money from him were like himself, persons of standing in the process of building up large agrarian holdings, or trying to keep them together. One was Lieutenant Fredrik Enander of Väversunda parish, who owned quite a few plots, some perhaps aquired with the the assistance of 400 Rdr borrowed from Petersson. Another was the future War Commissioner Lallerstedt, who borrowed 200 Rdr from the dean but had not yet embarked upon the career of plot acquisition which led him to become one of the major landholders in the region. Other borrowers were the tenants of the dean's farms in Trehörna, Fornåsa and Adelöv. A third group, the largest in terms of money, comprised his parishioners and neighbours in Örberga, Källstad, Herrestad and Väversunda. A fourth group, the largest in terms of number of loans, were crofters and soldiers, hands and widows, whose debts ranged up to 13 Rdr, sums which were negligible in relation to the debts of large farmers, but perhaps equally significant in relation to their household economy. All the substantial loans were mortgages registered in the local courts and were older, some stemming from the preceding decade. Smaller amounts were borrowed for shorter times.

In sum, in the 1820s small-scale peasants used little money from outside their own farms, and if they did it was to make the initial purchase of their farm, which was then run on traditional self-sufficient lines. The money they needed came mostly from friends, neighbours and relatives: farm purchases were funded from inside the traditional peasant economy. However, farmers who built up large holdings were dependent on outside funding, though some of them, usually the smaller ones, also skimmed the local agrarian money market. Most of the capital channelled into agriculture this way was of urban origin, created in trade, craft or mercantile activity; that from the Bank of Sweden came from outside the country. It was only on the plains of Dahl where mortgages were common in the 1820s. In southern Aska, and particularly on the shieldland in northern Aska, agriculture was still run on lines which were typical of peasant economy, and hence money from outside was not needed for investment in farm expansion. Even so, a Bank of Sweden mortgage to a Björketorp smallholder suggests either that farm expansion was just getting under way, or that there were other economic activities through which farmers on the shield made contacts with the outside world, which needed investment.

The 1850s

At the end of our study period mortgages were much more common. Only four out of twenty-five that were given or renewed in Dahl in 1855 were from private people; all the others were funded by Östgöta Hypoteksförening (Mortgage Bank). It was founded in 1845 and frequently borrowed money from from abroad, as in a loan of 10 million German Marks from Hambros Bank in London.[15] The private merchant bank established in Vadstena in 1857 was of no importance as a provider of mortgages in Dahl.

The number recorded in the sampled year of 1855 suggests that about 250 mortgages were registered or renewed in Dahl in the 1850s, double the total for the 1820s. For Aska, the mortgage registers of 1858 were studied for the large parishes of Motala and Västra Ny, where mortgages had by the 1850s become as common as in Dahl. However, of the twenty-nine given there in 1858, only eight were from Östgöta Hypoteksförening, one from the Bank of Sweden and one from Vadstena Enskilda Bank. All others were given by individual people, conforming to the pattern in Dahl in the 1820s. Many of the mortgagors were from Motala Town, such as brewers Rosendahl and Kjellman and trader Thamsten. The number of mortgages given in the sampled year suggests a total of 290 in the 1850s for the rural parts of Motala and Västra Ny (the town of Motala was excluded from the registers). These two parishes contained 253 plots in 1860, and therefore had a higher ratio of mortgages per plot than Dahl, with 250 mortgages and 316 plots.

In general mortgages were bigger in the 1850s than in the 1820s, not only due to the devaluation in the 1850s, but also because they became related to the market rather than the taxation values of plots. Thus, the total mortgage value for one year in Dahl hundred was 106,000 Rdr, about 700,000 Rdr for the decade allowing for the devaluation, at a time when the total agrarian taxation value of the hundred was slightly more than 600,000 Rdr (deflated value). The average sum mortgaged in 1855 in Dahl was 4,200 Rdr: that is, it represented a category I farm in taxation value. Whilst there were still farmers who kept their loans low or non-existent, others happily mortgaged large farms for all they were worth. For example, Gustaf Larsson had mortgages for 25,000 Rdr on his farm at Bårstad, Rogslösa, which had a taxation value of 9,800 Rdr. Although this might have been much less than the market value of his land, it does seem that Larsson had been borrowing up to, if not beyond, the limit of the security provided by his property.

The average size of mortgages in Motala and Västra Ny was, at 4,100 Rdr, almost the same as in Dahl. The total mortgage sum for 1858 was 119,000 Rdr, about 800,000 Rdr for the whole decade, whilst the tax value of agricultural property in the two parishes in 1860 was less than 400,000 Rdr. Thus, northern Aska had closed up on Dahl in the extent of mortgage funding. Our figures in chapter 3 show that the value of category I farms there increased by a considerably smaller amount than in Dahl over the period, which might suggest that other than purely agricultural activities were affecting the need for capital in the shieldland.

The number of inventories available for the 1850s allows us to deal separately with persons of standing and freehold farmers on the plains and shieldland. The picture given by the inventories of Dahl farmers is slightly different from that in the mortgage registers. Even though banks were important, particularly the Östgöta Mortgage Bank, some of the smaller freeholders still stuck to older patterns. This might be because the people who died were on average older than the total farmer population, and in consequence both stuck to more traditional habits and had had time to pay off most of their mortgages. But this seems to be only part of the explanation because as in the 1820s small farmers generally had small and locally sourced debts, whilst it was large farmers who borrowed heavily from banks and wealthy townspeople.

Sven Swärd, a category I freeholder of Kasta village in Örberga parish had debts of 10,000 Rdr and total assets of 14,700 Rdr. In spite of the relatively small difference between his credit and debit and his own large borrowings, Swärd seems to have operated as a small-scale moneylender. Twenty-five people owed him money, though not on such a grand scale as the person-of-standing moneylenders of the 1820s. Swärd was owed 1,370 Rdr by Johan Larsson of Herrestad and and 1,200 Rdr by his own father-in-law; all other debts were between 10 and 260 Rdr, even though most of them were formal mortgages; as in the 1820s, moneylenders preferred security guarding their money, however small the debt. Swärd's own debts were few but large. The biggest was 5,300 Rdr owed to Östgöta Mortgage Bank, the second largest was 1,850 Rdr owed to a freeholder for a plot of land. Swärd had another three debts, to the estate of his father-in-law and to two local freeholders, amounting in total to 2,600 Rdr.

At his death, Anders Larsson of Broby, Strå left a fortune of 80,700 Rdr, which was very large compared with run-of-the-mill category I farmers like Swärd. He owed 47,900 Rdr, including a mortgage to Östgöta Mortgage Bank for 25,000 Rdr – more than the tax value of his entire landholding.

Judged by these sums, farming by the 1850s had become much more demanding of capital than in the 1820s. The largest farms were becoming ever larger, and enclosure, reclamation and the introduction of new equipment meant that each acre of commercially farmed land required more capital. The large expansive farmers at the end of our period had bigger debts in proportion to their wealth than their predecessors; each loan was much larger, and local moneylenders had been replaced by banks, which accounted for between 30 and 60 per cent of each farmer's total debt.[16] Indebtedness had also become more common among smaller farmers in Dahl by the 1850s. Lena Maria Persdotter, who died in 1855 leaving a category III farmer husband and small children in Ullnäs, Örberga, had total assets worth 7,200 Rdr and debts of 6,100 Rdr, more than the tax value of the land. Local farmers had lent most of the money, including one mortgage of 2,300 Rdr and another of almost 1,000 Rdr. Örberga Parish Fund gave the third largest mortgage of almost 700 Rdr. The situation was similar on the plain in southern Aska in the 1850s. Inventories of many small farmers listed substantial debts, but most were locally sourced and few were from institutions, whilst larger farmers more commonly had mortgages from banks. Carl Johan Johansson, who died in 1852 at an active age as a category II farmer in Fivelstad parish, owed 3,400 Rdr against assets of 2,600 Rdr. Eight loans from local farmers accounted for almost all of the debt; only 375 Rdr from Fivelstad Forest Common Fund came from institutions. Also in 1851, the category III freehold farmer Anders Magnusson of Hagebyhöga village lost his wife at an active age. His total debts were 2,700 Rdr and his assets 2,720 Rdr. Johansson had a mortgage for 400 Rdr with the Bank of Sweden and another with the Östgöta Mortgage Bank.

These examples show that it is difficult to generalize about the sources of plainland farmer's credit from inventories in the 1850s in the way that was possible for the 1820s: some smaller farmers now had mortgages with institutions. Otherwise, even though all sorts of money-lending institutions existed, small farmers on the plains still mainly borrowed from their relatives, neighbours and parish funds. Perhaps they mistrusted the institutions or got better rates with private loans,[17] but all farmers on the plains generally had a higher proportion of their debt in mortgages in the 1850s than in the 1820s. The higher levels of debt relative to assets registered in the inventories of large farmers in the 1850s seems to indicate the higher cost of farm expansion and improvement as the size of the largest plainland farms increased. Despite their large debts, their inventories usually showed a healthy surplus of assets. Their investment in expansion seems, therefore, to have been profitable. It was different for the small farmers of categories II and III, whose holdings cannot have exceeded their own subsistence requirements by much in normal harvest years. As we have seen, in the 1850s their debts could exceed their assets, whilst in the 1820s their proportionate indebtedness had been less than that of large farmers. What this meant in terms of the changing nature of their household economies is hard to say. It could be that the two small farmers who died young were traditional peasant farmers who had simply not yet been able to pay off the initial cost of acquiring their holding with loans from relatives and neighbours. However, it might equally well be that small farmers were suffering the indebtedness that inevitably comes in a partially commercialized farming system, borrowing to cover losses in bad years as a prelude to losing their land to larger farms.

Category I farmers' debts were not only of a different magnitude and proportion, and for a different purpose, but increasingly funded from outside farming. The capital inflow into agriculture from elsewhere increased greatly between the 1820s and 1850s. Whether the ultimate source of institutional capital was in the towns, abroad or the deposits of other farmers is difficult to tell; it was certainly channelled into farming on the plains through the bank in Vadstena. We will examine the ledgers of Vadstena Enskilda Bank later, but first we will see if the economies of freehold and tenant farmers on the shieldland of northern Aska were moving in the same way as on the plains, and where persons of standing got capital to expand their estates on both the plain and shield.

The mortgage registers show that farmers in the shieldland of northern Aska still had non-bank mortgages to a larger extent than those on the plains in the 1850s. In the 1820s these shieldland farmers in general had less debt when they died than those on the plains. This seems to have persisted into the 1850s for the smaller farmers. Few of the nine inventories of category II and category III farmers examined from northern Aska in the 1850s contained substantial debt, in absolute terms or in relation to the value of assets. None of them had a mortgage with a bank. One of these was Anders Kastbom, a freeholder with a distillery at Västgöteby in Motala parish. The value of his assets was 8,388 Rdr when he died in the mid-1850s, but his total debt was only 381 Rdr. As in the 1820s, tenant farmers seem to have had larger debts in relation to their assets. When tenant's wife Greta Persdotter of Ishultet, a category III farm in Västra Ny parish, died in 1858 the household owed 234 Rdr and the total value of its belongings was only 740 Rdr. The individual debts were all small, the largest, for 66 Rdr, was owed to a local farmer and three of the ten debts were unsettled credit purchases for items of consumption, such as 3 Rdr for straw or 9.67 Rdr to a trader in Motala for half a barrel of herrings. The reason for the small debts of small farmers on the shield could be that they were still self-sufficient peasants, not yet troubled by the market, who stuck to the traditional attitude that money debts were in principle wrong, financing any land purchases required over a longer period through savings.

Persons of standing and estate owners, irrespective of whether they lived on the plains or the shield, were like the large freeholders of Dahl in having large debts. District Judge Duberg of Lindenäs estate in Motala parish, whose property was worth 186,000 Rdr at his death, had debts of 63,000 Rdr. However, he was also owed 38,000 Rdr, and only 5,650 Rdr of his large debts came from banks and other institutions; most was owed to other persons of standing in Östergötland. Whereas Dahl freeholders borrowed from towns, Duberg pushed money in the opposite direction: 9,500 Rdr was owed to him by inhabitants of Motala *köping*. When War Commissioner Lallerstedt of Kungs-Starby died in 1851 his fortune was assessed at 312,000 Rdr and his debts amounted to 193,000 Rdr. More than a third of his mortgages were from banks and institutions; much of his property was outside our study area, some as far away as Stockholm, and his large mortgage and institutional debts mirrored these widespread economic contacts.

Many other persons of standing had large debts, sometimes larger than the total value of their assets. When the wife of auditor Planck of Sonnorp in Styra parish died in the 1850s, debts, mainly unsettled purchases of plots, amounted to almost 60,000 Rdr, while his assets were worth 32,000 Rdr. About 10 per cent of his debt was owed to Östgöta Enskilda Bank and another 800 Rdr to local parish funds. The largest part was owed to large landowners in parishes scattered over Östergötland. If Duberg and Lallerstedt show how estate owners could expand their holdings with even more alacrity than freehold farmers, Planck shows that such speculation could end in expensive failure – something we did not see in the inventories of large freeholders on the plains.

Debt and investment among the landless

From what we have seen so far, we would expect semi-landless crofters and completely landless hands, maids, cottagers and labourers of Dahl and Aska hundreds to have had little debt, and not to have borrowed from banks and institutions as they could not provide any security. Investments in agriculture on crofts must have been small, and to some extent funded by the landowners, who frequently lent haulage animals and implements to their crofters. We calculated debt and credit balances for the inventories of twenty-six crofters in the 1820s and forty-eight in the 1850s. The average asset value per crofter in the entire region was only about 200 Rdr and the average debt about 50 Rdr. Soldiers from Dahl were the wealthiest crofter group in the 1820s, owning about 400 Rdr on average and owing only 34 Rdr, while ordinary crofters in Dahl had assets worth 71 Rdr and debts amounting to 62 Rdr. The wealthiest crofter by far was a Dahl soldier who owned 950 Rdr of arable land as well as his croft. The poorest was an ordinary crofter in northern Aska who only had 25 Rdr's worth of assets. The largest debt amounted to 359 Rdr, owed by a soldier in southern Aska with assets of 462 Rdr. The small debts crofters owed were invariably to local people. One or two owed a few *Riksdaler* to the moneylender Dean Petersson of Örberga, and crofters on estates often owed money to their landlords, though rarely more than 10–20 Rdr. The soldier with the largest debt owed money mainly to another soldier (270 Rdr), and the remainder to local farmers and crofters. The sums involved grew over time: in the 1850s crofters on average owned assets to the value of almost 300 Rdr and their debts averaged 100 Rdr. The largest belonged to three *förpantning*-crofters in northern Aska, whose average assets were 1,146 Rdr and debts 46 Rdr. The same type of crofter on the plains was poorer, with assets of only 300–400 Rdr. Their debts were still very much like they had been in the 1820s, owed to local farmers and crofters, and to landlords by estate crofters.

Thus, the semi-landless crofters' properties and debts changed little over the period. However, on the plains two or three crofters did manage to acquire plots of arable land, thereby joining the category III farmers (with whom they were grouped in chapters 3 and 4), and crofters also occasionally appeared as creditors in the inventories of (invariably small) farmers. This shows the fluidity of the boundary between the crofting and small freeholding populations. In fact they were often related, and it was not uncommon for small freehold farmers' sons to begin their married life as crofters, eventually to inherit land of their own. Crofts might thus be the first stepping-stone to a farm for peasant farmers' sons, and it is not surprising that some crofters increased their land and wealth during the course of their lives.

The banks and farm capital

Only the ledgers of the Vadstena Enskilda Bank, which opened in 1857, were available for inspection. After our period ended, branches were opened in Motala, Askersund, Skänninge and Borensberg, and Vadstena Enskilda Bank became the dominant merchant bank in western Östergötland until 1879, when a series of unfortunate deals forced it into bankruptcy. The original 281 shareholders invested a capital of 250,000 Rdr subdivided into 2,500 shares of 100 Rdr each, with financial guarantees of three times their shareholdings. The total capital backing of 750,000 Rdr was bigger than the total tax value of agricultural land in Dahl, and Vadstena Enskilda Bank was one of the largest limited companies in Östergötland at the time in terms of share capital.[18] Its initial share capital was large enough to get it rolling without large loans from elsewhere, unlike the Östgöta Mortgage Bank which borrowed from outside the country.[19] Its main earnings were from credit notes and interest on loans. Credit notes amounted to 2,500,000 Rdr in 1860, while the total amount of money lent was 1,400,000 Rdr. Deposits were of lesser importance and mainly by share- and credit-holders, although the Östgöta Mortgage Bank often deposited sums of between 20,000 Rdr and 50,000 Rdr with the Vadstena bank. It is thus difficult to discover the eventual sources of the bank's funds, apart from the original shareholder's deposits and guarantees. The Vadstena bank came to be called '*Bondbanken*' (the farmers' bank) locally, due to its reputedly large number of shareholders and customers in farming. Was it, in fact, mainly a channel between the wallets of different farmers in the area, or did capital flow through it to link urban craft and trade to agriculture, or industry to other sectors?

The size of the bank ledgers, like the other sources we used, forced us to limit ourselves to a few aspects of the bank's activities and the initial six years of its existence. We tried to explore three things in particular, to complement our findings from inventories and mortgage registers. Firstly, who invested money in the bank as shareholders and how large a proportion of the capital came from the agrarian and non-agrarian sectors? Secondly, who borrowed money from the bank and how much of this went into agriculture in our region? Finally, who used the bank as depositors, and how many of them came from agriculture?

Of the original 281 shareholders, 197 definitely came from agriculture, and the occupational designations of 86 suggest that they were from other economic sectors. However, 20 of the 86 'non-agricultural' shareholders were state servants, such as vicars, district judges and captains in the army, who were all at the time provided with farms on Crown land as the main part of their state income. The largest numbers of shares were bought by a factory-owner in Vadstena (34), an apothecary from Vadstena (49), two of our enterprising farmers from Dahl, Gustaf Larsson from Bårstad, Rogslösa (40) and Anders Persson of Östra Starby, St Per (36), a trader from Linköping (50), a distant relative of the Lallerstedts in Stockholm (42), and local estate owners Staaf and Åhl (42 and 45 shares respectively). In general, the non-agrarian people bought more shares each than the farmers and, hence, their minority was less marked when shares are counted instead of shareholders: they owned 918 shares in 1857 rather than the 714 which they would have possessed if their holdings had been of average size. Of the 196 farmers who owned 1,582 shares, the 71 who came from our region owned 508. They were not evenly distributed across it (see table 5.21): ten came from northern Aska, while thirty owning 282 shares came from southern Aska, and thirty owners of 182 shares came from Dahl. Local persons of standing and people from Motala Town and Vadstena owned almost 500 shares, so that in total nearly 40 per cent of the bank was owned from our region. Quite a high percentage of the enterprising commercial farmers of the plains were shareholders. Of the twenty-seven Herrestad farms in the late 1850s, for example, five were shareholders in the local bank, but the largest farmer there was not among them. Representation from individual villages and hamlets in southern Aska, and particularly in more distant northern Aska, was smaller.

The short-term credit register and the loans ledger for 1857–63 show that farmers were even more active as borrowers. Quite a few people had credit accounts with the bank, with limits ranging between the 900 Rdr of a farmer and the 30,000 Rdr allowed to Motala Engineering Works. Because a majority of the account holders used their credit for only short periods of time, there were few outstanding credits at the year's end, when all accounts must be settled to avoid the imposition of a higher rate of interest. In Herrestad parish, seven farmers had credit accounts, among them Anders Samuelsson and Per Johan Larsson, both peasants with large expanding holdings. In 1860 Larsson had used none of his credit limit of 2,000 Rdr, Samuelsson half of his 1,000 Rdr. Neither was a shareholder, but four of the other five account holders from Herrestad were.

Seven farmers from Motala parish, as well as many traders and industrialists from Motala Town, had accounts. Only two were ordinary freehold

Table 6.2. *Money loans[a] (in Rdr) taken out from Vadstena Enskilda Bank 1857–64 by residents of Herrestad and Motala*

Herrestad	
Farmer P. J. Larsson	45,900
Farmer J. P. Larsson	3,000
Farmer N. Larsson	300
Farmer L. J. Larsson	2,250
Farmer E. G. Johansson	3,000
Farmer M. Pharmansson	1,250
Farmer N. Persson	650
Farmer A. J. Petersson	2,150
Farmer J.A. Holmberg	7,500
Farmer P. A. Johansson	5,000
Farmer L. E. Larsson	36,470
Farmer M. Andersson	25,900
Total	133,370
Motala	
Miller J. R. Johansson	30,110
Farmer S. M. Liderström	1,000
Farmer P. Persson	2,500
Total	33,610

Note:
[a] The totals are summed loans. The amounts owed at any one time were smaller because many loans were renewed quite frequently.
Source: Vadstena Enskilda Bank Registers, Landsarkivet, Vadstena.

farmers: the others were estate owners or part-industrialists, like M. F. Berndes of Säter whose estate on the shield contained a mill, a saw and an oilworks plus Kathrinefors Ironworks on Motala Ström. His credit limit at the bank was no more than 3,000 Rdr, nonetheless.

Other bank lending was in the form of loans for three months and upwards, mainly for two to six months. Table 6.2 shows that some farmers used this loan facility much more than others. Of the people from Herrestad on the Dahl plain, Styra in southern Aska, and Motala on the shield who borrowed money in this way in 1858, none from Styra were farmers, while Herrestad was represented by three people, two shareholders and Per Johan Larsson, who took out two loans, one for 3,000 Rdr and one for 2,250 Rdr, both for three months. One of the Herrestad shareholders borrowed 2,000 and 1,000 Rdr for three months. Nineteen short-term loans were given to people from Motala in 1858, but only four were to pure farmers. The bank was not important in the development of farming on the shield, where its funds flowed into secondary economic activities.

Conclusion

In the 1820s a significant amount of borrowed money was involved in the purchase of farmland in our region: the farm system was not maintained by the inheritance of peasant children. Many freeholders had debts representing borrowed money rather than unsettled credit transactions. Strikingly, tenant farmers were more indebted than freeholders, and in southern Aska and the northern shieldland freeholder indebtedness was far less than in Dahl. Farmers with large holdings were most indebted, absolutely and in proportion to the value of their assets, than small ones, suggesting that money was being borrowed to buy farmland rather than to cover farming losses in bad years.

The flows of cash into farmland were of two kinds. First, many little often unsecured sums lent by kin and neighbours to set up small holdings, which could represent the peasant practice of helping to set up newly weds on farms of their own. Secondly, large loans were made on mortgage to farmers who were actively expanding already substantial holdings. This occurred almost exclusively on the plains of Dahl where farm expansion and on-farm processing activities were developing strongly. Ploughed-back profits were not sufficient to fund this: or, perhaps rather, reliable future profits could be safely anticipated through borrowing. Money flowed into farm expansion in Dahl from two sources. First, the more desirable mortgages granted by the Bank of Sweden, which borrowed from abroad to lend to farmers. Secondly, mortgages granted by a few local persons of standing whose capital often originated in urban craft or trade were in aggregate much more significant. Ordinary freehold farmers rarely lent money in large individual sums; perhaps their own holdings offered greater potential returns than interest on loans. The persons of standing who lent money were usually relatively old, whilst those who were still building their estates borrowed even more than the large freeholders of Dahl. Their funds often came from much longer distances, through a widespread network of people of similar social standing. This pattern persisted throughout the period.

The intricate nexus of local flows of small sums focussed on small farms survived into the 1850s, implying that the setting-up of peasant children on their own holdings at marriage continued. The survival of peasantry is also suggested by the existence of young hands with money to lend, and of crofters who went on to acquire land of their own as they progressed through the life cycle. However, the indebtedness of small farmers generally was much greater in the 1850s than in the 1820s. The high proportion who were now heavily in debt suggests that they were suffering the inevitable fate of small farmers in an increasingly commercialized agricultural economy. That large farmers had larger farms in the last decade of the period than in the 1820s, were absolutely much more indebted, but less so in proportion to their asset values, suggest that they were reaping the benefits which accrue equally inevitably to large holdings in this kind of system. As in the 1820s, the fact that commercial farmers tended to be heavily indebted, rather than to lend money or deposit it in banks, can only mean that the profits available from investing in the production and processing of farm output were greater than the 5 or 6 per cent which could be got in interest. As we would expect from chapters 3 and 4, these trends were almost confined to Dahl. There was still little borrowing or lending in the 1850s by farmers on the northern shieldland, where small farms remained absolutely and relatively numerous. In southern Aska, where farm expansion was less pronounced than in Dahl, farmers more often lent to the bank than they borrowed from it, although in smaller unit sums than the farmers of Dahl borrowed.

Borrowing was even more commonly on mortgage in the 1850s than in the 1820s. However, mortgages were now overwhelmingly taken out from institutions, and capital was drawn from greater distances as the mortgage bank in particular sucked money from abroad. Investment by direct lending to farmers by persons of standing became much less common as people with money to spare entrusted it to the intermediary institutions of the mortgage and merchant banks; indeed, the funds of the latter were drawn mainly from within the region, particularly from the savings of townsfolk and rural persons of standing who had previously lent on mortgage. These institutions were obviously indispensable to the acceleration of arable farm expansion and development in Dahl, and to its initiation to a lesser extent elsewhere in the region, during the last two decades of our period. The direct investment of urban-derived capital was also involved in the beginnings of the process in Dahl at the start of our period. On the shield there was more bank investment in non-agricultural activities than in farming. At the start of the period surplus capital *from* the urban secondary sector helped to fund the development of farming; by the end, there were much bigger investment oportunities in the towns themselves, but funds flowed even more freely *through* the new urban tertiary sector.

7
Flows of labour

Setting the scene

Introduction

How did labour flow between the different sectors of the regional economy as they changed in different ways? Could townsfolk develop their activities independently from peasants, crofters and cottagers, or either of these separately from the industrial development in Motala based on long-standing milling and iron-making? Of course not; but we have said little about this so far. We have looked at the structures of livelihood positions and how they developed in different sectors of the economy without much reference to the *people* who occupied those positions. How did the life-paths of people change as the ways in which they derived their livelihoods altered? Were the widows and spinsters of Vadstena, for example, survivors of dead farmers; were the industrial workers of Motala former metal handicraft or ironworks employees; was the menial population of Vadstena drawn from peasant children dispossessed as farms expanded, and was the proletarian labour force of agriculture drawn from the same source? It is important, too, as with capital, not to consider our region as an isolated island. Labour in its agriculture, industry and trade might just as well have been recruited from the same economic sectors elsewhere as from different sectors locally.

Flows of people within, into and out of an area have not often been considered as an integral part of the process of economic and social change. Similarly, although we know from specialist studies of migration that rates of population movement have been high and occurred over long distances for centuries,[1] such studies have rarely taken the changing structures of livelihood positions in the areas of in- and out-migration into account. However, we would expect connections between the subregions and economic sectors of our region, and between them and places outside it, through flows of workers as well as of capital and goods.

334

Migration as a response to agrarian change?

Chapter 4 showed that the agrarian population of Aska and Dahl grew during the early nineteenth century, and that this was at least partly due to natural demographic change. It is generally considered that areas of large population growth were likely to be areas of out-migration. However, if we assume that the natural population growth in our region was connected with local agricultural and industrial change, migration can be viewed in another way. The strong demand for proletarian labour created by the amalgamation of farms, the intensification of agricultural production, urban economic change and industrial growth did not give rise to high agricultural wages, as we have seen. This could either be because we have underestimated local population growth, or because there was a source of labour outside the region big and mobile enough to prevent a rise in local wages when demand for labour grew. Higher wages in agriculture in the northern shieldland than on the southern plains could imply a number of things: the supply of extra labour might be greater in the south, from local natural increase or migration from neighbouring Småland where population growth was rapid at the time; or agricultural and industrial labour demand might have increased faster in the north. In fact, the commonness of long-distance migration in Sweden[2] suggests that Motala would not be perceived as much farther away than Dahl by someone in northern Småland, and higher wages in rapidly growing industry would tend to push up wages in agriculture in the north.[3] The pattern of flows could of course have been much more complex than differential rates of inward movement due to variations in labour demand reflected in wage levels. Flows of people *inside* the region from agriculture to town or industry or between peasant and commercial farming could be compensated by in-migration from other regions, or different rates of out-migration from different sectors of the region could be compensated by flows between the sectors within it.

The high quality of the sources of information on the composition of the population, and, from the nineteenth century, migration, mean that these topics have been more fully investigated in Sweden than elsewhere. However, because population flows are rarely linked to changes in structures of livelihood positions, this work does not provide completely clear expectations about what would happen to population flows in response to the changes we have described. The turnover of labour was very high on a number of estates in South and Central Sweden, and although mobility was generally much lower among crofters and qualified labour, all categories of employees contained some people who were more or less mobile than others.[4] Between 1861 and 1870 unmarried hands and maids had annual turnover rates of between 25 and 50 per cent,[5] while married hands had turnover rates of 20 per cent and crofters only 10 per cent.[6] Seasonal labour also had a high turnover rate, over 50 per cent in one case, and betweeen 20 and 30 per cent on average. These findings from estate ledgers suggest that the number of positions for hired labour on large farms are underestimated in our calculations because many seasonal workers were not officially registered,[7] and that the total expansion of the labour force in our region was higher than the 80 per cent revealed by taxation registers. As in many other studies of migration, the highly migratory tended to be younger and occupy positions which had been a stage in the peasant life-cycle pattern, such as hands and maids, while the less migratory older labour worked as married hands or crofters. The average age among crofters for example, was between forty and forty-five years, that of maids and hands was between twenty and thirty, and married hands were in their late thirties. Seasonal workers were of about the same mean age as crofters; they were frequently migrants, who might be crofters or cottagers in their home parishes. According to these results, agricultural labourers began their careers in a series of hand or maid positions, passed into the married hand phase, still with many moves between positions, and reached a more stable stage as crofter or crofter's wife from the age of about forty. In the parts of our region where crofts became fewer, such careers must have become rarer and, according to these findings, the total mobility of agricultural labour must have increased as this occurred. It is reasonable to expect, too, that mobility across the boundary between agriculture and industry would increase in these circumstances. But in the areas concerned, where towns and industrial districts were distant from the studied parishes, migration to urban or industrial destinations was only about 10 per cent of the total and the outflow from towns to the countryside was not much less. On the other hand, urban migration from parishes adjacent to towns was between 27 and 36 per cent of the total outflow, and rural parishes next to towns had an inmigration surplus from more remote rural areas and a deficit in relation to the towns, suggesting a stepwise flow to towns.

This stepwise migration hypothesis, which cannot be properly tested without longitudinal analysis of migrant life-paths, implies that migration distances were short. We know that this was actually so in Asby parish in southern Östergötland,[8] where 78 per cent of in-migrants came from less than 17 km away, and 90 per cent from within 34 km, irrespective of the social group to which they belonged. The pattern of out-migration was similar: 77 per cent of migrants moved within 17 km, 88 per cent within 34 km. A significant change in this pattern only occurred in the 1890s, when more people started to move longer distances. There was little migration to towns from Asby, even though a growing new town existed nearby. Between 77 per cent and 91 per cent of the unskilled workers who moved into a number of parishes elsewhere in Central and Southern Sweden had travelled less than 34 km.[9] Qualified labour and persons of standing moved further: in two parishes the proportion coming from within 34 km was just above 40 per cent. Their information maps were obviously quite different from those of the landless. Only inter-parish moves were included in the calculation of these migration distances. The dominance of local moves suggests that intra-parishional migration must have been relatively very common (depending, of course, on the sizes of parishes). In Martinius' sample of parishes, the proportion of intra-parish migrations varied between 13 per cent and 56 per cent, so that about 90 per cent of all moves originating in them were to destinations within 34 km.

So far we have seen that rural migration was very local; that the proportion of people who moved to town was small, and that there seems to have been a stepwise flow of migrants to the towns consisting of agricultural workers, who thus crossed the town–country boundary.[10] The information on which these conclusions are based relates only to the single moves which led to people's current places of residence: statements about the paths they traced during the course of their lifetimes are purely inferential. Highly abbreviated versions of migrants' life-paths derived by adding information on their parishes of birth yield results which are not completely consistent with these inferences.[11] Martinius did find that migrants' parishes of birth were in general more remote than the origins or destinations of single moves, but the differences were relatively small: between 80 and 95 per cent of proletarians were born within 34 km of where they lived at the time observation. This does not suggest that stepwise migration was occurring towards towns. Furthermore, the distribution of the parishes of birth of migrants to parishes next to towns were no different from those of parishes further away.

Åsunda hundred, north of Lake Mälaren and dominated by estates, had low rates of natural population growth compared with other parts of South and Central Sweden, and out-migration to the Stockholm area, 70 km away. It was similar to our region in that during the nineteenth century crofters and tenant farmers were replaced by married hands. This change towards a more proletarian structure of livelihood positions increased mobility as the 'formal and practical restrictions on the size of the married hand households made it difficult to maintain un-productive family members'.[12] Despite the increased migration this brought and the strong attraction of Stockholm, movement was almost as spatially restricted as in other parts of Southern Sweden. Between 1846 and 1870 30 per cent of migration was internal to the hundred, and about 80 per cent of the external migration was within 30 km. Stockholm and other provinces attracted between 15 and 20 per cent of the out-migrants and accounted for about 10 per cent of total in-migration, though towns nearer than Stockholm attracted more of the urban-bound migrants. The birth-parishes of the landless population were mainly nearby: 92 per cent of a sample born in 1804–06 originated less than 30 km away, and 88 per cent of a sample born thirty years later.

In sum, the landless rural population of Sweden in the nineteenth century moved frequently in search of better livelihood positions. A high degree of mobility had existed amongst the young before capitalist agriculture; after it, high mobility persisted into older stages of the life cycle with the disappearance of the crofts and tenant farms on which they had been able to fix a firm footing. Predominantly they moved within the same parish or across just one or two parish boundaries. The hypothesis that migration was stepwise towards towns and industries is supported by the fact that parishes next to towns had migration deficits in that direction and surpluses in rural directions. But landless people's parishes of birth were not much more distant than their average migration distances, which suggests that migration was mainly circular. Rooted by property, farmers and crofters were much less mobile than the landless, but while a spell as a juvenile maid or hand before marriage was conventional in the peasant life cycle, a high rate of mobility was also common among young future farmers.

Although they mainly aimed to explain changes in social mobility, we can use the results of longitudinal cohort studies to provide further background to our enquiry. Only about 6 per cent in each of two sample cohorts from Åsunda hundred ever left their home region.[13] After thirty years, 52 per cent of the members of the older cohort and 41 per cent of the younger one still lived in Åsunda. The movements of the later cohort covered a larger spatial field. Migration rates of married farmhands were very high, those of peasants much lower; the landless accounted for 80 per cent or more of all migrations, crofters for less than 20 per cent. Social mobility over life-paths was predominantly downward, in particular among the crofters but also among peasants. The only way out of a downward spiral in the more and more concentrated capitalist farming system was to move elsewhere or leave agriculture, but in fact few moved to towns or other rural areas.

Although it was on plainland, agriculture in Skåning hundred in Västergötland did not change in the same way as in western Östergötland: it comprised mainly smallholdings, which grew in number and proportion of total land value while large and intermediate farms became fewer and less valuable. The absence of large-scale capitalist farming was reflected in the life-paths of Skåning's inhabitants: 40 per cent of a cohort of men born 1810–12 were sons of farmers and 42 per cent sons of crofters, while farmhands, craftsmen and foremen on large farms only accounted for 10 per cent of the total.[14] Again, a majority of the cohort did not move outside a very limited area or agriculture during their entire lives. Twenty-eight persons (about 5 per cent) moved at some time to the nearby towns of Skara and Lidköping, but most of these moves were parts of highly migratory careers and most moved back into the countryside. The work they took up in town seems to have been as hands or unskilled labourers. Only twelve of the local urban migrants remained in town, four of them moving to higher echelons of the urban economy as apprentices or labourers. In addition, one or two people moved directly to Stockholm or Göteborg, but in all only 7 per cent of the cohort at any one time lived in a town and at most 3 per cent stayed on there. Instead, notwithstanding the multiplication of small farms and the decline of large ones, the Skåning cohort underwent agrarian proletarianization: at the age of fifty, 305 of the 392 survivors were landless.

Spatial mobility was obviously intensive in the countryside of nineteenth-century Sweden; it was largely circulatory, and for many people it was associated with downward social mobility. This was more marked, but not exclusive to, areas where agriculture was becoming most capitalist. The rate of spatial mobility remained high until the landless were well on in age as they searched for better positions. Very few rural people tried their luck in towns and those who did were not a social or educational élite, but rather the opposite.[15] This was reflected in the kind of positions they succeeded in getting, even after a lengthy stay in town. Was it the case, therefore, that only menial jobs were left in towns for migrants from the countryside, who returned as soon as they could, because specialist urban craft and mercantile positions were filled by people born in towns? Although this was perhaps feasible in staid and stable Vadstena, the speed of its growth means that it cannot have been so in Motala Town.

Recruitment to towns and factory industry

Nearly 40 per cent of twenty-year-old men living in the industrial town of Örebro in 1894 had been born there, and 35 per cent in Örebro province; of the 1895 cohort, 34 per cent had been born in Örebro town and 40 per cent in the province.[16] Between 2 and 3 per cent of each cohort had been born in Stockholm. The rest came from other Swedish provinces. The proportion born in rural areas was between 40 and 50 per cent. Only a small proportion of Örebro's rurally born returned to the countryside; a majority stayed in towns of at least the same size. About 70 per cent of the people born in Örebro province, and about 80 per cent of those born further away, moved on from Örebro. Those born in the nearby countryside were more stable than those from other towns: about 35 per cent of them stayed in Örebro and about 30 per cent moved only within the Örebro province. It is possible to distinguish two groups in terms of migration behaviour. The urban-born moved over long distances and were highly mobile. Their mental maps of alternative livelihood positions consisted of urban nodes with no intervening rural opportunities. The rural-born, on the other hand, generally moved to the nearest town and stayed there or returned to the local countryside. There was some exchange between these two groups. About 10 per cent of the rural-born of Örebro province moved on to Stockholm and 25 per cent to other parts of Sweden. Thus, there was an inflow of highly mobile urban people to Örebro, but the main migration stream was of local rural people, most of whom were employed in menial tasks to begin with. Some of these local in-migrants advanced within the urban secondary sector, some moved back to the countryside, and some stayed as proletarian workmen.

Three obvious questions are raised when these findings are put together with
the overwhelmingly rural destinations of rural out-migrants and set in the
context of the massively rural and agricultural basis of the Swedish population
at the time. First, how could towns expand at all? To begin with, all our rural
examples are from fertile regions. When a number of such areas are added
together, the number of potential town migrants grows. Moreover, urban-
bound migration flows might have been greater from shieldland areas where
large scale intensification of agriculture did not occur, population growth
rates were often higher, and non-agricultural supplementary occupations were
common. Secondly, were towns perceived by all country folk as alternatives to
a hand or maid position at yet another farm, or did some kind of mental
barrier, such as the survival of customary peasant attitudes to towns, prevent
the majority of rural folk from going there? Thirdly, and not unrelated to the
last question, were skills in secondary production needed by rural recruits to
towns, or could they be taken straight from the plough and the rake?

Swedish industrial development did not occur only, or even mainly, in
traditional towns. Sites of abundant water power in the forest grew rapidly to
become new industrial towns, as in Motala. Where did their industrial workers
and other inhabitants come from? The literature discussed so far revealed very
small migration flows to such places, but they could hardly have been
sustained by rural labour which was seasonal, and therefore unregistered and
invisible to scrutiny. Neither could they have been peopled from older towns –
which would have been emptied in no time. Traditional theories of industrialization suggest that labour was recruited to the factories spawned by
technological innovation in the nineteenth century from completely unskilled,
(necessarily) mainly rural people. No special skills were needed to work the
machines, and it was possible to go straight from the plough to the lathe or
loom. More recently, this view has been contested. Industrialization has been
shown to have been a much more gradual process, often passing through a
distinctive stage of proto-industrialization, and the origin of workers in new
formal industry has been ascribed to local rural craft and pre-industrial mass-
production. Proto-industrial development was regionally self-contained: capital and entrepreneurship came from putters-out and labour from domestic
workers already active in the region concerned.[17] However, we have already
seen that both capital and labour were highly mobile, certain kinds of both
over long distances: investment, proto-industrial and already skilled industrial
labour might well have moved outside their home areas to distant new
opportunities which were suited to their aptitudes. In our region, this would
mean that people who had acquired secondary skills in homecrafts or in plant
like mills and ironworks were attracted to the new expanding industry over
long and short distances, while those who had worked all their lives within
local agriculture were unwilling to change the source of their livelihood to

industry, even if it needed minimal spatial dislocation and brought higher wages.[18] This hypothesis cannot be tested simply by counting the urban and rural origins of migrants or allocating them to distance bands; we need to trace them over at least part of their occupational careers, discover whether formal industry existed in their areas of origin, and penetrate the hidden economy of homecraft production there.

Little work of this kind exists, but we do know that according to birth registers about one-third of the industrial workers in Norrköping in the nineteenth century were children of peasants and crofters and that a further 10 to 20 per cent came from the homes of farmhands and farm labourers.[19] Agrarian backgrounds were most prevalent during the 1840s; by the 1870s children of peasants and crofters were far fewer and those of urban and rural proletarians were most common. Only between 17 and 35 per cent of the workers in the various textile and engineering factories originated in the town itself, but between 54 and 65 per cent had been born in Östergötland and only between 9 and 23 per cent elsewhere in Sweden. There were fewer locals and Norrköping-born labourers at the engineering works than in the textile mills, which is consistent with the emphasis recently given to the importance of the pre-existing pool of skilled labour at ironworks and domestic forges for the emergence of industry generally, particularly engineering, in the late nineteenth century.[20] Traditionally, a limited group of skilled labourers moved between early industrial plants like ironworks, mines, papermills and textile factories, after initial recruitment through a system similar to that which had been typical of urban guilds: sons of skilled workers were trained as master hammersmiths, paper makers and the like. Proto-industrial activities, on the other hand, associated with ever-smaller scale farming on subdivided holdings and population growth, maintained diverse sources of livelihood and fixed people in their home areas by removing the need for children to migrate in search of work. According to this model of development, factory industries grew within such areas; established by merchant putters-out, they sucked labour from domestic looms and forges as the process was completed.[21]

Reliance on each of these sources of labour would produce a characteristic spatial pattern of movement to early large-scale industry. The recruitment of 'green' agricultural workers would give rise to smooth distance-decay patterns of inflow, locally confined and distorted only by variations in rural population density. If the source was already proletarianized farm labour, the pattern would be further biased towards sources on the arable plains. Recruitment from proto-industrial households would also show smooth distance-decay patterns, in this case biased towards the shieldlands rather than the plains. If recruitment to factories was largely from early workshops, ironworks and other plant-based industry, the spatial patterns of in-migration and workers' birth-places would comprise nodes, largely, again, in shieldland regions but, continuing the pre-industrial tradition, many of them at considerable distances away. This would explain the low figures of urban and industrial migration found from areas on plains.

It is clear, then, that during the early phase of Swedish industrialization, labour was largely recruited from the local countryside, before the rate of its self-reproduction grew. But we do not know how the careers of the early industrial workers were formed: whether they gradually became industrialized through prior employment in craft workshops or manufactories, or if their parental homes, though formally registered as agrarian, were involved in homecrafts. It is possible to draw inferences about this from the geographical patterns of workers' places of birth and last residence, but to do it rigorously requires detailed studies of their whole life-careers.

Flows of labour in Aska and Dahl, 1810–60

Introduction

How far were these expectations based on existing research, plus our conjectures about its blind-spots, fulfilled by the flows of people through the changing structures of livelihood positions in Aska and Dahl? Our analysis is based on two samples of data drawn from *husförhörslängder*[22] for rural Herrestad on the plain and Västra Ny on the shield, and the towns of Vadstena and Motala. The first set of data represents information on the places of birth, last residence and next residence of everyone except dependent children who lived in the selected areas, in their own or others' households, whilst the registers for 1855–60 were in use. These data are not so easy to interpret as they might seem. The registration of a new household on every change of address in town and every move between neighbouring farms means both multiple-counting of inter-parish movers and that many people have places in the same parish recorded as their last, current and next residence. Where they lived between their place of birth and their parish of current residence, and the dates of their entry into and/or exit from it, and thus the length of time they stayed in it, cannot be discovered. The abbreviated life-paths therefore yield underestimates of people's inter-parish mobility because they were very mobile over shorter distances.[23]

The second set of data does not suffer from this drawback. It comprises detailed life-path reconstructions for samples of people drawn from the registers of 1855 and traced forward and backward through the registers of all the other parishes in which they lived between birth and death. Some individuals passed out of observation because they could not be discovered in the registers of one of the parishes to or from which they were recorded, elsewhere, as having moved; but it was in fact possible to reconstruct complete life-paths for most of the people in the samples. There is, of course, a difference between the movements normally registered in migration research, as for example between parishes, and the pattern described here, where every registered move is included. However, although for some they only meant the distance of a few fields or doors along a street, many moves within parishes could be just as long as movements between parishes, and short moves between adjacent farms or streets could be much more significant in a person's life-path development if they were associated with changing from hand to farmer or journeyman to master than moves over many miles which involved no change of livelihood position.[24]

Flows through the countryside

Evidence from abbreviated life-paths
The expansion of non-agrarian production in Vadstena and Motala Town must have required strong inflows of labour. As the vast majority of the population of our region lived in the countryside, we would expect this to be the major source of those flows. We will examine whether the proletarianized sector of rural society did, indeed, provide these recruits. Furthermore, we need to look at how far the traditional kinds of flows in peasant society – such as the movement of young males and females to be hands and maids off their home farms before marriage and inheritance – continued and were meshed in with the patterns associated with the penetration of capital into agriculture and secondary production. How was this traditional flow of peasant society transformed into one which produced a permanently wage-earning farm labour force as the scale of farm production expanded? If agrarian labour moved on to secondary production who, then, did the necessary work on the growing farms of the region? What happened to the farmers who failed to hold onto their land in the capitalist transformation of the plains? What became of the crofters who had to leave their crofts on the reorganization of farming villages and estates at enclosure in the 1840s and 1850s?

Table 7.1 shows the relationship between the number of livelihood positions of different kinds registered in 1860 and the numbers of people who flowed through them between 1855 and 1860.[25] In both these rural parishes the rate of throughflow of people occupying positions as male hands and female maids, which traditionally came in the life-cycle stage between childhood and household formation, considerably exceeded any other. However, differences between the two parishes across almost all the positions imply that mobility was considerably greater in the capitalist farming system of the plain than in the mixed small pastoral farm/estate/industrializing economy of the shield. At first sight, it is surprising that the parish where the economy was developing more rapidly and where total population was increasing faster had the lower rate of population throughflow. One of the reasons for this must be that Herrestad was much smaller in area than Västra Ny, so that moves of an

Table 7.1. *Numbers of agrarian livelihood positions as recorded in the* mantalslängder *of 1860, and numbers of people occupying them between 1855 and 1860 as recorded in the abbreviated life-paths taken from the* husförhörslängder

	Herrestad			Västra Ny		
	Abbreviated path	Livelihood position	Ratio A/L	Abbreviated path	Livelihood position	Ratio A/L
Farmer	49	27	1.8	120	79	1.5
Crofter	24	12	2.0	187	114	1.6
Cottager	34	18	1.9	66	83	0.8
Hand	320	55	5.8	632	164	3.9
Labourer	8	5	1.6	24	16	1.5
Maid	278	54	5.1	757	118	6.4
'Unproductive'	36	42	0.9	151	166	0.9
Other	10	2	5.0	93	28	3.3
Total	759	215	3.5	2,030	768	2.6

identical short distance would cross the boundary of the former but not of the latter parish. Even so, it is probably the case that capitalist farming on the plains generated much more turnover amongst wholly proletarian labouring positions than the combination of part-time industrial employment linked with small-scale agriculture and forest cottage-holding in the north. The higher proportion of farmhands who were married in Västra Ny than in Herrestad (29 per cent compared with 23 per cent) also suggests that this group at least was more residentially stable on the shield than on the plains. Labourers were hands who had progressed into middle age and usually lived in houses of their own rather than with their employers, so the much lower rate of flow through labourers' than hands' positions is unsurprising.

Maids were the female equivalent of hands, living in the households of those who employed them. Like hands, they tended to be young, but the areas over which they were recruited were even more constricted than those on figure 7.1.[26] The flow of maids through their livelihood positions was the most rapid of any group, yielding a complete turnover annually (see table 7.1).[27] Like hands (see figure 7.2) they were cascading through Västra Ny into the rapidly expanding industrial and urban area along Motala Ström, and through Herrestad on the southern plains to Vadstena. Mainly in Vadstena, the maids included a small number of adult women who headed their own households and had therefore moved their positions relative to juvenile maids in the same way as labourers had relative to hands. It is unfortunate for our purpose that it is not possible to distinguish these older independent maids in the abbreviated life-paths.

The maps of places of birth and last residence of farmers between 1855 and 1860 on figure 7.3 show that all were overwhelmingly locally recruited. Also, the more productive resources that were attached to livelihood positions, the more spatially restricted the area from which the occupants came. Freeholders and tenant farmers must to a large extent have inherited their resources, or capital to invest in land or equipment and stock for leased farms. This implies a low degree of upward social mobility, which is not surprising considering that in the southern part of the region, exemplified by Herrestad, a rapid concentration of landholding and tenure was taking place. Many small farmers were forced to leave their positions by farm amalgamation and the tenurial turmoil that came with enclosure.[28] The similarity between the parishes of birth of hands and crofters[29] implies upward mobility between these two positions as erstwhile hands progressed through the life cycle; the further away a hand came from, the less likely he would be to come across a croft to work. It seems that the system was not violently unstable in terms of landholding. Although a number of farmers dropped out, more substantial farm holdings were retained inside particular families. Hence, the extremely localized places of birth and last residence of the farmers themselves shown on figure 7.3, even if before taking over their farms they would probably have worked temporarily as hands. This reminiscence of a peasant society was especially marked in Herrestad, in spite of the rapid capitalization of agriculture there. The cohort of hands who were embarking on a proletarian career were supplemented by this traditional flow which simply marked a short stage in the life-path of a peasant farmer. Of course, as the number of farmers declined and the number of wage labourers increased, so this component would make up a smaller and smaller part of the total flow of hands. In general, for Västra Ny, where economic expansion was more rapid, all the maps of recruitment show wider catchment areas, even if turnover was less intense in nearly all positions there. This might be because positions for hands required more specialist skills on estates and in industry, or because they were more attractive and therefore drew people over longer distances.

Crofts virtually disappeared as the number of farmhand positions, especially for married hands, expanded on the plains. The decline of crofts from nineteen to twelve as hands and labourers increased from thirty-nine to fifty-five in Herrestad between 1810 and 1860 implies a shift of erstwhile crofters and their sons into farmhand positions. In Västra Ny the number of hands trebled over the fifty years, but the number of crofts increased from 108 in 1810 to 114 in 1860, so that there must have been an inflow of new hands from further away. This is reflected in the proportion of hands born in neighbouring counties shown on figure 7.1(a).

The outflows of the three groups also differed, even if in general they were

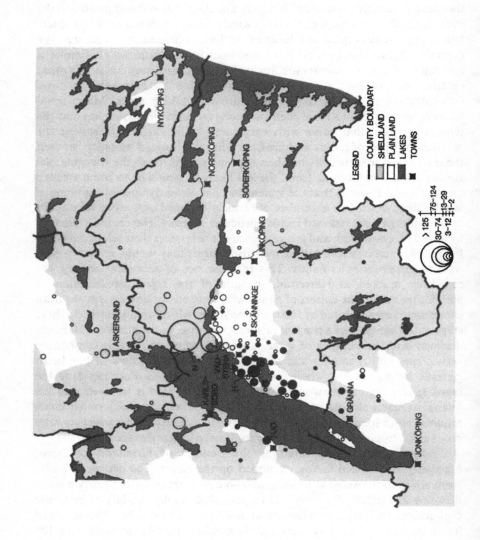

LEGEND

COUNTY BOUNDARY
SHIELDLAND
PLAIN LAND
LAKES
TOWNS

> 125
75-124
30-74
13-29
3-12
1-2

NYKÖPING
NORRKÖPING
SÖDERKÖPING
LINKÖPING
SKÄNNINGE
ASKERSUND
VAD-
STENA
KARLS-
BORG
GRÄNNA
JÖNKÖPING

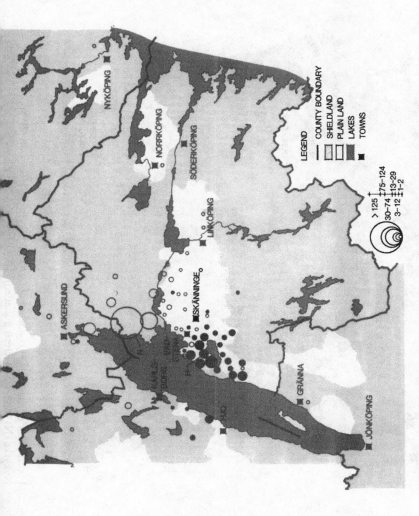

Figure 7.1. (a) Parishes of birth of farmhands living in Herrestad (H) and Nykyrke (N) parishes 1850–60 (Herrestad shaded). (b) Parishes of last residence of farmhands living in Herrestad (H) and Nykyrke (N) 1850–60 (Herrestad shaded).

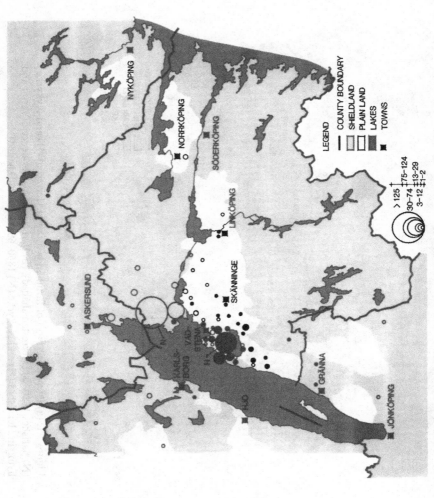

Figure 7.2. Parishes of next residence of farmhands living in Herrestad (H) and Nykyrke (N) 1850–60 (Herrestad shaded).

quite local and Västra Ny again displayed a slightly less local pattern. Very few freeholders moved out at all during the eight years studied. A few more crofters did, those in Herrestad presumably evicted on enclosure. Their destinations were mainly within the parish, whilst the Västra Ny crofters, even though they more than maintained their numbers, more frequently moved longer distances, outside the parish and the county. Finally, 5 per cent of Herrestad's hands and 3 per cent of Västra Ny's moved to towns. This adds a new dimension to their mobility, although it was of little significance in the overall pattern of flows of the cohort resident in rural areas in 1855–60. Västra Ny hands showed a strong bias towards neighbouring Motala, although the abbreviated life-paths do not reveal if any of them went directly into industry there.

Evidence from full lifelines
We know enough from the abbreviated lifelines to decide if the small samples of full lifelines are atypical in some respects. They will be used to illustrate, and verify or refute, what the abbreviated paths have suggested. Detailed lifelines can be depicted and analysed in a number of ways, from qualitative verbal descriptions of individual lines to highly schematic and quantitative tables or diagrams treating one or two dimensions of all the lines at once. We have done the latter, and will discuss the aggregate spatial and social mobility character- istics of each sample of lifelines only in general terms. A rural cohort was sampled for the reconstruction of complete lifelines of freeholders, crofters and hands. To examine differences between freeholders with farms of different sizes, some wealthier farmers from Väversunda parish were sampled together with small farmers from Orlunda where farms were massively subdivided further north on the plain, and Motala on the shield – eighteen farmers in all. In addition to the farmers, eleven crofters and eleven hands from Väversunda and Motala were sampled.

Although the information on figure 7.3 is very generalized and the spatial categories are highly aggregated, it reveals much more about the massive residential mobility shown by the abbreviated life-paths. Many of the moves, although by no means all of them, occurred between leaving the parental home, which on average occurred at nineteen years and eight months but ranged from twelve to twenty-four years of age, and the age of thirty. However, moves whilst living with parents or after marriage and the setting up of a family home were only rather fewer. Despite the frequency of movement, people rarely moved very far. Only 1.8 per cent of the years registered on the full lifelines were spent outside Östergötland, and 7.4 per cent in other parts of the province than the hundreds of Aska and Dahl, where over 90 per cent of the total lifetimes of the sampled individuals were spent. Of the 2,655 years contained in the lifelines (an average of 66 per individual) only 44, or 1.7 per cent, were spent in towns. This minuscule urban experience was accumulated

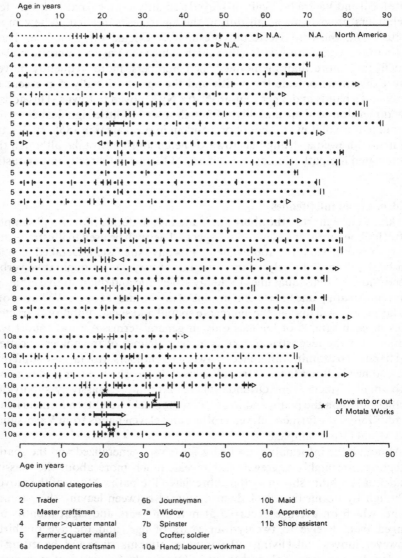

Figure 7.3. Lifelines of the rural cohort. Each line represents the life-path of an individual, with years from birth marked on the horizontal axis, tracing each change of residence from birth to death or the time at which they fell out of observation. The lines are grouped according to the occupations of the people concerned, designated by the codes at the beginning of the lines. A change of residence is marked by a vertical peck across the lifeline. How the line is drawn designates whether the place of residence is urban or rural (solid or dotted), and within Aska and Dahl, elsewhere in Östergötland, elsewhere in Sweden (fine, medium or bold thickness).

by two farmers, one of whom spent two and the other three years in towns during lifetimes of sixty-nine and eighty-eight years, and by six labourers, hands and maids. Only one of the towns visited was outside Östergötland, where one of the maids lived from the age of twenty-three. Two large farmers, eschewing the traditional peasant lifecycle stage as a hand off the parental holding, were born, lived for seventy years or more and died on their farms without ever moving elsewhere; another moved just once at the age of twenty-two. At the other extreme, two other farmers migrated to North America, one at forty-eight and the other at fifty-eight years of age; the latter had already changed his place of residence ten times before crossing the Atlantic. The six large farmers for whom full life-paths were reconstructed accumulated eighteen changes of residence between them and moved on average once for every twenty-two years of their lives; the twelve small farmers and eleven crofters moved every eleven years on average, and the eleven labourers, hands and maids every six years. One hand or maid lived in seventeen places during his/her fifty-seven year lifetime, two others in fourteen.

We have already seen that the birth-parishes of all farmers in both plain and shield were concentrated in tiny areas around the place where they lived their productive lives. This was also true of the farmers whose full lifelines were reconstructed. Only one of the Väversunda group was born outside the county, in the neighbouring one, and only two of the eighteen were born outside Aska and Dahl. Small differences can be distinguished between the three subsets: three of the five from Väversunda and Motala were born in their current home parishes, but only two of the eight Orlunda smallholders were. This indicates a greater reliance on inheritance among the Väversunda and Motala property owners, which is easily explicable when their social origins are considered. Four out of five of the Motala and Väversunda farmers came from freeholding or tenant farmer families, but none of the Orlunda farmers had a freeholding father, though five came from tenant farms. In all, thirteen of the eighteen farmers had farmer fathers. All five others were crofters' children: none had purely proletarian origins.

Only two of the eleven crofters had fathers who held land in any capacity, in each case as tenants. Thus, even if some social circulation was occurring, the son of a crofter had a much smaller chance to acquire land than the son of a farmer. At the same time, the crofters' sons had a better chance of becoming crofters themselves, or even tenants, than hands' sons, none of whom became a freeholder, tenant or crofter. Upward social mobility was therefore limited; obviously, it was much easier to go downward. Among the fathers of the eleven hands were two farmers and five crofters. More than half of the hands had farm origins, which would seem to destine them for a better position later in their life cycles. No less than eleven of the eighteen farmers had worked as hands before getting a farm of their own. The peasant pattern of life-cycle mobility persisted into our period, despite the growth of capitalist agriculture.

Nine of the eleven crofters had previously been hands so that the eleven hands might also be expected eventually to attain slightly better positions. However, only two of them actually became a crofter or a freeholder, irrespective of whether they came from the plains or the forest. Five, all from the plains, remained agricultural hands, whilst all the hands from the shield moved into secondary production. Three went to Motala Works, one was employed on the Göta Canal and one in craft work in Motala *köping*. The shieldland hands also came from slightly more prosperous homes than those from the plains: three plains hands had fathers who were proletarian labourers, two who were crofters, while one was illegitimate; one of their northern counterparts came from the home of a freeholder, three from crofts and one was illegitimate. Evidently, changes had occurred in society since the freeholders and crofters of 1855–60 had been hands.

Changing times are also amply reflected in the fates of our cohort after they were sampled. In particular, the freeholders of the plain went through a difficult period: the two who emigrated to North America had lost their farms in the 1860s and presumably hoped to get land under the Homestead Act; another ended up as a married hand, and only two ended their days as retired farmers. Only three of the thirteen smaller freeholders from Orlunda and Motala lost their farms, which is consistent with large farms on the plains growing at the expense of intermediate rather than small farms, and small farms on the shield maintaining their large number and proportional predominance. Similarly, the number of crofts declined dramatically on the plains between 1810 and 1860, but not on the shield. Not surprisingly, the plains crofters of our cohort became the victims of the takeover of the arable of their crofts by the farmers. None of them retired as a crofter, but as cottagers or married hands. Only one of the five crofters on the shield suffered this fate.

We can conclude that the sample of full lifelines is representative of the population from which it was drawn. Most of the changes and patterns described summarily earlier have been given fuller substance. We have also learned that the growing proletarian agrarian population was recruited from freeholders and crofters (and also, of course, to a growing extent, internally). There was little willingness to leave agrarian life, even to alternative opportunities close by, particularly among people from the plains: none in our cohort left agrarian life, although some moved across the Atlantic to stay in it. Agricultural positions were more stable on the shield, but strong population growth forced many superfluous sons of freeholders, crofters and cottagers out of their paternal livelihoods. They moved into the proliferating agrarian proletarian positions on the plain, or to industrial positions on the shield, perhaps because they had youthful experience of proto-industrial part-time production in their homes, or part-time employment in local ironworks, sawmills and other forest industries.

Table 7.2. *Numbers of livelihood positions in Vadstena as recorded in the* mantalslängder *of 1860, and numbers of people occupying them between 1855 and 1860 as recorded in the abbreviated life-paths taken from the* husförhörslängder

	Abbreviated paths	Livelihood position	Ratio A/L
Adult professionals, craft and services	310	102	2.3
Independent craftsmen, labourers and farmers	351	178	2.3
Apprentices, journeymen and shop assistants	391	100	3.9
Hands	335	67	5.0
Maids	637	190	3.4
'Unproductive'	384	101	3.8
Totals	2,408	738	3.2

Flows through Vadstena

Evidence from abbreviated life-paths
Comparison of tables 7.1 and 7.2 shows that the through-flow of people was greater in Vadstena than the countryside in all livelihood positions. Adults in basic urban occupations had a greater rate of turnover than rural farmers, and even the adult menial group had a higher ratio of occupants to positions in the town than the combined hands and labourers of the countryside, a large majority of whom were juvenile hands. Maids were not so mobile as in rural areas or as some other urban groups and many more were adult. In town there were more positions for older unmarried women without any resources of their own. The apprentices, journeymen and adult menial workers employed by craftsmen and merchants were much more mobile than all rural groups except the maids – even than the rural hands.[30]

We will concentrate on male townsfolk and maids, for whom occupations are given in the sources we have used for residential mobility. It is impossible to discover how many of the 'unproductive' households headed by women were producing homecraft wares in significant quantities, and how many were truly unproductive except through the examination of inventories, which are available for only the small proportion of the 1855–60 population which died. In the abbreviated life-paths we must therefore limit ourselves to male-headed households which can be categorized into different sectors of the labour force through the occupations given in the tax and church registers. There are, however, five widow and spinster heads of household in the urban sample for whom we have drawn full lifelines.

Figure 7.4 reveals clear distribution patterns in the parishes of birth of the

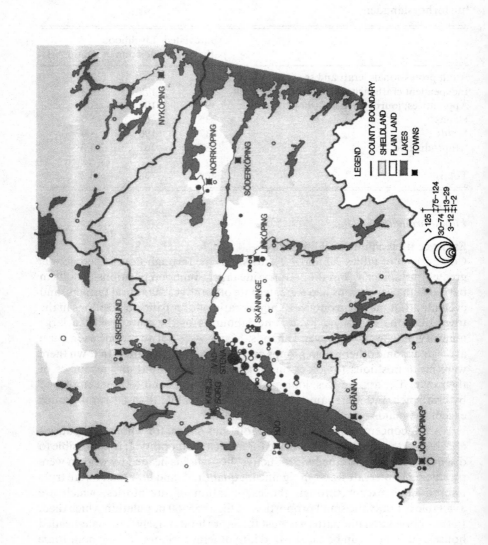

LEGEND

——	COUNTY BOUNDARY
☐	SHIELDLAND
☐	PLAIN LAND
■	LAKES
■	TOWNS

> 125
75–124
30–74
13–29
3–12
1–2

NYKÖPING

NORRKÖPING

SÖDERKÖPING

LINKÖPING

SKÄNNINGE

ASKERSUND

VAD
STENA

KARLS
BORG

GRÄNNA

JÖNKÖPING

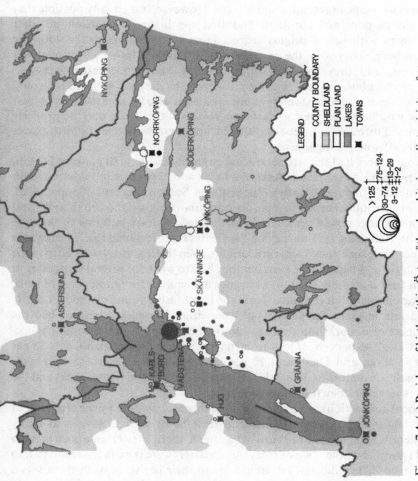

Figure 7.4. (a) Parishes of birth within Östergötland and its immediate vicinity of journeymen and apprentices living in Vadstena 1850–60 (apprentices shaded). (b) Parishes of last residence within Östergötland and its immediate vicinity of journeymen and apprentices living in Vadstena 1850–60 (apprentices shaded).

traditional craft component of Vadstena's economy, although not in the way that might be expected of a 'closed shop' urban activity. More than 50 per cent of all urban craftsmen were born in the countryside;[31] more than 60 per cent of journeymen and apprentices, in training to occupy craft positions later in their lives. Rural–urban links were obviously close, and they were growing more intense if the increasingly rural origin of the more junior craft occupations reflects a change in recruitment patterns. However, it is equally possible that the shares remained constant, and that the higher proportion of adult craftsmen with urban origins reflects greater professional success for the urban-born. They were the ones who could inherit a father's or a relative's business, and they had the urban contacts and hence easier access to independent businesses of their own, accepted by the local guilds. This would give a degree of spatial stability to the craftsmen who became masters, at least those who inherited their workshops. This hypothesis can only be properly examined through fully reconstituted lifelines, but the pattern of birth-places of all craftsmen is suggestive.

Apprentices had the strongest local connections, with 80 per cent born in Östergötland, 56 per cent in Aska and Dahl, and 23 per cent in Vadstena itself. Journeymen had much further-flung origins. Some 65 per cent of apprentices, and in turn qualified craftsmen, had agrarian origins: the urban and rural worlds of this region were closely connected, at least in terms of flows of labour. Reflecting the 'journeying' part of their professional role, journeymen did not have such close local rural origins, even though many were born in the countryside. To qualify as a master they had to work for several businesses to learn their craft fully. Indeed, it was not uncommon for journeymen to work on the continent for a while to learn the latest fashions and methods. Locally born journeymen who stayed put were either less enterprising or had the prospect of a workshop through kin or their parents' professional connections. In all 45 per cent of journeymen were born in Östergötland, only 25 per cent in Aska and Dahl, and 12 per cent in Vadstena itself, so they were normally adventurous in search of prospects outside their home areas and the places in which they trained as apprentices.

Local connections were a continuing influence, judging by the tendency of craft masters to return eventually to their home region. Almost as many masters as apprentices were born in Vadstena, though fewer were born in Aska and Dahl as a whole. Two of every three masters were born in Östergötland, so that even if they did not return exactly to their parish of birth there was a strong tendency to come back to their home area after a spell further away during the journeyman phase of their life cycle.

Independent craftsmen, who increased in number as guild control became less strict, were recruited differently. Half of them were born in Vadstena or the local hundreds, 75 per cent in Östergötland. They were more rurally recruited than master craftsmen, but less so than apprentices and journeymen.

Obviously, they did not organize their lives according to guild rules, although they may have begun as apprentices but left the system before the journeyman stage, either due to a lack of skill or a concern to profit immediately from the training they had already received.

The trading community of the town also contained distinct categories. Their recruitment was more exclusively urban and geographically wide-ranging than that of craftsmen, even though 40 per cent of them were born in the countryside. When shop assistants (juveniles in training to be merchants), shopkeepers and traders are taken together, the urban bias is even more marked. Three out of every five shop assistants were urban born and, judging from their places of last residence, spent most of their active careers in towns. Unlike the craftsmen, there was not a recruitment flow from the countryside, through the local town, then on to more distant parts of the urban system before eventual return. Rather, most people within trade were urban from the beginning, with only minority reinforcement from the local countryside.

These patterns, with flows of people into Vadstena's crafts and trade originating in the countryside or other towns, and then largely remaining within the urban system, are fully evident on figure 7.4. The initial moves from country to town were very local: most of the craftsmen and traders of Vadstena between 1855 and 1860 had been born in the rural parishes of Aska and Dahl. This seems to imply that it was in the towns where things happened and hence, where underemployed countryfolk went for training or general experience of the wider world, never to look back. But although Vadstena's population was overwhelmingly local and rural in origin, the analysis of flows in the countryside showed that very few of the mobile country dwellers ended up in or passed through the town: well over 90 per cent of them moved around solely within the countryside. As far as the aggregate pattern of mobility in Aska and Dahl was concerned, the net flow of population to the town was insignificant. The process of rural–urban migration apparent in the Vadstena figures alone is a misleading picture of the flows in the regional system as a whole.

Most of the townsfolk were not trained or in training for a skilled urban occupation, but hands, labourers or workmen, sometimes for craftsmen or merchants, sometimes with a more tenuous connection to a specialist urban occupation. Rural birthplaces were completely dominant in this group: 87 per cent of the menials came from the countryside, a large majority from Aska and Dahl and most of the remainder from elsewhere in Östergötland (see figure 7.5(a)).[32] The expansion of both crafts and unqualified labouring positions drew in people mainly from the narrowly local rural population, but the two groups had very different career patterns. Whereas nearly all craftsmen stayed on in towns, locally or more remote, unqualified people retained their close contact with the local countryside. Some 25 per cent of their moves after coming to Vadstena were back to Aska and Dahl hundreds or other parts of

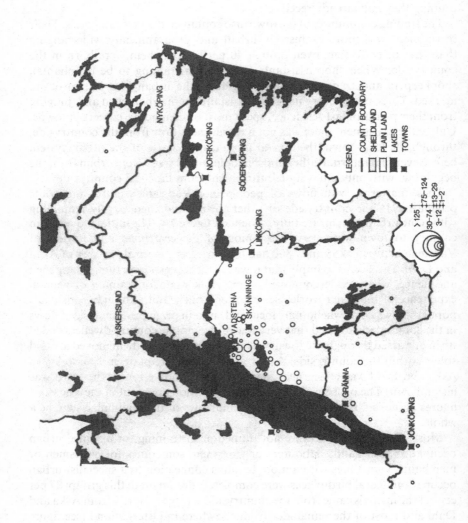

LEGEND

COUNTY BOUNDARY
SHIELDLAND
PLAIN LAND
LAKES
TOWNS

> 125
75–124
30–74
13–29
3–12
1–2

NYKÖPING

NORRKÖPING

SÖDERKÖPING

LINKÖPING

ASKERSUND

VADSTENA
SKÄNNINGE

GRÄNNA

JÖNKÖPING

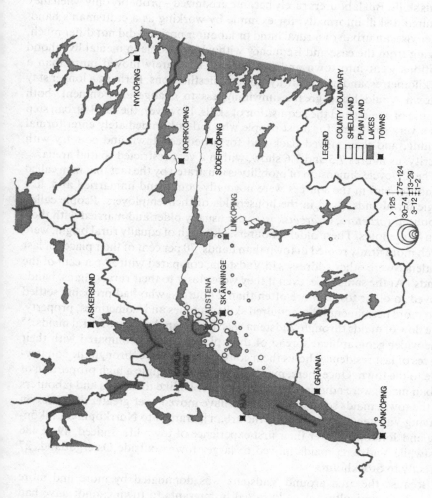

Figure 7.5. (a) Parishes of birth of hands living in Vadstena 1850–60. (b) Parishes of last residence of hands living in Vadstena 1850–60.

LEGEND

COUNTY BOUNDARY
SHIELDLAND
PLAIN LAND
LAKES
TOWNS

> 125
75–124
30–74
13–29
3–12
–2

NYKÖPING

NORRKÖPING

SÖDERKÖPING

LINKÖPING

ASKERSUND

VADSTENA

SKÄNNINGE

KARLS-
BORG

GRÄNNA

HJO

JÖNKÖPING

Östergötland, and almost 70 per cent were to other addresses within the town itself. The labourers' flow was thus completely different from that of the craftsmen. The step from rural to urban was a definite one for the future craftsman: through a gradual specialization for work in specifically urban livelihood positions he was unable to return to the countryside unless he gave up his skills. But labourers rarely became urbanized – probably only when they acquired a skill informally, for example by working as a craftsman's hand. Otherwise, an urban or a rural hand or labourer position did not differ much, judging from the ease and frequency with which people in menial livelihood positions went into town and out again. They rarely moved more than a few kilometres, and nearly always to rural destinations. Perhaps a longer stay in town would bring greater unwillingness to change environment, both because of ageing and the acquisition of skills. However, the rural–urban step was a very small one for most people who did not immediately enter formal training, and they moved back and forth between town and country with alacrity, as figures 7.5 and 7.6 show, within a very restricted spatial arena.

The life-cycle dimension of mobility is illustrated by the fact that men called 'hand' (*dräng*) in the sources were normally young and unmarried and, like single farmhands, lived in the households of their employers. People called 'labourer' (*arbetare*, or *arbetskarl*) were usually older and married with their own households. These older labourers, although of equally rural origin, were much more firmly rooted in town than hands: 70 per cent of their places of last residence were other addresses in Vadstena, compared with 37 per cent of the hands'. At the same time, even if they were closer to their rural origins, hands moved to other towns more often than labourers, who had probably settled into more permanent jobs, acquired skills, families and, sometimes, property. The flow of maids through Vadstena was not so rapid as that of rural maids.[33] The wider geographical spread of their places of birth compared with their places of last residence shows them converging stepwise through the countryside to the town. Once there, many of them stayed; quite a high proportion of urban maids were adult rather than juvenile. Unlike the hands and labourers of the town, maids seem quite often to have moved over greater distances on leaving Vadstena. Some went up the urban hierarchy to Norrköping, Linköping and Jönköping after their first experience of town life. Indeed, 49 of 486 erstwhile Vadstena maids moved to larger towns outside Östergötland, 27 directly to Stockholm.

Because the area around Vadstena was dominated by more and more capitalistic agriculture, locally raised in-migrants to town cannot have had training or experience in tasks of an urban nature through involvement in proto-industry or forest industries. All training for skilled non-agrarian occupations had to take place in town, but hands and labourers who stayed in town doing menial tasks did not receive craft or industrial training. As unqualified labour, it was as easy to move back to the countryside as within

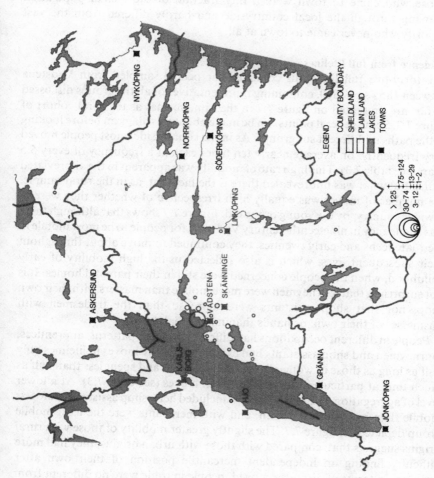

Figure 7.6. Parishes of next residence of hands living in Vadstena 1850-60.

LEGEND

COUNTY BOUNDARY
SHIELDLAND
PLAIN LAND
LAKES
TOWNS

>125
75-124
30-74
13-29
3-12
1-2

NYKÖPING

NORRKÖPING

SÖDERKÖPING

ASKERSUND

LINKÖPING

VADSTENA

SKÄNNINGE

KARLS-BORG

GRÄNNA

JÖNKÖPING

town, because the tasks involved were simple and similar in both cases. Moves to other towns would not mean any improvement in livelihood without further training, as it would for journeymen who could go to a more profitable business higher up the urban hierarchy. In consequence, there was very little migration to other towns or over long distances by urban menial workers. Those who came to town were a tiny fraction of the menial population coursing through the local countryside and hardly differed from the vast majority who never came to town at all.

Evidence from full lifelines
The forty-five full life-paths compiled for people sampled from Vadstena between 1855 and 1860, containing representatives of all the groups discussed so far, are presented on figure 7.7 in the same format as the rural cohort of figure 7.3. Some general points can be made about this diagram before looking at the paths of different subgroups. As in the countryside, most people moved very frequently, on average nearly ten times each at a frequency of every 5.5 years (see table 7.3). This high rate of mobility was apparent in the abbreviated paths, but what was not revealed there is the fact that, as in the rural sample, the number of moves was equally high irrespective of whether they were in town or country, by the young or the old. Figure 7.7 shows that although there was a tendency in nineteenth-century Vadstena for people to be most mobile in their late teens and early twenties, they continued to move about throughout their subsequent lives, which is also reflected in the high mobility of early childhood, when the people concerned were still in their parental homes. It is not surprising that journeymen were more mobile than masters with their own shop, nor that shop assistants were more so than the tradesmen with businesses of their own, or hands than married labourers.

People in different occupations had different mobility patterns: apprentices, journeymen and shop assistants had experienced more moves in lifelines only half as long as those of women heads of household and spent less than half as much time at particular addresses between moves (see table 7.3). At a lower level of aggregation, judging by the few included here, shop assistants were less mobile than hands or craftsmen, and with merchants were the least mobile group depicted on figure 7.7. The slightly greater mobility of those with rural origins suggests that, compared with those with urban origins, they had more difficulty finding an independent mercantile position of their own after training. In terms of distances moved, people in trade were no different from craftsmen. Moving from Stockholm to Vadstena or from Vadstena to Landskrona in the south of Sweden could obviously be achieved with ease by those in the higher echelons of the urban economy. In parallel with their less frequent movement behaviour, traders and shop assistants were also a socially stable group. Although only one had a father who was a tradesman himself, none had an agricultural origin: in four cases out of five the fathers were urban

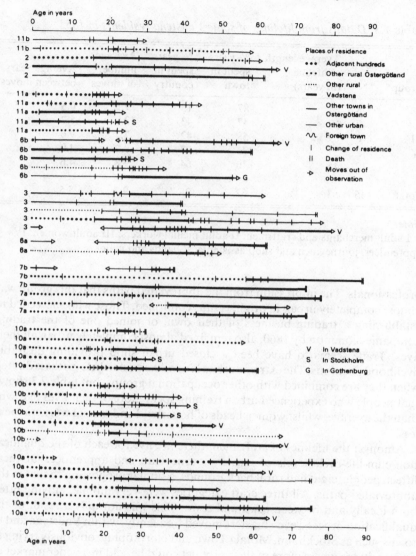

Figure 7.7. Lifelines of the cohort from Vadstena. A key to the numerical occupational codes to left of the lifelines is given on figure 7.3. A wavy line represents a foreign place of residence.

364   *Peasantry to capitalism*

Table 7.3. *Details from lifelines of some Vadstena residents in 1855*

Group*	N	Average length of lifeline (in years)	% years spent in town	% years spent in country	Average number of moves	Average years between moves
I	8	64	87	13	9.5	6.8
II	7	67	57	43	12.9	5.2
III	6	69	58	42	7.2	9.6
IV	12	37	73	27	9.2	4.0
V	12	45	36	64	9.8	4.6
Total	45	53	62	38	9.7	5.5

Note:
* I adult merchants and craftsmen; II adult menial workers; III adult women; IV apprentices, journeymen and shop assistants; V hands and maids.

professionals. The impression from the abbreviated paths that this group was almost completely urban in origin is thus confirmed. Those who succeeded in establishing a trading business of their own, or joined one of the trading companies common by 1860, also remained almost completely urban all their lives. Trade seems to have been a closed urban shop providing desirable livelihood positions. The extreme urban bias of the traders' life-paths is diluted when they are combined with other occupational groups, but table 7.3 shows that people who experienced urban training spent more of their lives in towns than the average, whilst women heads of household and menial workers spent less.

Amongst the lifelines of craftsmen, there are five for each of the qualification-cum-life-cycle levels of master, journeyman and apprentice. Of these fifteen people, eight had rural backgrounds, a similar proportion to that in the abbreviated paths. All three craft categories were very mobile, moving often both locally and between town and town. This was not only a result of the qualification process: even masters moved frequently, within Vadstena and to towns such as Stockholm, Motala Town and Norrköping, obviously trying to find positions where more of their product could be sold in a wider market or against less competition. The rules of the parochial registration system allowed out-migrants who were looking for a job they were not already aware of at their place of departure to be recorded as moving to a 'destination unknown'. It is impossible to trace these people continuously over space and time unless they are found in the registers of other places by luck. This is why apprentices could fall out of observation when they became journeymen and moved out of their previous position, and why journeymen not yet qualified as masters sometimes did the same when they moved to another master's shop.

The journeymen and apprentices of 1855–60 seem to have followed paths similar to those traced in the earlier stages of the lifelines of those who were masters by that time, though their origins were rather more humble and less based in secondary production. All the apprentices came from the countryside, from homes of crofters, agricultural labourers and hands. As in the abbreviated paths, they all originated in Aska and Dahl. The journeymen and masters had similar social origins: three out of five in each group came from towns and the fathers of three out of five were involved in secondary production. Those who came from the countryside had fathers who occupied slightly better positions than those of apprentices. This verifies the conclusion drawn from the abbreviated paths: that it was easier to succeed if of urban origins, with a father already involved in a specialist urban trade. It shows, too, that most of the expansion in urban production came from an inflow of apprentices from very humble rural proletarian rather than farming homes. The great increase in their numbers relative to craftsmen make it likely that they would eventually be employed as qualified wage workers in larger craft workshops, not become masters of their own. Thus, the craft life-paths not only reflect a life-cycle development, but also a historical process. The apprentices were all from the local countryside, while the older journeymen and masters had much more widespread and more urban origins. This was not only a result of their further education and search process for a proper position, but also of a change in recruitment patterns as more and more apprentices and journeymen became hired labourers rather than trainees for succession to urban workshops.

The independent craftsmen, of whom there were only two in our sample of lifelines, seem to have been journeymen and apprentices who failed to qualify, but after the abolition of guild monopolies succeeded in establishing a small craft shop. Their background and life-paths were similar to those of apprentices. This may explain the differences between craftsmen of different skill levels in the late 1850s: apprentices who did not fully qualify became independent craftsmen, leaving a dominant group with urban origins to go on to become master craftsmen.

The five hands in our urban cohort were all born in rural Östergötland near to Vadstena. Their life-paths began in varied circumstances. Two had fathers who were freeholders, the others farm labourers and cottagers, and four began their careers as farmhands. After passing through a couple of these positions, the move to town occurred between the ages of eighteen and twenty-five. No rationale is apparent behind the move to town of these hands whilst hundreds of others stayed in the countryside: none of them had worked in the secondary economy before moving to Vadstena, and as far as we can see from the sources, neither had their parents.[34] It is not possible, either, to see exactly what urban hands first did for a living in town, but judging from their future occupations they were employed in menial tasks by traders and craftsmen. Three of the five stayed urban and two returned to farming, reflecting the significant return flow of urban migrants in the abbreviated paths and the short social distance between town and country for the vast majority of people in the region. The three who stayed in towns followed different careers: two stayed in Vadstena as married labourers in joinery and hauling, and the third moved on to Motala Works and then Stockholm as an industrial labourer. The two who returned to agriculture were sons of farmers, although they did not have a farm awaiting them. One ended up as a wandering country book pedlar; the other died as a hand on a farm outside Stockholm.

The seven adult labourers in our sample, one step further into an urban proletarian life cycle, also all had rural origins: six came from crofts and cottages and one was the son of a tenant farmer. All left home in their teens to become farm hands, and six moved to town at about the same age as the hands. Without exception, they became urban hands in identical positions to those of the sampled hands, and as they were encountered later in their life cycles in more established positions, they remained labourers in Vadstena, except for one who later moved to another town. Their final positions were normally identical to those occupied when they were sampled, although four gradually specialized in secondary production as building, haulage, joinery and carpentry labourers. Some of the menial workers had honorary titles like night watchman and bell-ringer, indicating establishment in urban society despite rural origins. In chapter 5 we saw that they usually worked as, and were no wealthier than, labourers.

What were the differences between craft apprentices, who were locally recruited in the countryside, but attempting to advance socially and normally moving on in the urban hierarchy, and the hands and labourers, most of whom stayed in Vadstena or moved back to the local countryside? Their origins were similar: almost all came from crofts or cottages; but all the apprentices moved directly from home to town. No apprentice started as a farm hand, and all of them were in town well before the age of eighteen. The decisions, initiatives and contacts causing this are unclear, but this age difference on first entering town was an important aspect of the modernization process.

Some of the women heads of household described simply as widows or spinsters were wealthy. Others earned their keep by producing or dealing in lace and other textile goods. Yet others were aged and completely or almost destitute. It would therefore be dangerous to generalize from the small sample of female head-of-households' lifelines depicted as on figure 7.7 (7a and 7b) and on table 7.3. Nonetheless, it is striking that they were on average both the oldest and least mobile group of household heads, and that all except one, who was born in Vadstena, came to town well into their adult years, generally from farm or vicarage homes. All but one stayed until they died. In accord with the conclusions we drew from the abbreviated life-paths, the seven maids represented as 10b on figure 7.7 mainly stayed in Vadstena or left for another town, in two cases eventually Stockholm, which suggests that many more than the

Table 7.4. *Numbers of livelihood positions in selected industries in Motala Town recorded in the* mantalslängder *tax records of 1860, and numbers of people occupying them between 1855 and 1860 as recorded in the abbreviated life-paths taken from the* husförhörslängder

	Motala Engineering Works			Four other industries		
	Abbreviated paths	Livelihood position	Ratio A/L	Abbreviated paths	Livelihood position	Ratio A/L
White collar and skilled workers	52	132	0.4	18	9	2.0
Labourers and hands	1,095	497	2.2	87	42	2.1
Apprentices	6	36	0.2	2	0	—
Maids	198	30	6.6	21	9	2.3
Total	1,351	695	1.9	128	60	2.1

twenty-seven who moved there directly from Vadstena would ultimately end up there.

Flows through the early industries of Motala Town

Evidence from abbreviated life-paths

Table 7.4 shows the relationship between the number of available livelihood positions in 1860 and the number of people who occupied positions of those kinds during the second half of the 1850s at Motala Engineering Works, a smaller ironworks, a paper mill and two breweries. The source from which the abbreviated life-paths were constructed only categorizes a small minority of workers in detail, so that the various kinds of smiths, pattern makers, founders and so on who are listed in the tax records are hidden amongst a vast majority of 'workers' in the *husförhörslängder*. Little credence can therefore be given to turnover ratios for particular occupations on table 7.4. Moreover, even the total figures, especially those of Motala Works itself, are not directly comparable with those calculated earlier for Vadstena and the countryside because the labour force grew very rapidly through the 1850s, which will cause the actual amount of throughflow to be understated by the tabulated ratio. Even so, the impression that people tended to stay for much longer in these works than they generally did in positions on farms and in Vadstena is supported by more detailed analysis.

The mills and sawmills along Motala Ström employed fewer than sixty people in 1820. All the labour of Motala Works had to move in from

368   *Peasantry to capitalism*

Plate 5. Motala River in the late nineteenth century.

elsewhere. While it supplied only the limited needs of the Göta Canal, its labour force was in 1830 just over 100, and in 1840 about 170. However, by 1850 360 people were employed and by 1860 around 700. The skilled workers and foremen recruited from Britain were gradually superseded by Swedish-born employees, who were completely dominant by the 1850s. From our present point of view, the period after the initial British incursion is the more interesting because it was only when labour recruitment became linked with other population flows in Aska and Dahl that we can examine if there was a continuation from proto-industry in the region into large-scale industrial work. Although Motala Works itself did not develop out of proto-industry, its labour force might have done.

Motala Works needed specialist workers in both of its two main parts, the ironworks and the engineering works. The former produced raw materials for the latter and can almost be seen as a large-scale traditional ironworks; the mechanical works employed all sorts of skilled people, from joiners who built models of the planned products to coppersmiths and filers. The works therefore needed a wide range of skills, and it is unfortunate that it is impossible to distinguish them in the sources we used. However, it is clear that the pattern of birth-places of employees at Motala Works was quite different from that of any of the urban groups in Vadstena. They came from a much wider area (compare figures 7.4 and 7.5, (a) and (b), with 7.8 (a) and (b)), even if 51 per cent were born in Aska and Dahl, and 73 per cent in Östergötland province as a whole. They were also very much more rural in origin than craftsmen in Vadstena: only 1 per cent of the Motala labourers of the 1850s who originated in Östergötland were born in a town, though a further 7 per cent came from towns elsewhere in Sweden (compared with 20 per cent from rural areas beyond the provincial boundary). Obviously, traditional towns like Vadstena did not channel skilled labour through to places like Motala Works, and we have already seen that their menial workers mainly stayed on, moved to other towns or back into the agrarian countryside they had come from. Recruitment to growing industry must have flowed along more direct lines. It is striking that so many workers came from the immediate hinterland of Aska and Dahl. Of the 73 per cent of the Works' labourers who were born in Östergötland, much more than half (40 per cent of the grand total) were born in forested areas like northern Aska, with ironworks and other kinds of non-agrarian production: at places like Åtvid and Boxholm, where for centuries copper or iron had been worked at water-powered sites, and the Mjölby area, also with numerous water-powered mills and hammers. The Motala Works labourers born outside Östergötland came even more preponderantly from such areas, and for them, too, the threshold of full-time industrial work must have been low.

The places of last residence of employees at Motala Works reinforce this conclusion. Over 20 per cent came directly from towns, compared with the 1 per cent who were urban born; 10 per cent had moved within Motala Works itself; 40 per cent came in from forested rural areas in Östergötland, and 5 per cent from forested rural areas outside the province, so that only 25 per cent moved to the Works from purely agricultural arable areas. Those from towns did not conform with the pattern of journeymen or menial urban labourers in Vadstena: they mainly came from other towns with big engineering factories. They had obviously been trained in similar establishments and were now looking for similar or better jobs, and rapidly expanding Motala Works in the 1850s was a strong magnet for skilled industrial labour. The large Bergsund and Bolinder Works in Stockholm, and the Motala Works branches in Norrköping, Göteborg and Nyköping all provided recruits. They were the source of about 15 per cent of the total inflow, which can therefore be subdivided into four distinct groups: 30 per cent came from completely agrarian areas on the plains and were presumably unskilled; 45 per cent came from forested areas where ironworks, handicrafts, mills, sawmills and char-coal burning were important subsidiary means of support; 15 per cent had already been trained at other factories, and the remaining 10 per cent came from miscellaneous towns, from inside Motala Works, or from abroad.

Motala Works must have trained the completely unskilled and proto-industrial workers. We would expect most of them, like the apprentices and journeymen of Vadstena, to continue working in the industry in which they had been trained. The final dimension of their abbreviated paths verifies this

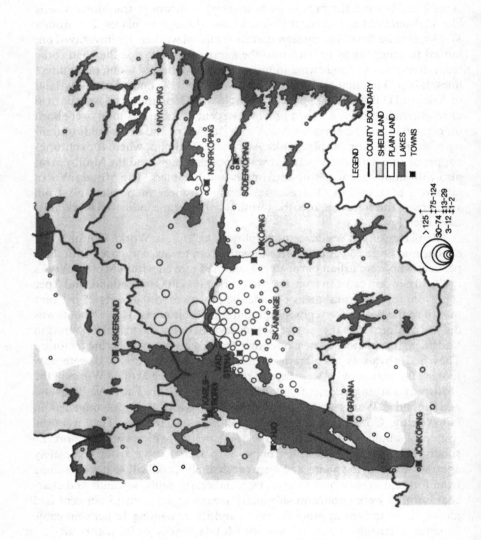

LEGEND

— COUNTY BOUNDARY

SHIELDLAND

PLAIN LAND

LAKES

■ TOWNS

> 125

75-124

30-74

13-29

3-12

1-2

NYKÖPING

NORRKÖPING

SÖDERKÖPING

LINKÖPING

ASKERSUND

VAD-
STENA

KARLS-
BORG

SKÄNNINGE

GRÄNNA

HJO

JÖNKÖPING

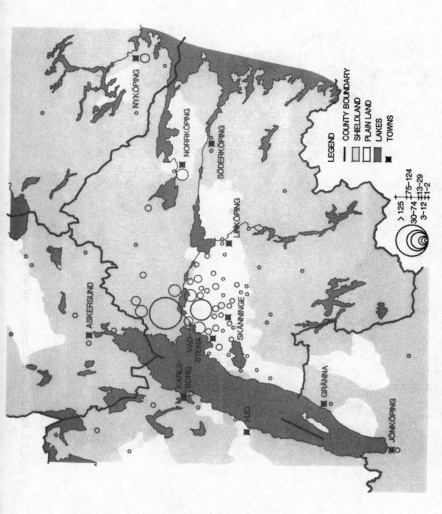

Figure 7.8. (a) Parishes of birth of workers at Motala Works 1850–60. (b) Parishes of last residence of workers at Motala Works 1850–60.

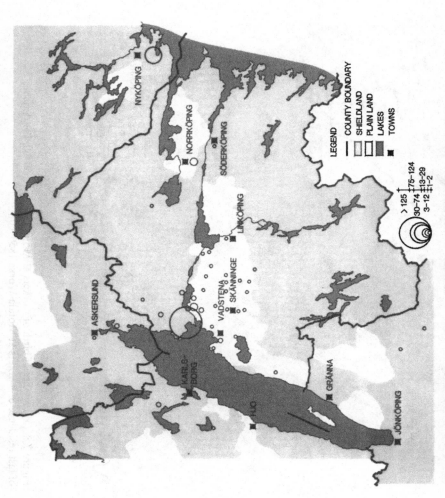

Figure 7.9. Parishes of next residence of workers at Motala Works 1850–60.

surmise, even though most of the places of next residence were actually in Motala Works itself (see figure 7.9), which provided housing and medical, social security and pensions schemes. Of the 25 per cent of the labourers who moved out, more than half went to other mechanical industries, while a slightly smaller group moved to areas similar to their places of last residence. Around 90 per cent of the labourers remained within industry. Motala Works thus functioned as a major industrial school, drawing its pupils mainly from areas where they would already have encountered non-agricultural production.

There were only fifty foremen and administrators at Motala Works. Because they had already to be highly qualified on arrival, we might expect them to come from places with similar industries or to have been trained at Motala itself. The places of birth pattern does not fit this picture. Almost 40 per cent of the foremen and administrators were born in towns, but only a few in places with mechanical industry (of which very little existed when the foremen were born). Local recruitment also occurred, although at 20 per cent it was much smaller than for the labourers. Moreover, few of this higher status group came from forested shieldland areas, and 54 per cent came from outside Östergötland. It seems that higher status employees were recruited about equally from the urban bourgeoisie and from labourers, though not to a great extent from Motala labourers. The places of last residence reinforce this picture but also show that some of this group did, in fact, move in from traditional ironworks communities.

Abbreviated life-paths were also reconstructed for everyone who worked between 1855 and 1860 at Katrinefors ironworks, Motala and Strömsnäs Breweries, and Duvedal Papermill. They give quite a different picture of worker recruitment than Motala Works, and there were also differences between them. Katrinefors drew its labour from parishes containing ironworks, presumably from the ironworks themselves. Only 46 per cent of its workers were born in Östergötland county, but 68 per cent were born in shieldland areas. Although at this stage no direct connection with proto-industrial families can be proved, the distribution of parishes of last residence shows that 85 per cent of the workers came to Katrinefors directly from ironworks communities, other shieldland areas, or the immediate vicinity of Motala.

The breweries had much stronger local connections. Of their workers 78 per cent were born in Östergötland and 88 per cent had lived in Östergötland immediately before coming to work at one of the breweries, implying a stepwise progression to Motala Town. More than half of the breweries' workers were born on the plains, mainly in Östergötland, and 7 per cent in towns. This reflects the tradition of non-industrial alcohol production, which was closely tied to grain farming or urban consumption. The persistence of old recruitment patterns into industrial brewing, and the fact that most tasks did not need much skill, brought in unqualified people from arable farming plus some skilled workers and foremen who had already worked in similar activities in towns.

The industrial paper mill seems to have had the most complex recruitment pattern apart from Motala Works, which in some ways it resembled. Skilled workers came from abroad and remote parts of Sweden (some from the traditional hand paper making areas in Småland), but others were from the local plains.

Differences between patterns of workers' origins and the destinations of those who moved provide a good index of the level of qualification and of the schooling process derived from employment in an industry (compare figures 7.8 and 7.9). Depending on the extent of skills acquired in childhood or the training received in a factory industry, out-migrants from new industrial areas could follow quite different patterns of migration and job-search. Few employees at Motala Works returned to the pattern of movement they exhibited before being employed there, which must mean that its schooling made workers almost completely dependent on the kind of work it provided. Not all industrial employment had this effect. Employees at Katrinefors, experienced in working at plants outside their homes before they came to Motala, had almost identical patterns of movement before and after working there. The brewery labourers were much more urbanized by industrial employment, whilst Duvedal Papermill employees stayed on to such an extent that it is impossible to say anything from the behaviour of the few who moved out. However, this itself is significant: its workers became much more stable, like those of Motala Works and unlike those of Katrinefors and the breweries. The varying transformative capacity of these industrial livelihood positions depended on the differing nature of the work done in them. Katrinefors Ironworks seems to have been an old-fashioned pre-industrial *bruk*, through which members of the traditional skilled and privileged smith community moved frequently in their customary way. The other Motala industries were newer-style factories which had to train most of their own workers in the necessary skills. In the process, workers became dependent on them for future industrial employment.

Evidence from full lifelines

Of the fifty-three industrial employees we sampled for the tracing of lifelines (see figure 7.10), thirty-five were from Motala Works and six or seven each from Katrinefors, one of the breweries and the Duvedal Papermill. The origins of workers, in terms not only of spatial position but also what fathers and mothers did for a living, are of particular interest: what had future industrial workers experienced over the years previous to their industrial debut, and can this explain why they moved from their home environment?

The spatial origins of the sampled Works' labourers have been grouped into

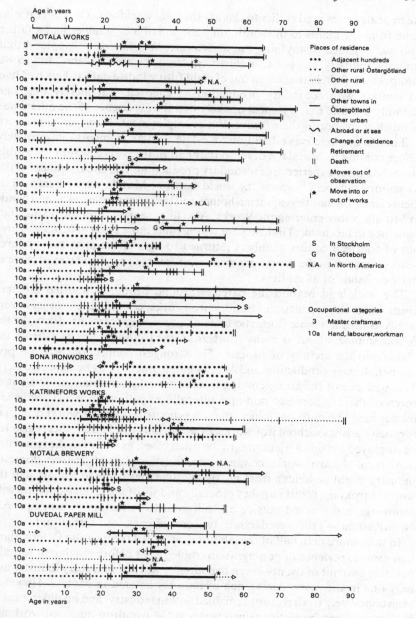

Figure 7.10. Lifelines of the cohort of employees at various works in Motala.

the five categories used earlier to classify the total workforce. Of the eight who came from the plains to the south, only the son of a Strå miller had a father who was engaged in anything but agriculture. We showed in chapter 4 that domestic industry was not entirely absent from the arable plains, and so we cannot say definitely that seven out of eight future industrial workers had had no contact with secondary production as children, but merely that one certainly did, and that it is most unlikely that the experience of the others was of much relevance to their future Works employment. However, three were mill hands and one was a hand on an estate with an ironworks immediately before coming to Motala Works, so that half of the workers from the arable plains had prior experience of secondary production. The circumstances of the six cohort members born on the shield north of Motala were similar: all of them came from landless agrarian homes and two were employed as mill hands in Motala before entering the Works. Also like the men from the plains, they remained in mechanical industry for a long time, most often in Motala. Three out of fourteen cohort members returned to agriculture after two or three years (contracts at the Works generally ran for three years) and stayed there as married hands or as crofters.

The shieldland beyond the parishes around Motala supported different kinds of craft production. Besides ironworks communities, others were involved in small-scale domestic craft or processing forest products. In the cohort sample, about as many (thirteen) came from these parts as from the plains and the vicinity of Motala. The strongest connection between pre-industrial mass production and Works employment is to be expected here. This was true of the group coming from ironworks communities: five out of seven of their fathers had non-agricultural occupations such as shoemaker, master hammersmith and joiner. But it was not true of the six workers born in forested parishes without ironworks, of whom only one had a father registered as employed outside agriculture. In the other cases, as well as in the remaining ones from the ironworks communities and the plain, part-time domestic industry might be hidden from us. By the time they came to Motala half the workers from the forest parishes generally and six of the seven from ones with ironworks had worked outside agriculture, two in factory industry proper. Notwithstanding these credentials, two moved back into agriculture.

In sum, more than half of the labourers recruited from within Östergötland had some experience of non-agrarian employment before moving to Motala, but only two out of twenty-seven had already worked in factory industry and only one in the particular branch of industry entered in Motala. However, even completely fresh recruits remained within industry, and no differences of background can be distinguished between the five drop-outs and workmen who stayed in Motala for thirty or forty years. The 15 per cent of the labour force that came from further away had obviously been trained elsewhere and were recruited to Motala because they were already skilled. Four of the five in the sampled cohort had prior experience of craft or industry, although apart from those from other ironworks communities their parental backgrounds were virtually identical to those of the locals, being almost totally agrarian.

As the largest engineering and shipbuilding enterprise in Sweden, and true to the intention of its founder, Motala Works was an important school for the Swedish heavy engineering industry in our period. Many of the locally recruited workmen who were trained as steam hammer- or puddle-smiths, as rolling mill operators or joiner-patternmakers, ended up in similar positions all over the country and as far away as Finland and Russia. We have already seen that the main flow from Motala Works went to other places where big mechanical works were developing. In the sample cohort, most of those who left Motala went to Stockholm,[35] Nyköping, Norrköping or Göteborg, all with engineering works owned by Motala or other strong business connections with it.

The differences in patterns of labour recruitment between Motala Works and other plant on Motala Ström revealed by the abbreviated life-paths are even more marked in the full life-lines. Four of the six Katrinefors workmen in the sample came from areas with ironworks; all of them stayed for only one or two years, before moving on to another ironworks in the neighbourhood, then another . . . Although Katrinefors was in scale more like a modern factory than the little ironworks scattered through the forests, its workers did not adopt the spatial stability characteristic of those at Motala Works. Like the abbreviated life-paths, the six workers' lifelines reconstructed for Strömsnäs Brewery all originated in agriculture on the plains in the vicinity of Motala. Before recruitment to Strömsnäs, only one had experienced anything but agricultural work, as a mill hand: the expertise required in the brewery was not very specialized and internal training was more important than previous experience. Most of the brewery workers remained, if not in brewing, in other secondary production. In this respect, they were more similar to those of Motala Works than those of Katrinefors. So, too, were the employees at Duvedal Papermill. Although recruitment was much less local, it was not of already skilled labour. Six of the seven cohort members had earlier experience of secondary production, one as a skilled paper maker's journeyman, two at Motala Works, one as a mill hand, and one as a carpenter in Vadstena. Their family backgrounds were agrarian in six cases, and their careers after the paper mill were no less homogeneous than those we have already seen: four stayed in paper making and all continued in some kind of secondary production. The average ten years of employment at the paper mill was long, as at Motala Works and the breweries. The industries recruiting fairly inexperienced labour could keep it for longer than that recruiting already skilled ironsmiths, who were well aware of the possibilities open to men with their trade.

Conclusion

It is clear, though perhaps surprising, that the patterns of population flows which underpinned the pre-industrial and apparently much less dynamic economic system that existed at the start of our period were little less intense and complicated than those related to the capitalist system of production that was emerging by the end of it. Even though the dramatic growth of industry on Motala Ström and the radical transformation of farming on the southern plain produced spates in the channels which carried people in and through the region, there was no major change in the shape of those channels to match that brought to capital flows by the development of banking. Indeed, we can recognize clear evolutionary steps between the patterns of population movement at the beginning and end of our period, none of which would have seemed particularly steep for most of those passing through them during the course of their life-paths.

Most people moved many times during their lives, tracing spatially complicated paths as they did so. We can recognize some characteristic paths which, in aggregate, sustained regularly patterned flows linking stations in the traditional peasant, urban and industrial economies. First, people without land or urban real estate coursed through the countryside from farm to farm, into town and out of it, moving only less often after marriage and family formation. They normally stayed in the subregion of their birth, although the aggregate distance moved during the course of their life could be very great. Secondly, and in the short term indistinguishable from this seething substratum, sons and daughters of peasant farmers usually left their family household to supplement the labour force of other family farms or of urban craftsmen's households before settling back into farming on inheritance and marriage. Of course, as farm population grew and as farms increased in size, an increasing proportion of this group would join with the first category and fail to complete the final settled step of the peasant sons' and daughters' life-paths. Thirdly, both these channels provided recruits to the urban economy where, as apprentices, they acquired a skill to fit them for craft positions. Their subsequent life-paths were very different from those of the previous two groups, as they moved over long distances from town to town to complete their training and search for a niche for a workshop of their own, which might well be back in the town where their training began. Fourthly, and somewhat distinct from the rest, the children of smiths at *bruk* in the shieldland had privileged access to the training their fathers could provide for succession to their protected and secure livelihoods. After it they moved frequently between *bruk* scattered widely through the forests before beginning, like peasant children, a rather less mobile lifestyle on marriage and family formation, perhaps at the *bruk* from which their father had retired. Although it is possible to generalize in this way, these patterns coalesced into a very intricate patternless whirl of movements.

The main change in this system by 1860 was the increase in all subregions of demand for labour which was more fully proletarianized than farmhands, journeymen and *bruk* smiths had been; labour which was related to employers more completely through the wage nexus and less closely through co-residence and the social ties binding the occupants of livelihood positions in one generation to their intended successors in the next. Parts of the demand for proletarian labour of all kinds was filled from areas where supply was richest, in the swiftly expanding wage labour-force of the increasingly capitalist farming system of the densely populated arable plains. This did not occur in the secondary sector only; recruitment was local in both town and country-side. It seems, too, that most movement to areas of high demand, not surprisingly, occurred in a stepwise fashion, just like the flows of the earlier period. At the same time, a proportion of the largely unskilled hands were absorbed into the expanding economies of Vadstena and Motala Town from where some moved off into a national labour market for craftsmen and industrial labour. Again, this pattern merely represents a continuation of that which had been characteristic of apprentices and journeymen in the earlier period.

Only younger people were willing or able to take this step from the primary sector to the secondary; although the number taking it increased massively as the economy was transformed, only youngish agricultural hands became 'green' industrial labour, or apprentices or hands in the town. Evicted crofters and peasants who had had to give up their farms found no alternative in the growth areas, either because of inherent conservatism, or because it was difficult (as it still is) for mature people to get gainful employment outside the sector where their skills were appropriate. The result was that part of the rural proletariat rapidly built up over the period was recruited from previously landed peasants and crofters, some of whom, particularly the older ones, tried to recreate their customary position abroad in the mass-emigration from Sweden to North America. In terms of position of origin, the base for all the flows of hands, labourers and apprentices, from the shieldland to the plains and to Motala, from the plains into Vadstena and Motala, and from more remote and less rapidly developing areas into our region, was the peasant and crofter groups and their already proletarianized offspring.

Finally, a much wider channel of movement developed by 1860 out of the traditional streams of smiths between *bruk*. There was a connection between areas with early rural ironworks, forest and proto-industry and expanding factory industry. Many of the new industrial workers represented in the abbreviated life-path data came from the shield, particularly from parishes with ironworks, and more than half of the industrial workers' life-paths began in or passed through plant or domestic manufacturing positions before their entry into factory industry. Even so, we must not forget that many workers were also recruited directly from agricultural tasks, in the same way as the urban hands and labourers: the connection between proto-industrial and early factory development was not so exclusive as is sometimes supposed.

Thus, a cross-sectional consideration of the patterns of population movement at the beginning and the end of our period would show stark contrasts. But these contrasts would not be present in the movements of the people whose life-paths spanned the transition. There was no sudden shift of migrational behaviour because the positions which were available at the end of the period as permanent sources of livelihood had had close equivalents at stages in the life cycle of the pre-industrial life-paths. Boys had always moved off the family farm or away from their home *bruk* in their youth; it was simply that by the end of the period many never returned. Moreover, their passage into a fully proletarian adulthood was blurred in all economic sectors. In agriculture the new married farmhand positions would normally involve hands living separately with their own families in farmhouses deserted as land was assembled into bigger holdings. Rural recruits into the urban proletariat still began their lives in the households of craftsmen with the title of apprentice and succeeded to that of journeyman. Migrants to full-time factory work from the small farms of the shieldland often began this apparently momentous transition by working full or part time at mills or in forest industries, whilst those from ironworking areas found strong echoes there of the security and paternalism of traditional *bruk*. In any event, almost everyone, from the plains or the shield, who made this change in lifestyle had experience from earliest childhood of domestic craft production in their own homes, whether for sale or for household self-provisioning.

Our analysis of the ways in which individual people moved through the economy as it was being transformed by capitalism has shown both how economic development affected people's lives and how their changing lives transformed the structure itself. Perhaps the longitudinal methods we have employed inevitably trace a very different picture from cross-sectional analysis and stage-wise theories of social and economic development. What appears clearly in historical hindsight as revolutionary change, and what was necessarily a product of massive aggregate changes in the eventual destinations of people's life-paths, meant no radical alteration of life-course for the vast majority of people as they launched themselves into it.

8

Conclusion: peasants and others in the development of capitalist economy

The cultivators of Aska and Dahl bore most of the distinguishing marks of peasantry at the beginning of the nineteenth century. Only 126 of their 810 farms were big enough to produce predominantly for the market, and most of them belonged to noblemen and persons of standing. The vast majority of peasants owned rather than rented their holdings, half of which were too small to provide more than a bare subsistence, if that. So were many crofts carved out of (or being used to extend) arable land and provide extra farm labour.

The livelihood positions supported by these stations were also typical of peasant economy. More than 80 per cent of the population lived on and from the farms. Of them, 61 per cent were members of families which worked farms or crofts; 29 per cent were non-family members of farm or croft households, and only 9 per cent lived in cottages without land. Occupancy of these positions was still closely related to stage in the life cycle. Farm families contained 44 per cent of the region's children. Crofters' children made up 43 per cent of that total, suggesting that crofters were more fecund and therefore younger on average than farmers. Only 4 per cent of children belonged to hands, maids and other non-members of farm families, many of whom were youngsters spending a spell off their parents' farm before returning to it directly or via a croft on marriage: only 17 per cent of living-in farmhands were married. Like most cottagers, non-family members of landed households were aged people, mainly single women, at the other extreme of the productive life cycle, sometimes destitute. Children were few, only 9 per cent of the total.

Farming projects were still diverse. Each holding had a few animals and grew a range of crops, and most were locked into a communal system of landholding and decision making, with strips scattered over open fields and access to commons. Even in the forested shieldland of northern Aska only 7 per cent of households were supported by non-agricultural pursuits, and most of those, including skilled forge-smiths' households, also farmed. Stores of cash passed down generations and between kinfolk and neighbours, marking and enabling the next stage in the life cycle of circulating hands and maids. The ancient port, administrative and marketing centre of Vadstena housed only 10 per cent of the population, many of whom farmed townland and holdings in nearby villages. The nearly 11,000 people living on 28 estates, 810 peasant farms, 710 crofts and 419 cottages supported only 94 specialist urban administrative, professional, dealing and craft positions – which by law were absent from the countryside.

All this is redolent of peasant economy. But things changed rapidly during the early nineteenth century: it was emphatically not true that 'in their ancient villages, time stood still; in their red moss grown cottages nothing happened which had not happened before; the children obeyed their parents and imitated them, and did the same things again which their parents had done before them'.[1] Everything became much more differentiated by 1860: in rural areas; between the countryside and the town, and between agriculture and industry. There were over a hundred more farms than in 1810 and the arable land had been extended by large-scale reclamation. But only the large farm size category had more land in 1860 than in 1810 and its share of the value of farmland increased from 47 per cent to 64 per cent. On average a category I farm was worth over 14,000 Rdr in 1860, possessed about 250 acres of land and produced a surplus of about 500 barrels of rye worth *c*. 2,500 Rdr per year. The number of crofts fell by 10 per cent, whereas the number of cottages more than doubled and the children they housed more than quadrupled. The number of living-in farmhands also doubled; the proportion of them who were married increased to over one-third, and the average number of children per married hand's wife increased from 0 to nearly 2. The average number of children of other women living-in with farm families increased from 0.2 to 0.6, whilst the total number of people living-in who were not hands, hands' wives or maids trebled. In 1810 only 11 per cent of rural children belonged to non-landed families; in 1860 43 per cent did so. The proletarianization of the farm labour force was not so neat and complete as is possible in theory (which would have been illegal), but these figures show that it got well under way between 1810 and 1860.

By 1860 most farmland was worked by people aiming to maximize their output of commodities for sale, including grain, potatoes, beef and milk products, but primarily liquor,[2] rather than goods for their own consumption. The engrossment of farms for this purpose was allied with the creation of integral holdings through enclosure, financed by heavy mortgages. Equipment, crops and labour-hiring practices were innovated to push the output of saleable commodities up and the margin between costs and receipts apart. The agents of these changes were individuated behaviourally as they became investors economically; separated from other farmers in their daily practices, and from the people who worked their land socially, spending earnings on 'Z goods'. Peasant farming was ceasing to be a way of life, and capitalist farming had become a means to the end of leaving the peasantry for the bourgeoisie.

Conclusion: the development of capitalist economy 383

In the process farmers spent more of their time and money in Vadstena, where the townsfolk withdrew almost completely from farming even their own townland. Although the occupational structure of the town was stable, changes in the relative wealth of different occupational groups, in the size and organization of businesses and households, in the status and origins of apprentices and journeymen, and in the town's topography and facilities, all suggest that the quickening capitalist pulse of farming redounded in and was facilitated by changes in the town. More obtrusively, and less directly connected with changes in peasant farming, Motala Town burst into growth along the Göta Canal at the junction of the plain and the shield. It was this independent industrial development, rather than multipliers from increasingly capitalist farming, which was responsible for the share of the region's population employed outside agriculture rising from 9 to 30 per cent between 1810 and 1860, and agriculture's share of its taxable wealth falling from 86 to 57 per cent.

However, it would be wrong to interpret what happened as the implantation of alien heavy industry into a region where subsistence farming was independently changing into capitalist enterprise. Maybe farming at the beginning of the period had, as in Sweden as a whole, approached natural economy. But as in peasantries generally, this implied an enormous amount of craft work for self-consumption in agrarian households. Where environmental conditions were propitious, as in the shieldland, peasant households could depend as much on secondary production as on subsistence farming. The symbiosis between industry and agriculture through labour was strongly reinforced by the operation of industrial plant by peasant millers and smiths, and especially by estate owners. The extent to which the heavy industry along Motala Ström was implanted into the region, and the extent to which it crystallized out of this traditional symbiosis is, in fact, a moot point.

The transformation, in agriculture as well as industry, was neither sudden nor wholesale. There were large capitalist farms in 1810 and small peasant farms in 1860; married hands and cottager labourers in 1810 and juvenile hands and maids destined after marriage for their own farms in 1860; 20 per cent of the population of one northern parish derived its livelihood outside agriculture in 1810. All statistical indicators of change moved slowly and incompletely. One reason for this was that the process itself was long drawn out, beginning before our period starts and continuing after it ends. Another reason is that the process of change inevitably gave rise to some equivocal statistical indicators, especially in terms of farm sizes because the search after maximum returns by capitalist erstwhile farmers and the continuation of peasantry could be symbiotic rather than mutually exclusive.[3] A further reason for the incompleteness of the changes indicated by our statistics is that although capitalist farming was becoming more pronounced in all parts of Aska and Dahl, its pace, direction and statistical impression were rather different in different subregions.

In some respects the statistical indicators of these differences are maddeningly equivocal: no subregion had all the strongest indicators of peasantry in 1810 or capitalist farming in 1860, and for no subregion do all the statistics change by equivalent amounts in the direction of capitalist farming over the period. The impression of a transition from peasantry to capitalist farming is strongest in Dahl, which contained the most fertile part of the plain and was nearest to Vadstena. In some respects, this appears to have been the most peasant-like subregion in 1810: married farmhands were much fewer there, and a higher proportion of its agrarian children belonged to farm and croft families, suggesting the smallest degree of labour proletarianization. The numbers of crofts, cottages, hands and maids per farm do not indicate that farming was differently organized in Dahl than in the other two subregions. However, already in 1810 large farms were relatively more numerous and farmed a higher proportion of the land in Dahl than in either part of Aska, and much less of its land belonged to farms in the smallest size category. Despite the equivocation of the statistics for 1810, all the changes between then and 1860 demonstrate a strong development of capitalist farming in Dahl. It is clear, too, that the peasantry itself was largely responsible for these changes: only in Dahl did the proportion of land in the hands of the peasantry increase over the period. Only in Dahl, too, did the total number of farms fall by 1860, when its large farms controlled the highest proportion of the land and its small farms the smallest in any subregion. The number of crofts also fell sharply, and the proportion of hands who were married rose from the lowest to the highest. The proportion of children belonging to landed families fell from the highest to the lowest, and the numbers of cottages, hands and maids per farm all increased much faster than elsewhere. Capitalist farming developed swiftly out of peasantry in Dahl, and by 1860 few peasant farms remained.

By then there was a marked contrast between Dahl and the other plainland subregion of southern Aska. In the latter the number of farms grew by 37 per cent and the proportion of small farms increased, although the amount of land they farmed fell. The number of crofts also increased, and the ratios of cottages, hands and maids per farm were less than half those of Dahl in 1860, whereas they had been the same in 1810. One-plot family holdings survived strongly in southern Aska. The expansion of estates over the period at the expense of peasant land in proportional terms, and the higher proportions of married farmhands and non-landed children in 1810 suggest that the estates and farms of nobles and persons of standing were responsible for most of the capitalist traits evident in southern Aska. In this subregion, which contained the vast majority of the region's farms and people, traditional peasantry survived strongly alongside more dynamic estate farming. The net effect was relatively little change towards capitalist agriculture.

The agricultural resources of the shieldland of northern Aska supported less than one-third as many farms as an equivalent area of plain. A much higher proportion was also in the hands of estates, which were larger on average than in either of the plainland regions, and their share of the land grew even more rapidly than in southern Aska, reaching 58 per cent by 1860. The forestry and haulage work provided to peasants by these large industrial/agricultural enterprises, the relative labour-intensiveness of forest pastoral farming combined with potato husbandry, the availability of resources for by-employments, and the fact that forest resources were undervalued for taxation compared with arable land, all combined to ensure the survival of a vast proportion of small peasant farms which required large amounts of labour. They comprised 70 per cent of northern Aska's farms in 1860; few farms belonging neither to estates nor persons of standing were in the largest size category. Despite this, the ratios of hands, maids and cottages to farms were higher than in Dahl, and that of crofts to farms was much higher. By 1860 the peasantry of southern Aska was still largely unaffected by capitalism; that of Dahl had transformed itself into agrarian capitalism, and that of northern Aska was intensifying through progressively greater dependence on industrial capitalism.

At this gross scale of spatial classification, fundamentally different processes do seem to have been occurring in different physical geographical regions: peasant ecotypes did exist. But they were neither homogeneous nor clearly marked-off from each other. On figures 8.1(a) and 8.1(b) we have delineated them in terms of parish-level data for four variables which give a reasonably clear impression of where a parish lay on the spectrum from subsistence peasantry to capitalist farming. On figure 8.1(a) parishes with lower than average scores on each of these variables, signifying a tendency towards peasant farming, are mapped. Parishes with much lower than average scores, and therefore the most extreme degree of peasantry, are distinguished by heavier shading. Only one parish (Strå in southern Aska) had very low scores on all four variables, signifying the most unequivocal existence of peasantry, and only eight had high scores on all four variables (and are therefore blank on the map), signifying the unequivocal existence of extremely capitalist farming relative to the situation in the region as a whole. Despite this degree of muddiness, the mapped scores are well patterned geographically. A large cluster of strongly capitalist parishes shows up clearly in Dahl, whilst estate-dominated Kristberg is distinctive in the north. The most peasant-like parishes were Västra Ny, furthest away from Motala Ström and the Göta Canal on the shieldland, and a swath of six adjacent parishes across the southern part of the plain in Aska. On figure 8.1(b) nine intermediate parishes stretched in bands across the middle of each hundred, but two separate areas of surviving peasant farming and two of capitalist farming show up quite clearly: estate-dominated capitalist farming in the extreme north-east; capitalist but erstwhile peasant farming in Dahl, fringed to the north by estate-dominated capitalist farming in St Per (Aska and Dahl); arable peasantry in the extreme south of Aska, and pastoral/rural-industrial peasantry in Västra Ny in the extreme north.

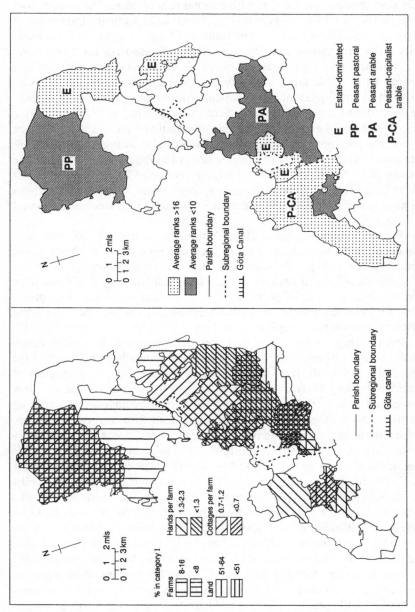

Figure 8.1. Peasant and capitalist farming parishes in Aska and Dahl in 1860: (a) Parishes with low proportions of land and farms in category I and low ratios of cottages and hands to farms; (b) Parishes ranking very high and very low on the sums of their ranks on the variables mapped on (a).

Parish data hide enormous variation on a smaller scale: arable villages existed in shieldland parishes, and forest patches survived in parishes on the most fertile part of the plain; the most peasant-like parishes contained large expanding farms of freeholders and persons of standing, often on land outside the ancient village open fields, and the most estate-dominated parishes had some small farms. Given this kaleidoscopic small-scale variation and the crudeness of the indexes used, the degree of geographical definition on the maps is quite remarkable. However, they do not show a simple and consistent relationship between the physical environment, the types of farming practised, and the ways they were progressing. Within each broad type of physical environment, two other factors were important, although again they did not have a uniform effect.

Firstly, land tenure. In both pastoral northern and arable southern Aska, parishes where freehold peasant land was predominant in 1810 saw the intensification of peasantry by 1860, whereas estate farming expanded in area and became more capitalist. Apart from in St Per, exactly the opposite happened in Dahl: capitalist farming developed directly out of peasant farming and freehold capitalist farming expanded at the expense of estates. Location relative to markets also had a strong effect, which again could yield opposite extremes. The most peasant-like parishes of southern Aska were no further from Vadstena than those in Dahl where peasant capitalism was burgeoning, and they lay on equally fertile plainland. Whereas nearness to Vadstena seems to have intensified either peasant or capitalist farming, closeness to Motala Town and the Göta Canal apparently did the opposite (see figure 8.1(b)). Perhaps farming was moving towards one extreme or the other there, too, but had had less time to change than around ancient Vadstena?

Different subregions were being propelled in different directions swiftly: our analysis spans only two generations. But none of the changes would have appeared to be severe to the people who lived through them. None required attitudes or behaviour which were particularly novel. Indeed, in a number of ways the more peasant-like the farming system was, the more it was suited to a quick and relatively smooth transition to capitalism. The expansion of holdings piecemeal when finances allowed was made easier by the existence of holdings which were heavily subdivided across plots scattered within and between the open fields of communal farmers, which meant that land could be traded in very small pieces. It was facilitated further by partible inheritance strategies coupled with the vagaries of family demography, which ensured that spare land was always available on some holdings and needed by others as they traversed generations. Certainly, partible inheritance was not, as is often suggested, a barrier to either the accumulation of large holdings across generations or the inculcation of individualistic acquisitive behaviour. It must have caused land to be split into numerous small easily affordable pieces, forced many of them onto the market, and driven peasants to seek the money needed to buy them by selling more commodities and/or borrowing.

Similarly, the existence of crofts, stemming from government insistence that all rural family households must, peasant-like, possess land, when first estates and then burgeoning peasant farms needed non-family labour, blurred the distinction between landholding and proletarian livelihoods. So did continuation of the peasant practice of evening out the labour surpluses and deficits of farms by circulating juvenile workers. Again, government insistence that all agrarian people must belong to landed households was complicit in this survival, ensuring that as more labour was required on larger farms it must be provided with at least the fiction of full subsistence plus accommodation. These labour contracts allowed farmhands to marry without land; whilst a few large farms were expanding at the expense of many small ones, they were less and less likely ever to get any. But farm expansion also made peasant farmsteads redundant and, because of its necessarily piecemeal nature, many of them were distant from the owners' home farms. Over one-quarter of married farmhands lived in one of these erstwhile farmsteads rather than with the farm family itself, providing their own subsistence as well as surpluses for the owner from the land attached to it, bringing up their children in households of which they were the ostensible head. The survival of peasant practices effectively disguised and mollified the effects of labour proletarianization.

The continued circulation of juveniles through livelihood positions off family farms, and of some married people through a number of crofts before settling on their own land, meant that peasants were used to moving about to work for others in the early stages of their life cycles. Many of their paths passed through mills, quarries and so on in the countryside, or through both apprentice and menial positions in their local town. At the beginning of the period, peasant sons provided most recruits to craft training: in adulthood, therefore, some peasant families had relatives in the local bourgeoisie. Many more peasant children occupied menial positions in town for a brief spell as they coursed around during their juvenile years. The speed of this circulation meant that many rural people spent some time in town, even though their stays were brief, which spread experience of urban marketing institutions and practices and bourgeois consumption patterns into the countryside. Because the distances covered by these movements were generally short this effect was heavily circumscribed spatially to within not much more than a dozen kilometres from the town. The retirement to rural retreats of wealthy urban widows, which was a feature of the early part of our period, took loan capital and, doubtless, urban consumer tastes, out into the countryside over a similarly restricted radius. The spatial mixing of bourgeoisie and peasantry was stirred further and spread much wider by the continuance of another aspect of peasant natural economy within the Swedish state structure unaltered into the nineteenth century. The part-payment of administrators, judges, church ministers, military officers and other state servants with the use of

Conclusion: the development of capitalist economy   389

farms on Crown land meant that few peasants, however remote their village, did not have a potential exemplar of capitalist farming in their midst, often operating amongst them in the open fields. Work by peasant sons as farmhands and daughters as maids in these bourgeois rural households just before setting up their own farm and home must have diffused non-peasant attributes into peasant economy and society.

Our findings about how the region changed are by no means self-evident in the sources we used. In fact, the taxation returns which were our major source have been used to yield quite different conclusions. The multiplication and shrinkage of plots of land, the rising difference between their market and taxation values, the fewness of living-out farm labourers, and the apparent stasis of the traditional urban economy indicated by population totals and occupational counts all suggest a predominance and continuation of peasantry. When rural population growth of 60 per cent in fifty years and increasingly stentorian government exhortations to modernize farming practices are added to this brew, they produce a strong whiff of peasant involution. This used to be the diagnosis of this period of Swedish history; besides being inherently plausible, it helped to explain the massive emigration and wholesale political and legal reforms which followed in the 1860s and 1870s.[4] However, the methods we used to collate and interrogate the sources show something very different in some parts of western Östergötland.

To dignify these ideas and methods with the title and terminology of time-geography might seem grandiloquent. In fact, we have used common parlance in preference to time-geographical terminology wherever possible; and surely, it might be asked, are not these aspects of the phenomena we have studied so obviously important that we would have studied them anyway, without reference to that literature? All we can say is that people who have studied these phenomena and analysed these sources without this methodological background have not done so: the tradition of historical geography has been to concentrate on elucidating source materials to let them 'speak for themselves', rather than using them to answer clearly specified a priori research questions.[5] Early on in our work, the need to discover more about the livelihood positions and life-paths of farmers, crofters and so on led us away from the taxation records with which we were intitially preoccupied to examine inventories and parish registers. Only research which is not source-material led can ramify in this way. More fundamentally, our deliberate intention to reconstruct stations as construed and defined in time-geography is what forced us to examine all the resources belonging to and activities carried out on farms, not just agriculture, and to bring together through nominal linkage all the land and other resources belonging to particular people. The concept of livelihood position forced us to examine everyone who was active in the countryside, rather than just members of analytically isolated social groups such as peasants or proletarians. The concept of life-path forced us to include all the stations, in agriculture and outside it, resource owning and proletarian, through which people might pass during the course of their lives. It was only because these things were done that we discovered the wide variety of subregional patterns of change and the gradual nature of the transition from peasant to capitalist economy which we believe are our most important findings.

The geographical variety shown by figure 8.1 demonstrates that no one of the models of the ways in which peasantry can change which we outlined in chapter 1 is appropriate to our region in the early nineteenth century. Of those processes, new feudalism is least so. Although there were some similarities between the traditional aristocratic estates on *frälse* land and the feudal estates of Eastern Europe, they were relatively few and their privileges had been greatly eroded before our period begins, and continued to be as it progresssed. The combination of a high proportion of land under noble estates with relatively rapid population growth on the north east of the plain along the southern shore of Lake Boren (see figure 8.2) might suggest that this model gets nearest to what was happening there, but this pair of parishes contained only 5 per cent of the region's total rural population. *A priori,* the secure tenure, political protection, and falling taxation of Swedish peasants would lead us to expect, from theories of peasant society, either the stable or involutionary version of peasant homeostasis. The first does seem to be appropriate to the six parishes in the bottom left-hand quadrant of the graph inset on figure 8.2, which ranked high on degree of peasantness in 1860 and low on rates of population growth between 1810 and 1860. They were generally remote from Vadstena and Motala Town. Isolation from ideas and economic forces coming from and through the towns seems to have allowed peasantry to continue to regulate population numbers and sustain its existence, as in theory it is strongly inclined to do. But this was so in parishes which contained only 16 per cent of the region's rural population in 1860, although it had been nearly 20 per cent in 1810.

Moreover, the demographic consequences of 'peace, inoculation and the potato' were not entirely absent from even those parishes. Although their population growth rates were low relative to that of the region as a whole, they nonetheless averaged 32 per cent over fifty years. This was dwarfed by the rate of over 75 per cent averaged by the parishes in the upper left-hand quadrant of the graph, whose high scores on peasant characteristics suggest that they were following the involutionary version of peasant homeostasis.[6] These parishes contained 45 per cent of the region's rural population in 1860, so that the most common experience in it does seem to fit Heckscher's thesis about what was happening in Sweden as a whole at the time. However, the exact nature of this process varied considerably over the long north–south spine of parishes which fall into this category. At the northern end in Motala and Västra Ny on the shield, colonization of forest land through potato cultivation was partly

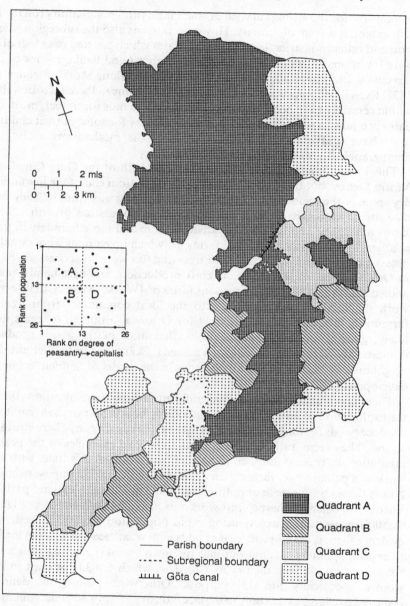

Figure 8.2. The relationship of rates of population growth, 1810–60 and the degree of peasantry or capitalist farming. Ranks along the horizontal scale are those used for figure 8.1(a).

responsible for the strong survival of small farms and population growth. To that extent, it was involutionary. However, this was also the subregion where iron and other industries, largely on the estates which covered over half of it, were traditionally important as providers of forest and haulage work to the peasantry, and where factory industry mushroomed along Motala Ström after 1832. Even if peasants were unlikely to work part-time in these factories, they would certainly have benefitted from the extra demands for timber, meat and dairy products, and the prospect of jobs in industry for some of their children must have weakened the force of some of the mechanisms leading to demographically stable homeostasis in peasantry.

This impact might well have spread to the south of the Göta Canal into Västra Stenby, but the group of parishes at the bottom end of the involutionary spine of the region in southern Aska were a long way from it. They were also at the extremes of both peasantness and population growth.[7] Small farms multiplied and land was subdivided more and more minutely in these southern Aska parishes. Of those farms for which inventories were sampled, bees, chickens, geese and goats, fruit trees and flax were most common by far in Orlunda, and so was domestic craft production, amongst hands, maids and cottagers as well as small peasant farmers. (Weak) evidence that textiles work might have been put out into the local community from wealthy farming homesteads also exists only for Orlunda, although it seems most likely, given Swedish law and custom, that most of the goods produced domestically were taken for sale in Vadstena, Skänninge or further afield by neighbours or kinfolk of their makers, as in the model of 'hidden' economic development.

There is no reliable evidence of significant proto-industrialization (that is, the exploitation of farm and cottage labour by bourgeois merchant capital to produce goods for long-distance sale) in these parishes, or anywhere else in the region. They seem, rather, to have been archetypal examples of the peasant involutionary process ascribed to Sweden as a whole at this time, with craft work as a palliative of, rather than a release from, agrarian impoverishment. Nor is there any evidence of putting-out in the northern shieldland parish of Västra Ny, where domestic craft work was also prevalent amongst a largely small-farm peasantry experiencing rapid population growth. Indeed, what evidence there is of this practice is at odds with what received wisdom leads us to expect. Spinning was put out into the town and surrounding countryside from the damask linen factory in Vadstena, which taught the craft to rural women specifically for this purpose. Otherwise, putting-out definitely occurred to a large extent only in the lace industry *within* Vadstena, spilling out into nearby rural parishes such as Örberga during our period. It was in plainland capitalist-farming Örberga, too, that large-scale production of *vadmal* cloth occurred, together with leather crafts and debts to urban merchants and tanners. However, leather working in particular was a feature of landless households, and seems to be an example of craft development due to urban capital battening onto proletarian households in a captalist farming area, rather than directly out of peasantry.[8] The later stages of the process of rural craft development were no more in line with the proto-industrial model than its inception: far from spawning factory industry, the urban linen works which brought much of it into being closed before our period ended.

It would be stretching the remit of terminology beyond the limit of its elasticity to claim that proto-industrialization was occurring anywhere in our region.[9] But homecrafts were important supplements to income in both the peasant and capitalist parishes where population was growing fastest. In southern Aska they were apparently practised by everyone – persons of standing with large farms, peasant farmers, crofters, cottagers and hands – as the only supplement to agriculture. In Västra Ny and other shieldland parishes homecrafts were equally ubiquitous, but probably less important than forest work for industries. Most of the parishes on the Dahl plain fell into the bottom right-hand quadrant of figure 8.2. Their peasantry had progressed to capitalist farming, and their average rate of population growth was barely 30 per cent, the lowest in the region. However, the parishes of Örberga and Rogslösa sit in the upper right-hand quadrant of the diagram, with equally capitalist farming to the rest of Dahl, but the sixth and seventh fastest-growing populations in the whole region. Domestic craft work, perhaps with putting-out or credit from Vadstena, was certainly common there. Moreover, around their churches and scattered along the post-road running through them, these two parishes contained rapidly emerging small village clusters of artisans, who served the increasingly professionalized farmers of the vicinity, following a pattern evident in arable England at the same time.[10] Perhaps, taking the parallel with England further, they were 'open parishes' into which the surplus and seasonal labour of neighbouring smaller, more integral and easily controlled parishes was decanted. This, rather than involuting peasant farming, was the matrix within which home crafts developed most strongly on the arable plain of Dahl.

Curiously, the industry which most nearly matched the specifications of proto-industrialization, growing out of a peasant home craft for domestic consumption, moving into larger-scale domestic production for long-distance markets, then into local factories, was one which is never considered in this light: distilling on the arable farms of the plain. But this was not really proto-industialization, either: it was associated with an increased commitment to commercial farming and expanding farm sizes, rather than their erosion, and with the control of population growth rates, rather than their explosion. Distilling is much better considered as a way in which the peasants of the plain satisfied their appetite for 'Z goods' and their aspiration after upward social mobility for their children, propelling the peasantry into capitalist farming and estate owners up the social echelon to which they already belonged. This did not come through the control of tenants by landlords. In Dahl it emerged amongst freeholding peasants, unchivvied by landlords; on estates it was the product of direct farming, not the manipulation of terms of tenancy, mainly by bourgeois persons of standing rather than traditional landowners.

Whilst the 'Z goods' model combined with a variant of the hidden economy model are appropriate to developments on the plains, the latter and what we might term the 'heavy industry' model of peasant transition best matched what happened on the shield. The peasantry of northern Aska were traditionally heavily involved in industrial by-employment, but by no means all of it was 'hidden' in their own nominally farming homesteads. Much of it was in large-scale plant owned and operated by ironworks estates, or in forest work or haulage for them. It was out of this non-peasant economic matrix that industrialization occurred after 1832. However conducive the customary patterns of land ownership, population growth, economic experience and social behaviour of the shieldland peasantry were to the emergence of heavy industry (and the significance of this should not be ignored), it was nonetheless from amongst local estate owners and entrepreneurs further afield that the impetus for heavy industrialization along Motala Ström and the Göta Canal came.

It is quite obvious that had we restricted ourselves to the study of peasant farmers themselves, we would have discovered little about the reasons for the ways in which their farming changed: on the ground itself, they lived and worked alongside and, at certain stages of their lives, for non-peasants; what they conceived of and did, or could not conceive of and do, was not innate within a hermetic society of their own but contingent on what other parts of society were conceiving of and doing. The cultural ether produced in other parts of society was channelled into the peasantry through churches and village meetings, and the political institutions through which they interacted with others bound them closely together. To that extent, our work bears out our most general introductory point: that peasantry and the way it changed cannot be understood separately from the other parts of society with which it co-existed. Each of the models of change in peasant society outlined in chapter 1 is, too, applicable to some extent somewhere in Aska and Dahl in the early nineteenth century; but none is appropriate to more than part of the region, or matches closely with any particular part. However, we have not been testing them: they were used as signposts to where empirical work might lead, rather than as targets for verification or refutation. Our objective was to make discoveries on an empirical journey, not to repaint the direction indicators we left at the start.

The discoveries we made were heavily dependent on where we started from. Having set out to seek geographical patterns and process in the transition from peasant to capitalist farming, it is unsurprising that we found them. Indeed, given the heavy variegation of these patterns, their unclear definition, and the inconsistency with which different variables were related within them, it might be argued that to claim that we did find geographical patterns comes from mesmerization by the signpost we saw at the start, rather than a clear sight of what we found. This would be to misconstrue what the signpost actually said: regional processes do not necessarily produce a series of sharply juxtaposed homogeneous slabs of territory. Indeed, they are hardly likely to do so because of the random effects of processess operating at a smaller scale, and because a number of regional processes operate simultaneously at different scales, each to some extent confounding the outcome of the others.

At the largest scale, agriculture and industry in Aska and Dahl as a whole, minus some of their marginal inland parts, shared the common experience of change stimulated by sudden and cheap linkage into a wider oecumene through the Göta Canal, which made it uniformly atypical of most of Sweden at the time. Development generated many flows of goods, people and capital within the region as what had been complicated but relatively autarkic peasant and estate economies in a heavily variegated physical environment became more specialized. The progressively diffentiating parts were bound together more closely through Vadstena and Motala Town and by inter-sectoral flows of materials, labour and capital which bypassed them. The greater concentration of the plains on arable farming, the shield on pastoralism supplemented by forest work, homecrafts and heavy industry, and the town on tertiary activites occurred as much because these complementary reciprocations were possible within the region as because of what was inherent in any particular part.

What happened to the parts at this subregional scale depended on the joint outcome of three variables, each of which was different in spatial patterning. The nature of the physical environment changed categorically from place to place: sharply along the fault bounding the plain and shield, but more gradually through the northern part of southern Aska. Distance from Vadstena and Motala Town varied continuously over space: steeply in this era of expensive land haulage, and at spatially highly irregular rates reflecting nearness to transportation routes. Land tenure varied more randomly and with a less clear pattern, although there was a general progression from estates in the north to peasant freehold in the south. The joint operation of these three different variables produced a pattern of recognizable subregions because they were not completely independent of each other. Physical geography influenced the other two: medieval Vadstena's location on the shore of Lake Vättern in the most fertile part of the plain was not fortuitous, and because they were created in this region after peasants had settled the best arable land, nor was the distribution of estates. Different subregional processes operating simultaneously, each separately intelligible, worked to produce subregional patterns, despite the confounding effect of smaller scale processes, and despite the fact that if they had been fully independent each might well have operated to mask the effects of the others.

The ghost at this geographical feast was Sweden itself, visible in all the scene-setting for our empirical work. Although our region did not develop because it was being carried along by a national economic entity growing as a whole, none of our findings about how peasant farming was changing and industry was developing would be intelligible without reference to the distinctive nature of the Swedish polity, Swedish social institutions and Swedish culture at the time. Ethnic homogeneity, the origin of the state in a war of liberation from Denmark in which peasants played a crucial role, and its dependence thenceforth on taxes and military conscripts from the peasantry, meant that the Swedish peasantry was much less different culturally, and more fully incorporated into the state structure than was usual elsewhere. The creation of an established national Lutheran church heightened this effect, as did the long survival of natural economic forms of remuneration in the administration of the state. Sweden's role in the world economy of the time, heavily dependent on the export of armaments and iron in other forms, combined with these influences and mercantilist dirigism to create a historical–geographical aspic in which peasantry was remarkably well preserved.[11] However, changes in the world economy during the Napoleonic era and the important role of the peasantry in the coup d'état of Gustav III meant that the mercantilist and other legal impediments to the development of capitalist farming were quickly and progressively removed: the bonds were loosed from a free peasantry which, although remarkably similar to the archetype of peasantry in its economic practices, was not culturally very dissimilar from the bourgeoisie. This national cultural and political context lay behind all the changes which occurred in the particular region we studied, and without reference to it the reasons for those changes are not intelligible.

Peasants everywhere do share many fundamental economic, social and cultural characteristics. But they do not exist in completely agrarian economies; their farming interacts with other aspects of their own household economies, and with the industries and urban activities of non-peasants scattered amongst them. This given, it is differences across environmental, economic and political spaces which account for how peasantries change in interaction amongst themselves and with others. Peasantry, in stasis or change, is necessarily a geographical process.

Notes

Preface

1 These sources are discussed in chapter 3.
2 Our preliminary plan is presented in Hoppe and Langton, *Countryside*.
3 During the last decade or so the beginning of development in Sweden has been pushed back in time. Winberg, Another route; Fridlizius, Agricultural productivity.
4 Arguments to support this view are set out in Langton and Hoppe, *Town*.
5 Pollard, *Conquest*; Jones, *Miracle*.
6 Gilbert, Regional geography; Pudup, Arguments; Cooke, *Modernity*; Entrikin, *Betweenness*; Soja, *Postmodern*; Johnston, Hauer and Hoekveld, *Regional geography*.
7 Entrikin, Issues; Giddens, Time; Gregory and Urry, *Social relations*; Soja, Spatiality; Marshall, Study.
8 For some historians this is a disadvantage. Coleman, Proto-industrialization; Houston and Snell, Proto-industrialization.

1 Introduction: peasants and economic development

1 Scott, *Moral economy*.
2 Ladurie, Peasants.
3 Shanin, *Defining peasants*, 229.
4 Wolf, *Europe*.
5 North and Thomas, *Rise*.
6 Blum, *End*, 5.
7 Lenin, *Development*; Chayanov, *Theory*.
8 Friedmann, World market; Friedmann, Economic analysis.
9 Macfarlane, Peasants.
10 Macfarlane, *Origins*.
11 Wallerstein, *World-System*; Massey, *Spatial divisions*; Harvey, *Condition*; Cooke, *Future*.
12 Wrigley, Corn yields, 107.
13 Crisenoy, Capitalism, 9.
14 Wrigley, Corn yields.

15 Galeski, Problems.
16 Shanin, *Peasants*, 1st edn 14–17; 2nd edn, 2–8; Scott, *Moral economy*; Wolf, *Peasants*.
17 Berger, *Pig earth*, 196.
18 Wolf, *Peasants*; Scott, *Moral economy*.
19 Ladurie, *Peasants*.
20 Berger, *Pig earth*; Pfeifer, Quality.
21 McCloskey, Economics; Scott, *Moral economy*.
22 In parts of nineteeth-century Ireland linen was made from flax crops to provide cash income for rent; elsewhere 'all the spring rents were made . . . of the pigs. The autumn rents would either be made of butter, or a little corn if they had it'. Beames, *Peasants*, 4.
23 Variations in the produce of peasant holdings were far less than differences within other orders of society. Blum, *Old order*, 4.
24 McGuire and Netting, Leveling.
25 Wolf, *Peasants*. In Ireland a priest 'charged between 10s and 3 guineas for marriages, according to the wealth of the farmer involved . . . the poorest paid nothing'. Beames, *Peasants*, 10.
26 Banaji, *Agrarian question*.
27 Seavoy, *Famine*; Boserup, *Conditions*.
28 Chayanov, *Peasant economy*, 108–9.
29 Wolf, *Peasants*.
30 Foster, Peasant society.
31 Chayanov, *Peasant economy*, 148–9.
32 Creighton, Family.
33 Hammel, The Zadruga.
34 Macfarlane, *Origins*, uses the existence of primogeniture and stem families to indicate that peasantry did not exist.
35 Goody, Inheritance.
36 Berkner, Stem family.
37 Chayanov, *Peasant economy*.
38 Goody, Inheritance.
39 Hanssen, Commonfolk, 94.
40 Reed, The peasantry; Reed, Nineteenth-century.
41 Blum, *Servitude*.
42 Hanssen, Commonfolk.
43 Bailey, Peasant view, 304.
44 Harvey, Money.
45 To pay a pedlar more than his trinket could be bought for elsewhere was to refuse to recognize his specialist dealing role in relation to the community. Hanssen, Commonfolk.
46 Remote, intermittently used, upland grazing and forests were often held in common. In Southern Europe so were chestnut groves, in Mexico surface water, and in Sweden coastal fishing grounds. Ladurie, *Languedoc*; Cleary, Patterns; P. M. Jones, Parish; Sheridan, *Dove*; Löfgren, Ecotypes.
47 The allocation of common fisheries randomly by lot to enable their private use at the appropriate season had the same effect through spreading good and bad luck fairly. Löfgren, Ecotypes.

48 Anderson, *Europe*.
49 Redfield, *Little community*.
50 Dobrowolski, Peasant culture.
51 Bailey, Peasant view, 303.
52 Berger, *Pig earth*, 202.
53 Popkin, *Rational peasant*; Kahn, Peasant ideologies.
54 Berger, *Pig earth*, 204–5.
55 Redfield, *Peasant society*, 23–39.
56 Crone, *Pre-industrial societies*, 15–16.
57 Wolf, *Peasants*.
58 Blum, *Old order*, 29.
59 In some Germanic states only the clergy and nobles were represented. Myers, *Parliaments*.
60 Wilkinson, Traditional concepts.
61 Kula, *Economic theory*.
62 Vries, *European economy*.
63 Kula, *Economic theory*.
64 Blum, *Old order*, 73–4.
65 More than half went in dues to lords with *dominium directum* over land, nearly 30 per cent to the state in taxes, and nearly 15 per cent to the church in tithes. Blum, *Old order*, 77.
66 Wrigley, *Population*, 108 and 111
67 These demographic controls would not apply if, as in Eastern Europe, labour dues as heavy as 360 days were required from peasant holdings, child labour was commandeered from them and paid by lords, and armies were manned by peasant conscripts. Blum, *Old order*, 54–7.
68 For the origin of this trait in Western Europe see Goody, *Development*.
69 Pre-mortem transfers might be ratified in contracts (known as *undantag* in Sweden) stipulating exactly the support to be given to parents. Whether they were enforceable was another matter. Zola, *Earth*.
70 Segalen, *Love*.
71 Wrigley, *Population*.
72 Peasant offspring were not pampered: in late eighteenth-century Sweden parents refused free medicine to spare their children a future which was 'miserable, devoted to the gain of others and slavish'. Gaunt, Pre-industrial, 185.
73 Ladurie, Peasants.
74 Salaman, *Potato*, 273–89; Beames, *Peasants*, 2–3.
75 Some 120 tenant farms in an area of Ireland became 160 in only four years through the subdivision of holdings. Conacre ground was leased by farmers for one season to people without their own land to grow potatoes. The proportion of the population producing its own subsistence increased, and early nineteenth-century Ireland had no full-time labouring population. Beames, *Peasants*, 2–9.
76 Wolf, *Peasants*, 20.
77 Arensberg, Old World.
78 Braudel, *Mediterranean*, 25–137.
79 Boserup, *Conditions*; Wolf, *Peasants*.
80 Arensberg, Old World; Anderson, Traditional Europe; Blum, *Old order*; Blum, Internal structure; Creighton, Family; Jones, Parish.

81 Braudel, *Mediterranean*; Cole, Anthopology; Netting, *Balancing*.
82 Anderson, Traditional Europe; Creighton, Family; Jones, Parish; Thirsk, Seventeenth-century; Underdown, *Revel*; Löfgren, Ecotypes.
83 Hefner, Culture problem, 555.
84 Homans, Explanation, 27; Kussmaul, *General view*.
85 Arensberg, Old World.
86 Stocklund, Succession.
87 Hefner, Culture problem; Orlove, Ecological anthropology; Redfield, *Communities*.
88 Harriss, *Capitalism*, 15.
89 Geertz, *Involution*, 12–37.
90 Homans, Explanation, 27.
91 Cole and Wolf, *Frontier*.
92 Cole, Anthropology.
93 Löfgren, Ecotypes, 101.
94 Arensberg, Old World; Langton and Hoppe, *Town*.
95 Hefner, Culture problem, 552.
96 Cole, Anthropology, 273.
97 For example, Braun, *Industrialisation*; Musgrove, *The North*, 75–91.
98 Cleary, Transhumance; Prince, Contrasts.
99 Nash, Markets, 162.
100 Wolf, *Peasants*, 41–50.
101 Kula, *Theory*.
102 Hodges, *Markets*, 43–6.
103 Wolf, *Peasants*; Blum, *Old order*.
104 Mintz, *Pratik*; Hanssen, Commonfolk.
105 Wolf, *Peasants*, 46.
106 Fairs were often outside towns and the control of their normal marketing institutions.
107 Hanssen, Commonfolk; Lucassen, *Migrant labour*, 88–92.
108 Wolf, *Peasants*, 46.
109 Lucassen, *Migrant labour*.
110 Löfgren, Historical perspectives.
111 Figes, Danilov, 123.
112 Banaji, *Agrarian question*, 3.
113 Chayanov, *Peasant economy*.
114 Arensberg, Old World; Anderson, Traditional Europe; Redfield, *Community*.
115 Blum, Nobility. Sometimes, as in France north of the Loire, the exemption attached to particular persons and in consequence whatever land they happened to own, and sometimes, as in France south of the Loire (and Sweden), to particular pieces of land, which could not be owned by non-nobles. Vries, *Economy*, 200.
116 Vries, *Economy*, 33–4.
117 Hilton, Introduction.
118 Our account of Eastern European serfdom and estate production is based on Blum, *Old order*; Blum, Servitude; Kula, *Theory*, and Pounds, *Europe*, 163–6.
119 In the early sixteenth century 140,000 tons of grain, chiefly rye, were exported annually through the Danish Sounds, and 60,000 head of cattle were sent

westwards from Denmark alone, with large supplements from Lithuania and Hungary. Vries, *Economy*, 45.

120 Feeny *et al.*, Tragedy.
121 Geertz, *Involution*.
122 Blum, *Old order*; Blum, *Forgotten past*; Pounds, *Europe*.
123 Goody, Introduction.
124 Kula, *Theory*.
125 Creighton, Family.
126 Hardin, Tragedy; Feeny *et al.*, Tragedy.
127 Dahlman, *Open fields*.
128 Østerud, *Structures*.
129 Brenner, Capitalism, 49.
130 Thirsk, Inheritance.
131 Specification of rents and tithes as set proportions of specified goods in kind gave landlords and the church a strong reason to keep the structure of peasant output unchanged, preventing the innovation of new crops. Share-cropping, common in Italy and Southern France, had the same effect. Vries, *Economy*, 41; Pounds, *Europe*, 171–2.
132 Aston and Philpin, *Brenner*; Butlin, *Transformation*; Glennie, Search; Hoyle, Tenure.
133 The effect on marketable surpluses was dramatic. A yield:seed ratio of 4:1 was common in Europe from medieval times to the nineteenth century, and 5:1 was exceptionally high. In England, 6:1 was achieved by the early seventeenth century and 10:1 by the late eighteenth. 'When the yield doubles, the persons fed nearly triples, and the nonagricultural population that could then be supported increases sevenfold'. Vries, *Economy*, 35–6.
134 A similar landlord–tenant system existed in Catalonia and parts of Northern France. Vries, *Economy*.
135 Kula, *Theory*.
136 Tribe, *Genealogies*, 71–93; Wilmot, *Business*.
137 Kula, *Theory*; Dodgshon, *Past*.
138 Beames, *Peasants*, 11.
139 Arensberg, Old World. As land fell into the hands of absentee landowners in Northern Italy in the seventeeth and eighteenth centuries advanced commercial agriculture was transformed into subsistence farming by share-cropping tenants on small holdings. Vries, *Economy*, 54–5; Pounds, *Europe*, 171–2.
140 Price, *Modernization*, 22.
141 Todd, *France*, 62–6.
142 Stone, *Crisis*; Mingay, *Gentry*.
143 Martin, *Feudalism*; Parker, Sovereignty.
144 Wrigley, *Continuity*.
145 Ranging from 10 per cent in France and 15 per cent in Italy as a whole through 25 per cent in Picardy and Laonnais to 56 per cent in Bavaria. Pounds, *Europe*, 161.
146 Campbell, Land.
147 Mingay, *Gentry*, 44–50.
148 Ridgeway, Mentalités, 177.
149 Goody, *Family*.

150 Ridgeway, Mentalités, 178.
151 Thirsk, Seventeenth-century agriculture; Ahlberger, Home industry.
152 Hanssen, Commonfolk, 97.
153 Vries, *Economy*, 32.
154 Hilton, Small town, 87.
155 Langton and Hoppe, *Town*, 6–13; Verhulst, Origins.
156 Jones, *Miracle*, 85–6.
157 Blum, *Old order*.
158 Kula, *Theory*; Johnson, *Peasant*.
159 Ports and capital cities 'very nearly monopolized the considerable urbanization' in Europe, 1600–1750. Vries, Patterns, 101; Dodgshon, *Past*; Vries, *European urbanization*.
160 Mui and Mui, *Shops*.
161 Shammas, *Consumer*, 17–51.
162 Howell, Stability.
163 Thirsk, *Economic policy*.
164 Polanyi, *Transformation*.
165 Hymer and Resnick, Model; Resnick, Decline; Vries, Peasant demand.
166 Tucker, *Instructions*, 245–6. The house of an English 'Peasant or Mechanic' contained three times the value of such wares in the mid-eighteenth century as in continental Europe.
167 Mui and Mui, *Shops*; Shammas, *Consumer*; Glennie, Industry.
168 Vries, *Economy*.
169 When English figures are removed from European totals, 'urbanization in north and west Europe recedes rather than advancing'. Wrigley, Urban growth, 176.
170 Shammas, *Consumer*, 7; Vries, *Economy*; Vries, *Urbanization*; Wrigley, Urban growth.
171 Hymer and Resnick, Model.
172 Vries, Peasant demand, 225.
173 Vries, *Urbanization*. The cities which became the *Randstad* contained as many people as London or Paris in 1650.
174 Vries, *Economy*, 95.
175 Mendels, Proto-industrialization; Butlin, Early industrialization; Hudson, Regional perspective; Berg, *Age*, 69–91; Vries, *Economy*; Langton and Hoppe, *Town*; Pollard, *Conquest*; Jones, *Miracle*; Kriedte, Medick and Schlumbohm, *Industrialization*; Berg, Hudson and Sonenscher, *Manufacturing*; Gregory, New; Thirsk, Industries.
176 Pollard, *Conquest*, 74; Jones, Environment; Jones, *Miracle*, 3–21.
177 Thirsk, Horn.
178 Vries, Peasant demand; Mendels, Agriculture.
179 Defoe, *Tour*, 465–8; Tupling, *Rossendale*; Swain, *Industry*.
180 Thirsk, Seventeenth-century; Langton and Hoppe, *Town*.
181 Court, Industrial organization, 241; Freeman, *Perspective*.
182 Mendels, Social mobility; Medick, Proto-industrial family; Medick, Structures; Levine, *Family formation*.
183 Tucker, *Instructions*, 245; Hudson, Proto-industrialization.
184 Mendels, Proto-industrialisation. Although the ever-cheapening supply of

domestic industrial labour made it competitive with factory industry well into the nineteenth century. Gregory, New, 355.

185 Britain still produced about 70 per cent of Europe's coal in 1840. Pounds, *Europe*, 334.

186 Access to commons was vital to households engaged in domestic industry in a number of English regions. Hudson, *Regions*, 21.

187 For example, Bettey, *Rural life*, 40–1.

188 Pfister, Appendix, 192.

189 Berg, *Manufactures*, 129–58; Hudson, *Regions*, 25–7; Gregory, New, 353.

190 Braun, Industrialisation; Gullickson, *Spinners*; Snell, *Annals*, 15–66.

191 Vries, *Economy*, 105–10.

192 Langton, Industry.

193 Kriedte, Medick and Sclumbohm, *Industrialisation*.

194 Braun, *Industrialisation*.

195 Franklin, *European peasantry*.

196 Johnson, Iron industry; Hyde, *Iron industry*.

197 Langton, Landowners.

198 Hyde, *Iron industry*, 9–10.

199 Ashton and Sykes, *Coal indus*try; Langton, *Geographical change*.

200 Pounds, *Europe*, 249.

201 Wolf, *Peasants*, 11.

202 Anderson, *Lineages*, 408.

203 Serfdom was legally abolished in Switzerland in 1525, but survived in Savoy until 1792, in Denmark until 1804, in Prussia until 1807, in Bavaria until 1808, in parts of the Austro-Hungarian Empire until 1848 and in other parts until 1861, in Russia until 1861 and in Romania and Poland until 1864. This reform was usually accompanied by the introduction of representative parliaments instead of assemblies of Estates. Anderson, *Lineages*, 24; Pounds, *Europe*, 167; Blum, *Old order*, 383–8.

204 Colinvaux, *Fates*; Anderson, *Lineages*; Coss, Feudalism.

205 In the sixteenth century there were only twenty-five years without large-scale military operations in Europe, and in the seventeenth century seven years. Anderson, *Lineages*, 33.

206 Myers, *Parliaments*, 9.

207 Except in the Swiss Confederation after it gained complete independence from Austria in 1393, in Britain after the Civil War of 1642–49, and in France after the Revolution of 1789.

208 Anderson, *Lineages*, 25–9. Roman Law's distinction between civil and public spheres in *jus* and *lex* allowed the principles of unconditional rights in private property and absolute royal power to co-exist, just as its distinction between property and possession allowed *dominium directum* and *dominium utile* over the same piece of land by different people.

209 Vries, *Economy*, 22.

210 Unwin, *Industrial organization*, 177–8.

211 Colinvaux, *Fates*, 160–2.

212 French military expenditure consumed over half the budget even in years of peace, and public debt was twice the state's annual income in the early eighteenth

century. In Holland it rose from 1 million to 140 million guilders 1579–1655. Colinvaux, *Fates*, 203.
213 Martin, *Feudalism*, 114.
214 Corrigan and Sayer, *Great arch*; Dodgshon, Space.
215 'The rise of protection . . . was the accidental consequence of taxation for revenue'. Thomas and McCloskey, Overseas trade, 93; Unwin, *Organisation*, 172–95.
216 Tucker, *Instructions*, 252.
217 The Statute of Monopolies (1624) limited the English Crown's ability to grant exclusive rights to individuals or companies through letters patent. Remnants of the system were used to allow private intellectual property rights to be held by inventors of new techniques. McLeod, *Inventing*, 10–39.
218 Thirsk, *Policies*, 133. The French Crown retained power to establish monopoly companies until 1789.

2 Conceptualizing regional change

1 Thompson, *Poverty*, 276.
2 The reconstruction of 'period pictures' linked by studies of 'vertical themes' was for a long time the conventional method of historical geography. Smith, Historical geography; Langton, Two traditions.
3 Samuelson, Analysis, 375.
4 Hägerstrand, Tidsanvändning; Hägerstrand, Diorama; Hägerstrand, Tidsgeografi.
5 Carlstein, *Time*; Pred, Of paths; Gregory, Time-geography.
6 Pred, Of paths; Carlstein, Sociology; Giddens, Time; Cloke, Philo and Sadler, *Human geography*, 93–131; Gregson, On duality.
7 Hägerstrand, Domain, 87.
8 Parkes and Thrift, *Time*; Sayer, Explanation; Törnqvist, On arenas.
9 Hägerstrand, Geography.
10 Hägerstrand, Survival, 9.
11 Hägerstrand, Tidsanvändning; Hägerstrand, Ecology; Hägerstrand, Survival.
12 Carlstein, *Time*; Hägerstrand, Tidsanvändning; Hägerstrand, People; Hägerstrand, Ecology; Hägerstrand, Survival.
13 Pred, Choreography.
14 Carlstein, *Time*; Hägerstrand, Tidsanvändning; Hägerstrand, Ecology; Parkes and Thrift, *Time*.
15 Hägerstrand, Tidsanvändning; Hägerstrand, People.
16 Hägerstrand, Geography, quoted in Pred, Choreography, 211
17 Pred, Social reproduction, 5.
18 Thrift and Pred, Time geography.
19 Pred, *Place*.
20 Pred, Production.
21 Hägerstrand, Tidsanvändning.
22 Baker, Perspective; Conzen, Historical geography; Paassen, Human geography.
23 These are of two kinds. First, those which are properties of the individual human beings, the 'capability', 'coupling' and 'authority' constraints which have been defined and schematized in the literature. Secondly, the scarcity of the resources of

time and space available to one individual constrains his or her behaviour in the short term. Pred, Choreography.
24 Carlstein, *Time*; Pred, Choreography; Pred, Production; Pred, *Place*; Thrift and Pred, Time geography.
25 Gregory, Suspended animation.
26 Carlstein, *Time*; Pred, *Place*.
27 Gregory, Suspended animation, 321.
28 Cannadine, Review, 67; Sayer, Explanation.
29 Carlstein, *Time*, 39.
30 Baker, Perspective; Buttimer, Grasping; Gregory, Suspended animation.
31 It was used by Hägerstrand to depict life biographies in past societies. As far as we know, it has received little attention since in the literature of time-geography, although it has been used in substantive studies of social and spatial change in Sweden. Hägerstrand, Survival; Gerger and Hoppe, *Education*.
32 Hoppe and Langton, *Town*.
33 *Livelihood activity system* might therefore be a better term, but it seems better to retain the expression already in use on the understanding that 'position' represents the full ensemble of resources assembled from the set of stations in a person's web of paths, not a point in terrestrial space.
34 Gerard, Entitation, 120–1; Langton, Potentialities.
35 Mendels, Proto-industrialisation; Mendels, Agriculture.
36 Carlstein, *Time*; Pred, *Place*.
37 The sampling procedures will be explained in the relevant chapters.

3 Agrarian land ownership and farm holdings

1 Swedish agricultural land fell into one of three categories of *jordnatur* according to its tax liabilities. Crown (*krono*) land paid no state tax, while land owned freehold by peasants (*skatte*) was burdened with land tax and other liabilities from which noble (*frälse*) land was exempted, to varying extents at different times. Carlsson, *Ståndssamhälle*.
2 Österberg, *Den gamla*.
3 We have tried to translate particular terms into something recognizably 'Western European', and to stress similarities between Sweden and elsewhere to aid comparison. *Bonde* was the word used to describe a cultivator of taxed land. It has become customary to translate this as 'peasant', but 'yeoman' would be equally suitable.
4 Revera, *Gods*; Lindegren, *Utskrivning*; Nilsson, Landbor.
5 Conditions in the trans-Baltic lands were generally different from those of the mother country and Finland, and we will deal only with those in Sweden proper.
6 Revera, *Gods*.
7 *Säterier* in Swedish. They were estates with special taxation privileges.
8 Revera, *Gods*.
9 *Indelningsverket* was the system by which the regular army was supported in the countryside as part of farm taxation. A certain number of farms were obliged to provide one soldier with a croft (*soldattorp*), as well as with his uniform, arms and – if it was a cavalry regiment – a horse. Crown farms were reserved for officers, who

were given them in lieu of salary, the size depending on rank. Guillemot, *Rask*; Kumm, *Indelt*.

10 Carlsson, *Ståndssamhälle*.

11 The Swedish *Riksdag* had a peasant Fourth Estate from its inception in the sixteenth century. Myers, *Parliaments*.

12 Herlitz, *Jordegendom*; Kyle, *Striden*.

13 Carlsson, *Ståndssamhälle*, 117ff.

14 Helmfrid, *Östergötland*, 234ff.

15 Jonsson, *Jordmagnater*, 57ff.

16 Magnusson, *Ty som*, 10ff.

17 Utterström, *Jordbrukets*; Ingers, *Bonden*; Montgomery, *Industrialismens*.

18 Wennberg, *Lantbebyggelsen*; Torbrand, *Johannishus*, 165ff.

19 Jonsson, *Jordmagnater*, 144ff.; Elgeskog, *Svensk*, 177ff.; Torbrand, *Johannishus*, 165ff.

20 Strict measures were taken in eighteenth-century Sweden to avoid vagrancy amongst the landless. Laws passed from 1747 onwards known as *tjänstehjonsstadgan* made it compulsory for landless people to be gainfully employed on a farm or in industry under annual contracts in which the employee admitted subservience to the employer who, for example, was allowed to exert physical punishment. These laws were abolished in 1833. The *försvarslöshetsförordningen* law made unemployment a criminal offence; an unemployed male could be conscripted into the army until 1824 and sentenced to between two and four years imprisonment until 1847. These statutes prohibited migration for vagrants: unless people had gainful employment at their destination, they were not allowed to move outside their parish. Utterström, *Jordbrukets*; Kyle, *Striden*, 147.

21 Elgeskog, *Svensk*, 141.

22 Lägnert, *Syd-och mellansvenska*; Hannerberg, *Svenskt*; Persson, Freehold farmers.

23 The medieval law in most of Central Sweden was that agricultural land must be organized along *solskifte* (sunshift) principles. The tofts of village farmsteads were proportional in width to their strips in the open fields. The name 'sunshift' refers to the fact that strips were arranged clockwise (i.e. in the direction of the sun's passage). This sequence was repeated in all fields and meadows of the village. Many villages and hamlets did not conform fully to these principles. Göransson, *Regelbundna*.

24 Olai, *Storskiftet*, 136ff., estimated that only 3 per cent of the arable in Ekebyborna parish, in our study area, consisted of colonizations of outlying land in 1700.

25 Torbrand, *Johannishus*, 253ff.

26 There was a 50 per cent increase in Närke between 1630 and 1700. Hannerberg, *Närkes*, 149ff.; Hannerberg, *Kumlaboken*.

27 Isacson, *Ekonomisk*, 70ff.

28 Martinius, *Jordbrukets*.

29 Bäck, *Bondeopposition*; Østerud, *Agrarian*.

30 G. Rydeberg, *Skatteköpen*.

31 Kyle, *Striden*.

32 An estimated 25 per cent of the tax purchases went to townsfolk, ironworks owners, and the like. Arpi, *Järnhanteringens*, 207ff; Nordström, *Relationer*, 26ff.

33 Kyle, *Striden*; Rydeberg, *Skatteköpen*.

34 Olai, *Storskiftet.*
35 Bäck, *Bondeopposition.*
36 Herlitz, *Jordegendom*; Kyle, *Striden.*
37 Carlsson, *Ståndssamhälle.*
38 Elgeskog, *Svensk torpbebyggelse.*
39 Forests were the source of firewood and building materials. Areas like the Östergötland plains with little or no forest had access to 'hundred commons' (*häradsallmänningar*) in places where forest was abundant.
40 In 1684 the limit was one-quarter of a *mantal*. With the *storskifte* enclosure regulations of 1747, subdivision was permitted down to one-sixth and one-eighth of a *mantal*, and in 1756 to below one-twelfth of a *mantal* in Värmland county. With the *laga skifte* enclosure regulations of 1827, farms could be subdivided as long as they had the capacity to feed one standard household. Heckscher, *Svenskt*, 212–14.
41 Winberg, *Folkökning*, 179ff.
42 Heckscher, *Industrialismen*, 4.
43 After its creation in the 1630s for cadastral mapping, the National Survey (*Lantmäteristyrelsen*) was important in controlling the organizational aspects of agriculture. Its county surveyors had to be consulted whenever field patterns were changed, boundaries were disputed, or farms were formally subdivided. It was quite natural for the *Lantmäteristyrelsen* to suggest changes in agricultural practices it saw as obstacles to improvement.
44 Hoppe, *Enclosure.*
45 Bäck, *Bondeopposition*, 198ff.
46 Frödin, *Skiftesväsendet.*
47 Bäck, *Bondeopposition*, 217ff.
48 Olai, *Storskiftet*; Olai, *Till vinnande.*
49 Bäck, *Bondeopposition*, 217ff.
50 Heckscher, *Sveriges ekonomiska*; Ingers, *Bonden*, 299ff.; Karlsson, *Mark*, 387ff.
51 Bäck, *Bondeopposition*, 212ff.
52 Hoppe, *Skiftesreformerna.*
53 Olai, *Storskiftet*; Olai, *Till vinnande.*
54 Bäck, *Bondeopposition.*
55 Hoppe, *Skiftesreformerna.*
56 Heckscher, *Svenskt*, 199ff. Later legislation considerably modified enclosure regulations, particularly the *laga skifte* laws of 1827 under which only two or three separate parcels of land were allowed per farm, and farmsteads were moved out from villages into the fields allocated to them. Helmfrid, *Storskifte.*
57 Kyle, *Striden.*
58 Olai, *Storskiftet*, 157ff. This also occurred at Väversunda in the south of our area as well as in Olai's research parish of Ekebyborna in the north. Hoppe, *Ventilation.*
59 Herlitz, *Jordegendom*; Herlitz, *Markegångspriser*; Isacson and Magnusson, *Proto-industrialisation.*
60 Heckscher, *Jordbrukets.*
61 Eriksson and Rogers, *Rural labor*, 32–4.
62 Magnusson, *Ty som*, 44ff.
63 Wennberg, *Lantbebyggelsen.*
64 Jonsson, *Jordmagnater*, 124ff. Jonsson gives examples of nineteeth-century crofts

in Södermanland with 20–30 acres of arable, more than many freehold farms.
65 On average 55 per cent of all grain (comprising rent paid in grain plus the production of the manor farm) was sold. Magnusson, *Ty som*, 111.
66 Helmfrid, *Östergötland*, 189ff.
67 Helmfrid, Storskifte.
68 Because differences in geology, land use and settlement did not correspond in detail with parish boundaries, it is necessary to refer to villages and hamlets at a sub-parish level.
69 Wannerdt, *Den svenska*; Lext, *Mantalsskrivningen*.
70 The tax values of farmland varied from year to year with agricultural produce prices because Swedish land taxes were still meant to represent given quantities of goods in kind (Jörberg, *Prices*, 8–16). Generally the taxation value was lower than its market value. In Östergötland in general during the early nineteenth century, the market value was 3–3.5 times larger than the taxation value.
71 Wohlin, *Den svenska*, 555ff.
72 Adamson, Kamerala; Persson, *Jorden*; Martinius, *Agrar*.
73 Hoppe, *Ventilation*.
74 The tax registers for Vadstena, also on the fertile plain, give the area as well as the value of holdings of townland arable. The average was 15.3 acres (7.55 hectares) per 1,000 Rdr. On the lowest-valued plot there were 30 acres, and on the highest-valued only 12 acres per 1,000 Rdr.
75 Bohman, Skiftesreformer.
76 Winberg, *Folkökning*, 244ff.
77 Lallerstedt's inventory of 1851 lists large properties in Västergötland and Stockholm provinces as well as Östergötland.
78 Heckscher, *Det Svenska*.
79 Adamson, Importance.

4 Agrarian livelihood positions, projects and production structures

1 The concept of peasant ecotype has been heavily used in Swedish research. See Hanssen, *Österlen*; Löfgren, Ecotypes; Löfgren, Perspectives; Gaunt, Natural resources; Winberg, *Folkökning*.
2 Erixon, *Gården*.
3 Löfgren, Family, 21.
4 Gaunt, Pre-industrial; Hanssen, Commonfolk.
5 Löfgren, Family, 25; Miller and Gerger, *Social change*, 94ff.
6 Ericsson and Rogers, *Rural labor*, 27.
7 Löfgren, Family.
8 Isacson, *Ekonomisk*, 96–105.
9 Winberg, *Folkökning*, 191ff.
10 Some 60 per cent of farm households in Örby parish in Sjuhäradsbygden were nuclear in 1792, but the percentage of extended families with two married couples (8 per cent) was larger than on the plains. Ahlberger, *Vävarfolket*, 101ff.
11 Winberg, *Folkökning*, 16; Hellspong and Löfgren, *Land*, 75.
12 Ericsson and Rogers, *Rural labor*, 65.
13 Elgeskog, *Svensk*, 116ff.

14 Jonsson, *Jordmagnater*, 100ff.
15 Magnusson, *Ty som*, 52ff.
16 Jonsson, *Jordmagnater*, 77ff.; Magnusson, *Ty som*, 87ff.
17 Mills, Lord; Holderness, Open.
18 Wohlin, *Den jordbruks-*, 26, 198 and 257.
19 Furuland, *Statarna*, 32.
20 Ericsson and Rogers, *Rural labor*, 35.
21 Elgeskog, *Svensk*, 269ff. and 385.
22 Sums of 100 Rdr were common for a forty-nine-year contract on ordinary crofts in Östergötland in the nineteenth century.
23 Jonsson, *Jordmagnater*, 52; Hoppe, *Enclosure*.
24 Adamson, Kamerala.
25 Hoppe, *Ventilation*, 174ff.
26 Winberg, *Folkökning*, 159ff.
27 Gadd, *Järn*, 85ff.
28 Hoppe, *Ventilation*, 188ff.
29 Hoppe, *Enclosure*; Hoppe, Skiftesreformerna.
30 The annual *stat* of one married hand in Östergötland in 1870 was 1,095 litres of unskimmed milk, 527 kilos of rye, 48 kilos of wheat, 104 kilos of barley, 155 kilos of oats and 110 kilos of potatoes plus 175 Rdr in cash. This was supplemented by the use of a potato patch and housing. Bagge, Lundberg and Svennilsson, *Wages*, 389.
31 Adamson, Kamerala.
32 Magnusson, *Ty som*, 82ff.
33 Carlsson, *Ståndssamhälle*.
34 Torbrand, *Johannishus*, 126; Wennberg, *Lantbebyggelsen*, 96.
35 Utterström, *Jordbrukets*, 781ff.
36 Ericsson and Rogers, *Rural labor*, 68.
37 Möller, *Godsen*, 61ff.
38 Jonsson, *Jordmagnater*, 76 and 78.
39 Marshall, *Travels*.
40 Guteland, *Biography*; Carlsson, Chronology.
41 Sundbärg, *Emigrations-*.
42 Winberg, Öst och väst.
43 Winberg, *Folkökning*.
44 *Ibid.*, 191ff.
45 Ericsson and Rogers, *Rural labor*.
46 Ahlberger, *Vävarfolket*, 116ff.
47 Jörberg, *Prices*.
48 Östergötland's wages were 20 per cent lower than the national average before 1840. Petersson, *Jordbrukets*; 1989, Herlitz, *Restadtegen*, 91.
49 Jörberg, *Prices*, 55.
50 Miller and Gerger, *Social change*.
51 Lundsjö, *Fattigdom*, 90 and 103.
52 Gerger, *Där nöden*.
53 Söderberg, *Agrar*.
54 Heckscher, Jordbrukets.

55 Hannerberg, *Närkes*; Olai, *Storskiftet*; Hoppe, Skiftesreformerna.
56 Olai, *Till vinnande*.
57 Hoppe, Skiftesreformerna.
58 According to Marshall, Swedish peasants attributed the innovation of new crops amongst the gentry to the mere pursuit of new fashions. He explained their own lack of innovation as a desire to cling to old fashions. Marshall, *Travels*.
59 Hannerberg, *Närkes*.
60 Dahl, *Torna*; Dahlman, *Open field*; Wennberg, *Lantbebyggelsen*.
61 Olai, *Storskiftet*, 100.
62 Arpi, Supply, 89; Stridsberg and Mattsson, *Skogens*; Bäck, *Bondeopposition*.
63 Olai, *Storskiftet*; Helmfrid, Storskifte.
64 Petersson, *Laga skifte*.
65 Winberg, *Folkökning*.
66 Olai, *Storskiftet*.
67 Isacson, *Ekonomisk*, 76–83.
68 Hannerberg, *Svenskt*, 23ff.
69 Slicher van Bath, *Agrarian history*, 250ff.
70 Lägnert, *Syd-och mellansvenska*, 157.
71 Regional statistics of Swedish agricultural production were compiled officially from 1858. For before then figures have to be compiled from randomly preserved descriptions of parishes or from farm inventories. As elsewhere, they are complicated sources to use. Gadd, *Järn*; Köll, *Tradition*.
72 Myrdal, *Medeltidens*, 64–5.
73 Slicher van Bath, *Agrarian history*, 239.
74 Gadd, *Järn*, 93ff.; Lägnert, *Syd-och mellansvenska*.
75 Gadd, *Järn*.
76 Petersson, *Jordbrukets*.
77 Gadd, *Järn*, 215.
78 Hannerberg, *Närkes*, 240; Petersson, *Jordbrukets*, 40.
79 Myrdal, *Medeltidens*, 66; Helmfrid, *Östergötland*, 31ff.
80 Petersson, *Jordbrukets*.
81 Gadd, *Järn*.
82 Petersson, *Jordbrukets*, 52.
83 Kuuse, *Från redskap*.
84 Myrdal, *Medeltidens*, 76–81.
85 Köll, *Tradition*, 89–90.
86 Petersson, *Jordbrukets*, 138.
87 Some 77 per cent of the farms only had ards. Köll, *Tradition*, 93.
88 Gadd, *Järn*, 157.
89 Petersson, *Jordbrukets*.
90 Myrdal, *Medeltidens*, 101.
91 Gadd, *Järn*, 163 and 169ff.
92 Köll, *Tradition*, 97.
93 Material on Dahl in this and immediately following paragraphs is from Petersson, *Jordbrukets*, 152–56.
94 Gadd, *Järn*, 172.
95 Myrdal, *Medeltidens*, 120.

96 Snell, *Annals.*
97 Söderbäck, *Rågöborna.*
98 Petersson, *Jordbrukets,* 156.
99 *Ibid.,* 157.
100 Köll, *Tradition,* 103.
101 Gadd, *Järn,* 178.
102 Köll, *Tradition,* 103–4.
103 Petersson, *Jordbrukets,* 158–9.
104 Gadd, *Järn,* 252–3 and 262.
105 Petersson, *Jordbrukets,* 162.
106 Hannerberg, *Svenskt,* 108.
107 Gadd, *Järn,* 117, 121 and 130.
108 Hoppe, Skiftesreformerna; Bäck, *Början.*
109 Köll, *Tradition,* 166ff.
110 Petersson, *Jordbrukets,* 40ff.
111 *Ibid.,* 67 and 71.
112 Helmfrid, *Östergötland,* 31ff.
113 Quote from a Riksdag speech of 1771, cited in Samuelsson, *Great power,* 77.
114 Attman, *Svensk;* Hildebrand, *Svensk;* Boethius, *Skogen.*
115 Nordström, *Relationer;* Hildebrand, *Svensk.*
116 Arpi, *Järnhanteringens;* Nordström, *Relationer.*
117 Helmfrid, *Tre sekler.*
118 Lake-ore alone made brittle finished goods and had to be mixed with mined iron. Hildebrand, *Svensk.*
119 Adamson, *De svenska;* Arpi, *Järnhanteringens;* Helmfrid, *Tre sekler.*
120 Borgegård, *Tjärhanteringen.*
121 Forsberg, *Gusums bruks;* Hannerberg, Den äldre.
122 Ahlberger, *Vävarfolket;* Boqvist, *Dolda;* Isacson and Magnusson, *Proto-industrialisation.*
123 Quoted in Samuelsson, *Great power,* 77.
124 Hoppe, Jordbruk.
125 Löfgren, Perspectives.
126 Isacson and Magnusson, *Proto-industrialisation,* 22–3.
127 *Ibid.,* 34.
128 Bergsten, *Östergötlands.*
129 Schön, *Från hantverk.*
130 Isacson and Magnusson, *Proto-industrialisation,* 26–7.
131 Odén, Domestic, 429.
132 *Ibid.*
133 Montgomery, *Rise,* 181.
134 Boqvist, *Dolda,* 136.
135 Isacson and Magnusson, *Proto-industrialisation,* 20.
136 Ljunggren, Handicrafts, 487.
137 Samuelsson, *Great power,* 150.
138 Brace, *Norse,* describes many still-prevalent folk costumes and their regional variations in 1857, particularly in Dalarne and Sweden north of Lake Mälaren generally. See also Ek, Booms.

139 Heckscher, *Economic history*, 232.
140 Brace, *Norse*, 141–2.
141 Ahlberger, *Vävarfolket.*
142 Isacson and Magnusson, *Proto-industrialisation*, 26.
143 Samuelsson, *Great power*, 152.
144 This privilege was granted on the incorporation of the province of Skåne into Sweden after its conquest from Denmark. People in these seven border parishes had come to depend on trade between the two countries. Borås and Ulricehamn were set up so that these traders could, as required by law, belong to towns, although they need not actually live in them. Dahl, Pedlars.
145 Hanssen, Commonfolk, 89.
146 Odén, Domestic, 495.
147 Passjournaler, 1849–50, BXIX. 2, Länsstyrelsens i Östergötland arkiv, Landarkivet, Vadstena.
148 Hoppe, *Ventilation.*
149 *Ibid.*
150 *Utjord* were outlying parts of village land belonging to farms in it.
151 The wages given for Aska only include the cash part of the wages, from which the total was assessed.
152 Petersson, *Jordbrukets.*
153 Gerger, *Där nöden*; Söderberg, *Agrar.*
154 Lundsjö, *Fattigdomen*; Söderberg, *Agrar.*
155 Gerger, *Där nöden.*
156 Lundsjö, *Fattigdomen.*
157 Winberg, *Folkökning*; Lithell, Breast feeding; Brändström, *Kärlekslösa.*
158 Ahlberger, *Vävarfolket.*
159 Olai, *Storskiftet*, 151–3.
160 *1858 års Finanskommittés Handlingar*, Östergötland, Riksarkivet, Stockholm.
161 *Ibid.*
162 Material on drainage in this and the next paragraph are from Möller, *Godsen*, 94ff., 105ff. and 110.
163 Petersson, *Jordbrukets*, 34.
164 Hoppe, *Ventilation.*
165 Petersson, *Jordbrukets*, 31.
166 *Ibid.*, 38.
167 *Ibid.*, 35.
168 Enclosure deed for Svälinge village, Herrestad parish, ÖLM i Östergötlands län, Linköping.
169 Enclosure deed for Övre Götala hamlet, Styra parish, ÖLM i Östergötlands län, Linköping.
170 Gadd, *Järn*, 173.
171 Material on farm equipment in Dahl is taken from Petersson, *Jordbrukets*, 156, 158 and 160.
172 Helmfrid, *Östergötland.*
173 Schön, *Från hantverk*, 24.
174 Ahlberger, *Vävarfolket*, 49ff.
175 Gadd, *Järn*, 123.

176 According to the 1858 Financial Commission Report, the number of pigs did not change in Dahl and Lysing hundreds from the 1830s, and grew slightly in Aska. The reduction in the number of pigs in Dahl shown by Petersson's figures must have been compensated by an increase in the the shieldland parts of Lysing hundred. Petersson, *Jordbrukets*.

177 Slicher van Bath, *Agrarian history*, 286; Persson, *Jorden*, 124ff.

178 Petersson, *Jordbrukets*, 174.

179 *1858 års Finanskommitté*, Östergötland, Riksarkivet, Stockholm.

180 From our own farm sizes and numbers, Petersson's model farms, and figures from the 1858 Financial Committee report. Petersson, *Jordbrukets*, 174.

181 Göta Canal accounts, Landsarkivet, Vadstena. Only very small samples from these voluminous and complex accounts could be examined in the limited time we could spend working on them. Complete transcriptions of ships and cargoes sailing through Motala canal depot from and to all the loading places in our region were taken for the months of May (when the canal opened for the year after freezing), June, September and October (when it closed) for 1835 and 1856.

182 *1858 års Finanskommitté*, Östergötland, Riksarkivet, Stockholm.

183 Thulin, Temperance, 740.

184 Heckscher, *Economic history*, 152.

185 Moberg, *Emigrants*, 9.

186 In the taxation registers of 1850 stills were ascribed to the particular plots on which they were located and to which distillation rights were attached. The numerator of the ratio is a count of these stills. The denominator is the number of separate farms in each parish given by our reconstitution of holdings. A single multiple-plot holding could have had a number of stills, so that the mapped ratio over-represents the proportion of farmers with stills.

187 Anders Larsson of Broby, Strå (who had stills at Kedevad in Strå and Ullevid in Örberga) as well as at Broby; Gustaf Lundberg of Nässja; Per Larsson of Storgården, Herrestad, and Carl Olofsson of Hageby.

188 Distilleries and liquor are listed in very few inventories, in each case for very large farms. Only stills (and dairies) housed in separate purpose-built structures seem to have been valued in their own right. Anders Larsson's distillery at Broby, Strå contained 4,449 *kannor* of liquor at his death in 1854; Lallerstedt's warehouse at Starby 24,854 *kannor* in 1851, and Anders Kastbom of Västgöteby was owed 207 Rdr for liquor at his death in 1852, which would have amounted to about 466 *kannor*.

189 The two brewhouses valued in Strå and Herrestad inventories were worth only 10 and 12 Rdr, that at Starby 16.5 Rdr.

190 The two looms plus accessories of an Orlunda housewife were valued at 1 Rdr 16 Sk; a single loom plus accessories belonging to a farmer in the same parish were worth 5 Rdr 16 Sk; the four looms, bobbins, three bobbin stools and a damaged loom at Starby were worth 20 Rdr.

191 Topelius, *Damastduktyg*.

192 *Ibid.*, 218.

193 Kerridge, *Textiles*.

194 Passjournaler 1849–57, BXIX 2, Länsstyrelsen i Östergötland, Landsarkivet, Vadstena.

195 Johansson, *Knyppling*, 16–17, records large lace production in the parishes next to Vadstena in the mid-nineteenth century.

5 Industrial and urban livelihoods

1 National tabulations of these locally gathered statistics were published annually as *Tabellverket* long before censuses were taken elsewhere in the nineteenth century.
2 Montgomery, *Rise*, 61 and 108; Corfield, *Impact*, 8.
3 Postan, Plague.
4 Heckscher, *Economic history*.
5 Montgomery, *Rise*, 107.
6 Samuelsson, *Great power*, 27, 31–2 and 92–3.
7 Boëthius, Swedish iron.
8 Hildebrand, Foreign markets; Ashton, *Iron*, 111n.
9 Boëthius, Swedish iron.
10 Montgomery, *Rise*, 10.
11 Söderlund, Impact.
12 Boëthius, Swedish iron.
13 Arpi, Swedish ironmasters.
14 Hildebrand, Foreign markets.
15 Rickman, *Swedish iron*, 17–18. Lake ore was still mixed with mined ore in Småland in the 1920s.
16 Seebass, *Bergslagen*.
17 De Geer, Sveriges.
18 Boëthius, Swedish iron; Friberg, Growth. The way the tax was calculated meant that by the late eighteenth century it represented only 4 to 6 per cent of the value of their pig iron.
19 Hildebrand, Foreign markets, 61.
20 Friberg, Growth.
21 Arpi, Supply.
22 Söderlund, Impact, 55–6.
23 Friberg, Growth.
24 Scott, *Sweden*, 450.
25 Montelius, Recruitment, 10–11.
26 Scott, *Sweden*, 450.
27 Food price inflation in the second half of the eighteenth century was accommodated by subsidizing the price of food supplied by *bruk* truck shops. These subsidies could be worth 90 per cent of nominal money wages at the end of the eighteenth century. Montelius, Recruitment.
28 Oak exports were forbidden and sawmilling was discouraged as wasteful of timber until the innovation of fine-bladed saws in the late eighteenth century, but Småland and Värmland supplied large exports of timber by that time. Tar making was important in Norrland. Layton, Industry, 208.
29 Boëthius, Swedish iron, 163–4.
30 These developed after arms making by peasants was banned in 1620. The large specialist plants into which it was forced were sold by the middle of the eighteenth

century, when some of them, such as Finspång, Bofors, Huskvarna and Åker, were among Sweden's most famous *bruk*. Samuelsson, *Great power*, 24–5.
31 Isacson and Magnusson, *Proto-industrialisation*.
32 Hörsell, *Borgare*.
33 Samuelsson, *Great power*, 84.
34 Heckscher *Economic history*, 183–7.
35 *Ibid.*, 85.
36 *Ibid.*, 78–9.
37 In the 1840s it took five days and eighteen pairs of oxen to get a load 80 km from the iron *bruk* and papermill at Lessebo to Kalmar for export. Gustavson, *Small giant*, 121.
38 Fridlizius, *Swedish corn*.
39 Tidander, Commerce, 935.
40 Layton, *Evolution*, 67–77.
41 Heckscher, *Economic history*, 72.
42 Layton, *Evolution*, 73.
43 Samuelsson, *Great power*, 23–5.
44 Heckscher, *Economic history*, 73–4.
45 Odén, Industries, 494.
46 Ljunggren, Handicrafts, 485. Prices were fixed by committees of a magistrate, a merchant, a broker and a guild artisan advised by valuers elected by each guild and place.
47 Montgomery, *Rise*, 34.
48 Stadin, *Småstäder*.
49 Hanssen, Commonfolk, 96. Master craftsmen contractually employed by the year by noblemen were exempt from the requirement to practise in towns. Ljunggren, Handicrafts, 485.
50 Stadin, *Småstäder*, 157.
51 Heckscher, *Economic history*, 192–3.
52 Bengtsson and Jörberg, Market.
53 Hanssen, Commonfolk, 89.
54 Stadin, *Småstäder*.
55 Sundbärg, *Sweden*, 526–7.
56 Berend and Ránki, *European periphery*.
57 Montgomery, *Rise*, 61.
58 Arosenius, Demography, 119.
59 Jörberg, Structural change, 94.
60 Scott, *Sweden*, 343.
61 Gustavson, *Small giant*, 80; Layton, *Evolution*, 70–1.
62 Ekman's invention was 'the most glorious page in Swedish iron handling and in all Swedish history'. Gustavson, *Small giant*, 79.
63 Boëthius, Swedish iron, 162.
64 Arpi, Supply.
65 Boëthius, Swedish iron, 155.
66 Gustavson, *Small giant*, 80.
67 Montgomery, *Rise*, 80.
68 Isacson and Magnusson, *Proto-industrialisation*.

69 Like *bruk, mekanisk verkstad* has no equivalent English term. We will use the term 'engineering works'. Literally, it means 'mechanical workshop', but has been translated in a dozen other ways, including machine shop, machine works, factory and mill. Gustavson, *Small giant*, 155.
70 Greater specialization came with heavier capitalization after the Limited Liability Act of 1848 and, especially, the demands of the railways after the 1860s.
71 Montgomery, *Rise*, 111.
72 Samuelsson, *Great power*, 151.
73 Brace, *Norse folk*, 141. Material on textiles is from Gustavson, *Small giant*; Montgomery, *Rise*; Samuelsson, *Great power*; Sellgren, Textiles; Jonsson, *Arbetsorganisationen*, and Ahlberger, *Vävarfolket*.
74 Arosenius, Demography, 126.
75 Jörberg, Structural change, 94.
76 Layton, Evolution, 69; Wik, *Norra*.
77 Larsson, Match industry; Berglund, *Industriarbetarklassens*.
78 Samuelsson, *Great power*, 152–3. The value of Swedish liquor output exceeded that of timber exports from 1861 to 1865.
79 Material in this paragraph is from Gustavson, *Small giant*.
80 *Ibid.*, 181.
81 Isacson and Magnusson, *Proto-industrialisation*, 135.
82 Brace, *Norse folk*, 147.
83 Scott, *Sweden*, 455.
84 The population of Göteborg was 12,804 in 1800 and 26,084 in 1850: those of Borås 1,774 and 2,733, Norrköping 9,089 and 16,916, Jönköping 2,684 and 6,005, and Eskilstuna 1,341 and 3,961. Arosenius, Demography, 207.
85 Gustavson, *Small giant*, 133.
86 Crafts, *Economic growth*.
87 Langton, Industrial revolution; Langton and Morris, Introduction; Turnbull, Canals.
88 Samuelsson, *Great power*, 149.
89 Montgomery, *Rise*, 31; Layton, Industry.
90 Heckscher, *Economic history*, 192.
91 Montgomery, *Rise*, 36.
92 Layton, Industry, 212.
93 Dahl, Pedlars, 169.
94 Layton, Industry; Layton, Handelsmans.
95 Heckscher, *Economic history*, 191–2.
96 Ek, Booms, 9. Not just gold and silver, but also such items as oak chests and bed covers were considered by peasants to be investments.
97 Ek, Booms, 11.
98 Brace, *Norse*, 214, 297 and 299.
99 Based on figures given in Arosenius, Demography, 207. The rate of urban growth along the Göta Canal and the lakes it connected would be even greater if Trollhättan and Motala, which did not receive urban status until after 1850, were included.
100 Hanssen, Commonfolk.

101 The others were a tutor, a footman, a housekeeper, a foreman and a nightwatchman.
102 The quarry was not valued because it belonged to the Crown. It was worked by convicts and leased by the Göta Canal Company from 1820.
103 Motala Ström, the only outlet of Lake Vättern, flows at 43 cubic metres per second. Its 15 metre fall between Lakes Vättern and Boren provided 1,156 horse power in the early twentieth century, and a hydro-electric power plant built then utilized 20,000 hp. Wallén, Hydrology.
104 Ek, Booms.
105 For example, in 1820 smiths Falk and Swartling worked the *knipphammare* forge at Motala owned by the former and H. Falk, of the large Mjölna mill in Strå, miller Molander and miller Andersson. Molander operated one of the peasant mills in 1810, but gave it up by 1820, when he was tenant of Duvedal mill and part-owner of the forge hammer.
106 The furnace was closed at the end of the seventeenth century, and in 1726 the *bruk* employed four hammersmiths and fifteen other workers at its forge. In 1748 its licensed output was about 69 tons, in the second rank amongst Östergötland's twenty-six iron *bruk*, four of which were about twice its size. Bergsten, *Östergötlands*.
107 A cadet branch of the house of that name. In the main line, Claes Grill was one of the wealthiest Stockholm merchants in the mid-eighteenth century, head of the Swedish East India Company, and the owner of large ironworks near Dannemora. Scott, *Sweden*, 348; Gustavson, *Small giant*, 194.
108 *Brukspatron* af Burén of Grytgöl owned half of Broby mill in Strå in 1830. Grytgöl ironworks, 10 km north-west of Finspång, was closed from 1806 to 1821. Bergsten, *Östergötlands*.
109 Karlsby estate, the site of Karlström's *bruk*, was on the edge of our region and probably also owned land immediately outside it.
110 Calculated from information in Tillhagen, *Järnet*
111 Sträng, *Vadstena*.
112 Set up in 1819 as the third workhouse in the kingdom to those at Karlskrona and Sveaborg, to house soldiers struck off regimental rolls, discharged but unreformed criminal prisoners and vagrants. Anyone without land or an urban workshop, or not in service to someone who had or formally employed in industry, was a vagrant. *Poor law*, 353–5.
113 Arosenius, Demography, 120.
114 Property values are absent from the *taxeringslängder* of Vadstena in 1810. Crown properties such as the castle, monastery and official residences of the vicar and others were never valued, and are excluded from this total.
115 One of them was Alderman Regnstrand, who lived in the castle where he held an official position. He was listed under St Per (Dahl), but said to be 'of Vadstena' in the taxation registers of other rural parishes where he owned property.
116 A widow's 156 Rdr arable and grass plots supported a cow and a pig and produced rye. Ten acres leased by Alderman dyer Hellström supported two horses, two cows, three sheep and two pigs, beside the pair of oxen required to plough for the rye. His animals and rye accounted for about one-sixth of his movable wealth. The

small townland holding of a Vadstena woman who owned and tended 3,000 Rdr's worth of land in Strå was the base from which both were farmed by three pairs of oxen to produce 300 Rdr's worth of rye which, with three horses, nine cows, a bull, two calves, four pigs and twenty sheep, were inventoried at her town home.

117 Occupants of property are listed in the *mantalslängder*, but it is sometimes impossible to decide whether families formed separate or common households, or whether single people lived in separate households, with families in the same property, or with each other. A count of households is therefore open to error. Baigent, Women.

118 Seven of them lived in the region, nine outside it, and nine at unknown locations.

119 Some allocations were arbitrary. For example, stone paviours were put with menial occupations rather than building crafts; town guards with menial rather than administative personnel; sextons with menial rather than professional activites.

120 The average tax tabulated for menial workers was as high as that of craftsmen because four of their five incomes were too small to be liable to tax.

121 People called aldermen with an additional occupation were classified under that occupation, those without in this category.

122 The Askersundian to whom the vast majority was owed was named Sundblad, as was an alderman–merchant of Vadstena in 1820.

123 Including master craftsmen, 'ex-craftsmen' and craftsmen's widows who paid tax on income from craft work, and one or two journeymen who paid tax on craft income or property, implying their independence from a master.

124 The town fisherman was classified as a menial worker.

125 Our account of the linen works is based on Topelius, *Damastduktyg*.

126 Heckscher, *Economic history*, 188–9.

127 Generally called *arbetskarlar*, though there were three *dagkarlar* (day-labourers).

128 Four boatmen are included in 'others', but no specialist land carriers were listed in taxation records.

129 Most spinning in Sweden was done in the households of 'the lower classes'. Heckscher, *Economic history*, 189.

130 Johansson, *Knyppling*, 8.

131 Topelius, *Damastduktyg*, 213.

132 Not evident on table 5.13 is the number of ex-soldiers amongst the retired males. (The Old Soldiers' Home closed before 1820.) The eight administrative and professional retired men included a former lieutenant-colonel, a former captain, a former quartermaster and two former sergeants, and seven of the ten former agricultural workers (so classified because they had crofts) were ex-soldiers.

133 In 1840, women made up 54 per cent of the Swedish population, but 57 per cent in Skänninge, 58 per cent in Linköping and Söderköping and 60 per cent in Vadstena. Arosenius, Demography, 125; *Mantalslängder, Sammandrag för Östergötland*, 1840.

134 The former was the daughter of Widow Tollbom, whose family was well known for making and dealing in lace. She was descended from a pastor at the Old Soldier's Home, and demonstrated lacemaking at the Paris World Exhibition of 1855. Johansson, *Knyppling*, 13–14.

135 The townland ascribed to brewers in table 5.9 belonged to brewers' widows following their former husbands' craft.

136 Parishioners must put up those who were poor but not in the poorhouse for a number of weeks per year, depending on the taxation value of their property, as instructed by the church vestry. *Poor Law*, 361–74.

137 Sometimes the word lodger was used, presumably for someone who paid rent. We did not distinguish between the two.

138 Nineteen of those tabulated, and all of those with children, lived in the linen factory where they were employed.

139 Some were listed as *gratialister* (i.e. state pensioners).

140 This was not true of merchant Wallberg, whose business income of 1,000 Rdr was the largest in the town. His household contained only himself, his wife and one maid. Any assistance he required in his business must have been obtained by paying proletarian menial and administrative workers.

141 Fahlbeck, Occupations, 503.

142 The settlements of Bergsäter, Bispmotala, Bondebacka, Duvedal (Motala and Vinnerstad), Holm, Hårstorp, Motala *köping*, Motala *utjordar*, Motala Village, Motala Engineering Works and Offerby.

143 Excluding inmates of institutions, whose inclusion gives an increase of 57 per cent to 2,081.

144 Qvarsell, Locked up.

145 The value of the bank business mapped on figure 5.6, which does not include the value of its building, was calculated from its tax payment, assuming the rate of 2.5 per cent paid by dealers on their profits.

146 They were entered in the records as 'carpenter workman', 'taxed as carpenter's journeymen', or 'practising carpenter', as distinct from 'master carpenter'.

147 If the 17,225 Rdr's worth of warehouses and ships are included, the dealers, to whom they nearly all belonged, account for 32 per cent compared with the 29 per cent of the unproductive group.

148 The income data on figure 5.7(d) comprises aggregations of the recorded incomes of members of each occupational group plus, for people for whom only a tax payment was recorded, multiples of taxes paid by the tax rate of the appropriate income band in each group.

149 The high income of the latter group on table 5.19 is due to possible mis-classification, discussed later.

150 Vicar Kinnander earned 5,520 Rdr a year, compared with his father's 1,995 Rdr as dean and vicar in 1820.

151 Perhaps salaries of officials and professionals cannot be equated with profits of dealers and craftsmen, which were more difficult to verify.

152 There were 2,500 shares in the bank, each valued at 100 Rdr. Shareholders also provided mortgage security to 3 times the value of their shares. Thus, there were 250,000 Rdr on deposit and security for a further 750,000 Rdr. 459,100 Rdr were out on loan.

153 Eight members of four trading companies were counted separately, although they paid tax jointly. Björk and Lagerdahl were categorized as officials because their earnings as harbour bailiffs exceeded their income as merchants.

154 Passjournaler 1849–57, BX1X 2, Landsarkivet, Vadstena. These journals were only examined for 1849 and 1850.

155 At her death in the 1840s, Widow Wetterstrand of Vadstena owed him for seven deliveries of fibre and thread made over the previous seven months; Mauritsson owed her for 4 ells of muslin and owned her loom. Many inventories list small debts to Mauritsson, but do not indicate how they had been incurred.

156 He was classified with craftsmen because he was taxed at the craft rate.

157 The Regnstrands built a theatre alongside the monastery garden in 1825. In 1850 Widow Regnstrand owned and ran a restaurant and kitchen in Castle Quarter, and her son Alderman Regnstrand owned rural farms, townland and a printing works and bookshop where the bank was opened between 1850 and 1860.

158 One of these weavers was the last producer of damask in the town in 1862. The original factory closed in 1843. Two former journeymen there opened a second one of about half the size in 1833 and continued it until 1839, when they split up, each to run his own works. Carl Magnus Pettersson's closed in 1851 and Petter Lundström's in 1859. Topelius, *Damastduktyg*.

159 They were not always easy to distinguish. Military officers were not ascribed salaries, and we can only assume that those not designated as 'former' or 'inactive' were still active. Similarly, we cannot know if two barons and a marshal of the court served in an official capacity or not. They were classified as 'unproductive' because they did not pay tax on income.

160 We do not know if lace was made by these women or by their maids.

161 Baigent, *Women*

162 Borsay and McInnes, *Debate*.

163 Under an Act of 1842 each parish must provide a school open to all children. Scott, *Sweden*, 352.

164 It was called a '*vadmalstamp*'.

165 They were bought from persons of standing and townsmen. This may have been related to the removal of prohibition on peasant ownership of mills in 1828, though peasants owned mills along Motala Ström at the start of the period.

166 A watermill at Väversunda was listed but not valued in the taxation register of 1850 and omitted from that of 1860.

167 Bergsten, *Östergötlands*.

168 It reopened in the 1860s and did not close permanently until 1900.

169 Gustavson, *Small giant*, 23.

170 Bring, *Göta Kanal*; Jonsson, *Arbetsorganisationen*.

171 Gustavson, *Small giant*, 94; Jonsson, *Arbetsorganisationen*.

172 Göta Canal accounts, Landsarkivet, Vadstena.

173 The discrepancy is because a convoy trapped by ice over winter left empty in the first days the canal was open.

174 Including fire bricks from Sweden's only collieries at Höganäs in Skåne, building bricks, sand, slaked lime, clay, sandstone and limestone uncut and as ashlars, bricks, tiles, spars, laths, joists, beams and boards from places around Lakes Vättern and Boren.

175 Bergsten, *Östergötlands*.

176 Bring, *Göta Kanal*, 184–5.

177 For Motala *köping* these were mainly got by multiplying taxes paid by 40, the tax rate being 2.5 per cent.
178 The 'works' was probably the Thamsten's tannery. Passjournaler 1849–57, BXIX 2, Landsarkivet, Vadstena.
179 These were employees resident at the works. Many of the half-dozen mechanics and scores of labourers living in the *köping* and riverside villages must also have worked there.
180 Bring, *Göta Kanal.*
181 Assuming a six-day working week and fifty-week year.
182 Given earlier as 400 Rdr due to the two-thirds deflation of values in 1860 in comparative statements.
183 John Fletcher came to the works from England with Fraser in 1822 and still worked there in 1860, as did two of his sons: one was a junior master mechanic married to a Swedish wife, another an apprentice. A Finn and a Norwegian also worked at the *verkstad* in 1860, but by then it had long been in the hands of a Swedish labour force. Jonsson, *Arbetsorganisationen.*
184 Katrinefors Ironworks had seventeen male workers; Rosendahl's Brewery and Smithy thirteen; Strömsnäs Brewery about the same, and Duvedal Papermill sixteen.

6 Investment and money in agriculture

1 These sums are based on inventory contents and values.
2 Bjurling, *Oxie*; Martinius, *Jordbrukets*; Isacson, *Ekonomisk.*
3 Banaji, Kautsky.
4 Heckscher, *Arbete.*
5 Köll, *Tradition*, 63ff.
6 Isacson, *Ekonomisk*, 140ff.
7 Isacson argues that the two largest farmers functioned as bankers for the parish inhabitants, who owed them more than 20,000 Rdr.
8 Ironworks owners in the area were thus owed substantial sums by the iron smelters of By, who produced pig-iron on a putting out basis for them.
9 The market value of farmland was set at 30–40 per cent higher than the tax value in the early 1840s, but about the same as the tax value in 1862, the second base year for of his study. Martinius, *Agrar.*
10 Martinius' boundary between large and middle-sized farms was 20,000 Rdr of taxation value, while our boundary between large farms and estates is 25,000 Rdr and that between large and middle-sized farms 3,500 Rdr. Martinius assumed that farms below 3,333 Rdr were pure family farms (which they rarely were in Östergötland) and that those above 20,000 Rdr mainly belonged to persons of standing, whom he wanted to put into one category (in Östergötland those owned by farms of less than 3,500 Rdr). Martinius, *Agrar*, 129–38.
11 Bank of Sweden mortgages were difficult to obtain, and only one-third of the taxation value of a property could be mortgaged. However, early redemption could not be forced, six months notice of early redemption rarely had to be given, the rate of interest of 4 to 5 per cent was normally slightly lower than that on private

mortgages, and the possible redemption period was, at up to twenty-eight years, longer than was usual on the private mortgage market, as well as being a fixed term. Nygren, *Svensk*.

12 Assuming that there were 115 mortgages in Dahl 1820–30 (10 × 70/6), and that their average was the same as that for the 70 sampled, the total mortgage sum would be 183,000 Rdr, or almost 31 per cent of the total taxation value of just over 595,000 Rdr. Although the tax value did not represent market value, few (if any) mortgages exceeded the tax value of plots, probably as a safety-measure for the moneylender.

13 Inventories have a number of shortcomings for these purposes. Indebtedness may have been smaller than inventories indicate because the values of land and other belongings were assessed at below market values, whilst debts were given at their face values. The undervaluation of assets could vary between 20 and 30 per cent. Gadd, *Järn*, 72ff.; Isacson, *Ekonomisk*, 223ff.; Martinius, *Agrar*, 121ff.

14 Bengtsson, *Glimtar*.

15 Nygren, *Svensk*.

16 The debt level on middle-sized farms in Västergötland was only about 30 per cent of total inventory value, while in our own study area farms of equivalent size generally had 50–60 per cent, using in both cases the taxation values of land. Martinius, *Agrar*, 132.

17 In Västergötland in the 1860s the average debt to institutions was between 3 and 7 per cent of total debt, so that even some of the small farmers of the southern Aska and Dahl plain were more advanced in this respect.

18 Only Motala Engineering Works, Holmen Works in Norrköping, The Göta Canal Company and Östergötlands Enskilda Bank were larger. *1858 års Finankommittés Handlingar*, Östergötland, Riksarkivet, Stockholm.

19 Nygren, *Svensk*.

7 Flows of labour

1 Åkerman, Demographic study.
2 *Ibid.*
3 Jörberg, *Prices*; Bagge, Lundberg and Svenilsson, *Wages*.
4 Martinius, *Agrar*, 22.
5 Thus between 25 and 50 per cent of the labour in these categories was replaced every year.
6 Martinius, *Agrar*, 22.
7 Parish registers yield lower figures than estate ledgers for seasonal labour, which obviously registered only in their home parishes. Martinius, *Agrar*, 23.
8 Hägerstrand, Landsbygdsbefolkning; Hägerstrand, Migration.
9 Martinius, *Agrar*, 50.
10 Whether they moved from one sector of the economy to another in the process is another matter because there was agriculture in Swedish towns in the nineteenth century.
11 Hägerstrand, Landbygdsbefolkning; Martinius, *Agrar*; Bergsten, *Östergötlands*.
12 Ericsson and Rogers, *Rural labor*, 186 and 189.
13 One cohort was born 1804–06 and one 1834–36. The members of each cohort were

traced from the age of twenty until they reached fifty. Eriksson and Rogers, *Rural labor*.

14 Martinius, *Peasant*, 9.
15 Hoppe, *Skolutbildning*.
16 Norman, *Från Bergslagen*.
17 Mendels, Proto-industrialization; Mendels, Social mobility; Kreidte, Medick and Schlumbohm, *Industrialisation*; Isacson and Magnusson, *Proto-industrialisation*.
18 Göransson, *Från familj*, 169.
19 *Ibid.*, 215ff.
20 Isacson and Magnusson, *Proto-industrialisation*.
21 Mendels, Social mobility; Levine, *Family*. It has also been argued that experience of homecraft production brought revulsion from the regimented life style of the factory. Bythell, *Handloom*; Thompson, *Making*.
22 Literally translated, *husförhörslängder* means 'household examination registers'. They were compiled for every Swedish parish, in principle from the seventeenth century. Their primary purpose was to record the results of annual checks on the religious knowledge of each member of the population. *Inter alia*, they record the name and address of every household head and the age, places of birth, last and next residence, relation to household head and occupation of each member of each household. They were up-dated continuously by church ministers. Kälvemark, The country.
23 People were registered more than once if they moved in and out several times, as well as if they moved around within a parish, and more mobile groups were over-represented. Thus, these data reflect *flows only* and cannot be equated with cross-sectional figures of population.
24 Hoppe, *Skolutbildning*.
25 Slight differences in the way people were classified in the two sources, especially in the allocation of individuals between cottager, unproductive, and other, are responsible for the number of cottager positions apparently exceeding the number of their occupants in Västra Ny.
26 Maps of the places of birth, last residence and next residence of rural maids are presented in Hoppe and Langton, *Flows*.
27 That is, six times as many maids were recorded in the two parishes over the six years as there were livelihood positions for maids in 1860.
28 Hoppe, *Ventilation*. This seems to have been a general feature of enclosure. Turner, *Parliamentary enclosure*.
29 Maps of the places of birth, last residence and next residence of crofters are presented in Hoppe and Langton, *Flows*.
30 The smallness of the town's area and the density of its housing caused more movement across and within its boundaries than in generally larger rural parishes.
31 Maps of the places of birth of adult craftsmen in Vadstena are presented in Hoppe and Langton, *Flows*.
32 Maps of the places of birth, last residence and next residence of Vadstena labourers are presented in Hoppe and Langton, *Flows*.
33 Maps of the places of birth, last residence and next residence of Vadstena maids are presented in Hoppe and Langton, *Flows*.

34 We cannot exclude the possibility of part-time textile production in these particular homes, with rural–urban contacts through putters-out giving an entrée to urban life.

35 One of our cohort of Motala Works workers became manager of the Bergsund Works in Stockholm in 1856.

8 Conclusion: peasants and others in the development of capitalist economy

1 Moberg, *Emigrants*, 61.

2 The restriction of farm distilling in 1855 might have been a reason for the difficulties of some of the farmers whose life-paths we traced into the 1860s and 1870s.

3 As inheritors of large capitalist holdings left farming they sold or leased out their land. They would get higher prices for small farms from people aspiring after a peasant way of life.

4 Heckscher, *Economic history*; Scott, *Sweden*; Herlitz, Capitalism; Langton, End.

5 Baker, Hamshere and Langton, Introduction; Hamshere, Data; Harley, Evidence.

6 Because only two categories were distinguished on each dimension, a few parishes had to be asssigned arbitrarily to one quadrant rather than another, but this does not affect the pattern on the map.

7 Bjälbo, Strå and Orlunda ranked 1,2 and 3 in terms of peasantness, and 1, 2 and 5 in terms of population growth rates.

8 Cf. Gullickson, *Spinners*.

9 Although no further than has already (inappropriately) been done for Sweden. Hoppe, Review.

10 Wrigley, Men.

11 Heckscher, Place.

Bibliography

Adamson, R., *De svenska järnbrukens storleksutveckling och avsättningsinriktning 1796–1860* (Meddelanden från Ekonomisk-historiska Institutionen, Göteborgs Universitet, 4, Göteborg, 1963)

The importance of the agrarian sector for industrialisation: some disparate reflections, Department of Economic History, Stockholm University, mimeo. (1980)

Kamerala källor rörande sysselsättning och bosättning, *Bebyggelsehist. Tidskr.*, 5 (1983), 54–62

Ahlberger, C., *Vävarfolket: hemindustri i Mark 1790–1850* (Institutet för lokalhistorisk forskning, skriftserie nr. 1, Göteborg, 1988)

Home industry and rural millenarianism in early nineteenth-century western Sweden, in Lundahl and Svensson, *Society*, 184–200

Åkerman, S., A demographic study of a pre-transition society, in S. Åkerman, C., Johansen and D. Gaunt (eds.), *Chance and change: social and economic studies in historical demography in the Baltic area* (Odense, Odense University Press, 1978), 33–48

Anderson, P., *Lineages of the absolutist state*, (London, New Left Books, 1974)

Anderson, R. T., *Traditional Europe: a study in anthropology and history* (Belmont, Calif., Wadsworth, 1971)

Arensberg, C. M., The Old World peoples: the place of European culture in world ethnography, *Anthropological Quarterly*, 36 (1963), 75–99

Arosenius, E., Demography, in Guinchard, *Sweden*, vol. I, 104–46

Arpi, G., *Järnhanteringens träkolsförsörjning 1830–1950* (Jernkontorets Bergshist. skriftserie, 14, Stockholm, 1951)

The supply with charcoal of the Swedish iron industry from 1830 to 1950, *Geografiska Annaler*, 35 (1953), 11–27

The Swedish Ironmasters' Association, *Scandinavian Economic History Review*, 8 (1960) 77–90

Ashton, T. S., *Iron and steel in the industrial revolution* (Manchester, Manchester University Press, 1924)

Ashton, T. S. and Sykes, J., *The coal industry of the eighteenth century* (2nd edn, Manchester, Manchester University Press, 1964)

Attman, A., *Svenskt järn och stål 1800–1914* (Jernkontorets Bergshistoriska skriftserie, 21, Stockholm, 1986)

426   Bibliography

Bäck, K., *Bondeopposition och bondeinflytande under frihetstiden: centralmakten och östgötabondernas reaktioner i näringspolitiska frågor* (Stockholm, LT:s Förlag, 1984)

Början till slutet (Borensberg, Noteria Förlag, 1992)

Baigent, E., Women and work in Vadstena, Sweden, 1830–1890, *Geografiska Annaler*, 71B (1989), 335–52

Bagge, G., Lundberg, E. and Svennilsson, I., *Wages in Sweden 1860–1930*, 2 vols. (London, P. S. King and Son, 1935)

Bailey, F. G. The peasant view of the bad life, in Shanin, *Peasants*, 1st edn, 299–321

Baker, A. R. H., An historico-geographical perspective on time and space, *Progress in Human Geography*, 5 (1981), 439–43

Baker, A. R. H. and Billinge M. (eds), *Period and place: research methods in historical geography* (Cambridge, Cambridge University Press, 1982)

Baker, A. R. H. and Gregory, D. J. (eds), *Explorations in historical geography* (Cambridge, Cambridge University Press, 1984)

Baker, A. R. H., Hamshere, J. D. and Langton, J., Introduction to *Geographical interpretations of historical sources* (Newton Abbot, David and Charles, 1970), 13–25

Banaji, J., Parts of K. Kautsky's *The Agrarian Question, Economy and Society*, 5 (1976), 3–49

Beames, M., *Peasants and power: the Whiteboy movements and their control in prefamine Ireland* (Brighton, Harvester, 1983)

Bedoire, F. and Hogdahl, L. Storbönder, *Bebyggelsehist. Tidskr.*, 5 (1983), 63–91

Bengtsson, C., *Glimtar från 1800-talets Vadstena*, Föreningen Gamla Vadstena, Småskriftserien 11 (Vadstena, Vadstena Affärstryck, 1965)

Bengtsson, T. and Jörberg, L., Market integration in Sweden during the 18th and 19th centuries, *Economy and History*, 18 (1975), 93–106

Berend, I. T. and Ránki, G., *The European periphery and industrialization 1780–1819* (Cambridge, Cambridge University Press, 1982)

Berg, M., *The age of manufactures: industry innovation and work in Britain 1700–1820* (London, Fontana, 1985)

Berg, M., Hudson P. and Sonenscher, M., *Manufacturing in town and country before the factory* (Cambridge, Cambridge University Press, 1983)

Berger, J., *Pig earth* (London, Chatto and Windus, 1985)

Berglund, B., *Industriarbetarklassens formering: arbete och teknisk förändring vid tre svenska fabriker under 1800-talet* (Meddelanden från Ekonomisk–historiska Institutionen, Göteborgs Universitet, Göteborg, 1982)

Bergsten, K. E., *Östergötlands bergslag* (Meddelanden från Lunds Universitets Geografiska Institutioner, Avhandlingar, 10, Gleerups, Lund, 1946)

Berkner, L. K., The stem family and the development cycle of the peasant household: an eighteenth-century Austrian example, *American Historical Review*, 77 (1972), 398–418

Bettey, J. H., *Rural life in Wessex 1500–1900* (Gloucester, Alan Sutton, 1987)

Bjurling, O. C., *Oxie härads sparbank, 1847–1947* (Malmö, Sparbanken, 1947)

Blum, J., The internal structure and polity of the European village community from the fifteenth to the nineteenth century, *Journal of Modern History*, 43 (1971), 541–75

The end of the old order in rural Europe (Princeton, NJ, Princeton University Press, 1978)

From servitude to freedom, in Blum, *Forgotten past*, 57–72

The nobility and the land, in Blum, *Forgotten past*, 33–48

The village and the family, in Blum, *Forgotten past*, 9–24

(ed.), *Our forgotten past: seven centuries of life on the land* (London, Thames and Hudson, 1982)

Boëthius, B., Swedish iron and steel, 1600–1955, *Scandinavian Economic History Review* 6 (1958), 144–75

Skogen och bygden (Stockholm, Thule, 1939)

Bohman, C., Skiftesreformer och jordbrukspolitik under 200 år som de avspeglas i Herrestad och Björka byar i Östergötland, Unpublished undergraduate dissertation, Department of Human Geography, Stockholm University, mimeo. (1983)

Boqvist, A., *Den dolda ekonomin: en etnologisk stidie av näringsstrukturen i Bollebygd 1850–1950* (Lund, Lund University Press, 1978)

Borgegård, L. E., *Tjärhanteringen i Västerbottens län under 1800–talets senare hälft* (Acta Regiae Societatis Skytteanae, 12, 1973)

Borsay, P. and McInnes, A., Debate: the emergence of a leisure town or an urban renaissance?, *Past and Present*, 126 (1990), 189–202

Boserup, E., *The conditions of agricultural growth* (Chicago, Aldine Press, 1965)

Brace, C. L., *The Norse folk: or visits to the homes of Norway and Sweden* (London, Richard Bentley, 1857)

Brändström, A., *'De kärlekslösa mödrarna'* (Acta Universitatis Umensis: Umeå Studies in the Humanities, 62, Umeå, 1984)

Braudel, F., *The Mediterranean and the Mediterranean world in the age of Philip II* (London, Fontana, 1975)

The identity of France, vol. I: *History and environment* (London, Collins, 1988)

Braun, R., *Industrialisation and everyday life* (Cambridge, Cambridge University Press, 1990)

Brenner, R., Agrarian class structure and economic development in pre-industrial Europe, *Past and Present*, 70 (1976), 29–73, reprinted in T. H. Aston and C. H. E. Philpin (eds), *The Brenner debate: agrarian class structure and economic development in pre-industrial Europe* (Cambridge, Cambridge University Press, 1985)

Bring, S. E., *Göta Kanals historia*, 2 vols. (Stockholm, Almqvist & Wiksell, 1930)

Burke, P. (ed.), *The new Cambridge modern history*, vol. XIII (Cambridge, Cambridge University Press, 1979)

Butlin, R. A., *The transformation of rural England c. 1580–1800: a study in historical geography* (Oxford, Oxford University Press, 1982)

Early industrialization in Europe: concepts and problems, *Geographical Journal*, 152 (1986), 1–8

Buttimer, A., Grasping the dynamism of the life-world, *Annals of the Association of American Geographers*, 66 (1976), 277–92

Bynum, W. F., Porter R. and Shepherd M. (eds.), *The anatomy of madness* (London, Tavistock, 1985)

Bythell, D., *The handloom weavers* (Cambridge, Cambridge University Press, 1969)

Campbell, B. M. S., Land, labour, livestock and productivity trends in English seignorial agriculture, 1208–1450, in B. M. S. Campbell and M. Overton (eds.), *Land, labour and livestock: historical studies in European agricultural productivity* (Manchester, Manchester University Press, 1991), 144–82

Cannadine, D., Review of periodical articles, *Urban History Yearbook*, 1982, 63–71

Carlestam, G. and Sollbe, B. (eds.), *Om tidens vidd och tingens ordning: texter av Torsten Hägerstrand* (Stockholm, Byggforskningsrådet, 1991)

Carlsson, S., *Ståndssamhälle och ståndspersoner, 1700–1865* (Lund, Gleerups, 1973)
Chronology and composition of Swedish emigration to America, in H. Runblom and H. Norman (eds.), *From Sweden to North America* (Studia Historica Upsaliensis, 74, Uppsala, 1976), 114–48

Carlstein, T., The sociology of structuration in time and space: a time-geographic assessment of Giddens' theory of structuration, *Svensk Geografisk Årsbok*, 57 (1981), 41–57
Time resources, society and ecology (London, Allen and Unwin, 1982)

Chayanov, A. V., *The theory of peasant economy* (Manchester, Manchester University Press, 1966)

Clarke, C. G. and Langton, J. (eds.), *Peasantry and progress: rural culture and the modern world*, School of Geography, University of Oxford Research Papers, 45 (1990)

Clarkson, L. A., *Proto-industrialization: the first phase of industrialization?* (London, Macmillan, 1985)

Cleary, M., Patterns of transhumance in Languedoc, *Geography*, 71 (1986), 25–33
Peasants, politics and producers: the organisation of agriculture in France since 1918 (Cambridge, Cambridge University Press, 1989)

Cloke, P., Philo C. and Sadler, D., *Approaching human geography: an introduction to contemporary theoretical debates* (London, Paul Chapman, 1991)

Cobbett, W., *Rural rides* (1830; reprint Harmondsworth, Penguin, 1967)

Cole, J. W., Anthropology comes part-way home: community studies in Europe, *Annual Review of Anthropology*, 6 (1977), 249–78

Cole J. and Wolf, E., *The hidden frontier: ecology and ethnicity in an Alpine valley* (New York, Academic Press, 1974)

Coleman, D. C., Proto-industrialization: a concept too many, *Economic History Review*, 2nd series, 36 (1983), 435–48

Colinvaux, P., *The fates of nations: a biological theory of history* (Harmondsworth, Penguin, 1980)

Conzen, M., Historical geography: changing spatial structures and social patterns of western cities, *Progress in Human Geography*, 7 (1983), 88–107

Cooke, P., *Back to the future: modernity, postmodernity and locality* (London, Unwin Hyman, 1990)

Corfield, P. J., *The impact of English towns 1700–1800* (Oxford, Oxford University Press, 1982)

Corfield, P. J. and Keene, D. (eds.), *Work in towns 850–1850* (Leicester, Leicester University Press, 1990)

Corrigan P. and Sayer, D. *The great arch: English state formation and cultural revolution* (Oxford, Blackwell, 1985)

Coss, P. R., Bastard feudalism revisited, *Past and Present*, 125 (1989), 27–35

Court, W. H. B., Industrial organization and economic progress in the eighteenth-century Midlands, in W. H. B. Court, *Scarcity and choice in history* (London, Edward Arnold, 1970), 235–49

Cox, K. R. and Golledge, R. G. (eds.), *Behavioral problems in geography revisited* (London, Methuen, 1981)

Crafts, N. S. B., *British economic growth during the industrial revolution* (Oxford, Clarendon Press, 1985)

Creighton, C., Family, property and relations of production in Western Europe, *Economy and Society*, 9 (1980), 129–67

de Crisenoy, C., Capitalism and agriculture, *Economy and Society*, 8 (1979), 9–25

Crone, P., *Pre-industrial societies* (Oxford, Blackwell, 1989)

Dahl, S., *Torna och Bara: studier i Skånes bebyggelse- och näringsgeografi före 1860* (Meddelanden från Lunds Universitets Geografiska Institutioner, 6, Lund, 1942) Travelling pedlars in nineteenth century Sweden, *Scandinavian Economic History Review*, 7 (1950), 167–78

Dahlman, C., *The open field system and beyond: a property rights analysis of an economic institution* (Cambridge, Cambridge University Press, 1980)

Darby, H. C., The draining of the English claylands, *Geographische Zeitschrift*, 52 (1964), 190–201

Daunton, M., 'Gentlemanly capitalism' and British industry 1820–1914, *Past and Present*, 122 (1989), 119–58

Davie, N., Chalk and cheese? 'Fielden' and 'forest' communities in early modern England, *Journal of Historical Sociology*, 4 (1991), 1–31

De Geer, S., Sveriges geografiska regioner, *Ymer*, 65 (1925), 393–415

Defoe, D., *A tour through the whole island of Great Britain* (1724–26; reprint Harmondsworth, Penguin, 1971)

Dobrowolski, K., Peasant traditional culture, in Shanin, *Peasants*, 1st edn, 277–98

Dodgshon, R. A., *The European past: social evolution and spatial order* (London, Macmillan, 1987) The changing evaluation of space, 1500–1914, in Dodgshon and Butlin, *Historical geography*, 2nd edn, 255–83

Dodgshon, R. A. and Butlin, R. A., *Historical geography of England and Wales* (London, Academic Press, 1978; 2nd edn, 1990)

Ek, G., Economic booms, innovations and the popular culture, *Economy and History*, 3 (1960), 3–37

Elgeskog, V., *Svensk torpbebyggelse från 1500-talet till laga skiftet* (Stockholm, LT:s Förlag, 1945)

Ellis, H. S. (ed.), *A survey of contemporary economics* (Philadelphia, Pa., Blakiston, 1948)

Entrikin, J. N., Philosophical issues in the scientific study of regions, in D. T. Herbert and R. J. Johnston (eds.), *Geography and the urban environment*, 4 (Chichester, Wiley, 1981), 1–27 *The betweenness of place: towards a geography of modernity* (London, Macmillan, 1991)

Eriksson, G. A., Advance and retreat of charcoal iron industry and rural settlement in Bergslagen, *Geografiska Annaler*, 42 (1960), 267–84

Eriksson, I. and Rogers, J., *Rural labor and population change: social and demographic developments in east-central Sweden during the nineteenth century* (Studia Historica Upsaliensia, 100, Uppsala, 1978)

Erixon, S., *Gården och familjen* (Etnologiska studier tillägnade Nils Edvard Hammarstedt 3.3.1921. Föreningen för svensk kulturhistoria, böcker no. 2, Stockholm, 1921)

Eskeröd, A., Jordskiftena och lantbrukets utveckling 1809–1914, in S. Carlsson (ed.), *Bonden i svensk historia*, 3 (Stockholm, Lantbruksförbundets Tidskrifts AB, 1956), 11–116

Everitt, A., Country, county and town; patterns of regional evolution in England, *Transactions of the Royal Historical Society*, 5th series, 29 (1979), 79–108

Fahlbeck, P., Occupations and industries: a general survey, in Sundbärg, *Sweden*, 497–510

Feeny, D., Berkes, F., McCay B. J. and Acheson, J. M., The tragedy of the commons: twenty-two years later, *Human Ecology*, 18 (1990), 1–19

Figes, O. and Danilov V. P., On the analytical distinction between peasants and farmers, in Shanin, *Peasants*, 2nd edn, 121–4

Fisher, J. (ed.), *Essays in the economic and social history of Tudor and Stuart England in honour of R. H. Tawney* (Cambridge, Cambridge University Press, 1961)

Floud, R. and McCloskey, D. N. (eds.), *The economic history of Britain since 1700*, vol. I: *1700–1860* (Cambridge, Cambridge University Press, 1981)

Forsberg, K., *Gusums bruks historia, 1653–1953* (Gusum, Gusums bruks AB, 1953)

Foster, G. M., Peasant society and the image of the limited good, *American Anthropology*, 67 (1965), 293–315

Fox, H. S. A. and Butlin, R. A. (eds.), *Change in the countryside*, Institute of British Geographers Special Publication, 10 (1979)

Freeman, M. J., *A perspective on the geography of English internal trade during the industrial revolution*, School of Geography, University of Oxford Research Papers, 29 (1982)

Friberg, N., The growth of population and its economic–geographical background in a mining district of central Sweden 1650–1750, *Geografiska Annaler*, 38 (1956), 395–437

Fridlizius, G., *Swedish corn exports in the free trade era* (Samhällsvetenskapliga studier, 14, Lund, 1957)

Agricultural productivity, trade and urban growth during the phase of commercialization of the Swedish economy, in A. D. van der Woude, A. Hayami and J. de Vries (eds.), *Urbanization in history: a process of dynamic interactions* (Oxford, Clarendon Press, 1990), 113–33

Friedmann, H., World market, state, and family farm: social bases of household production in the era of wage labor, *Comparative Studies of Society and History*, 20 (1978), 545–86

Economic analysis of the Postbellum South: regional economies and world markets, *Comparative Studies of Society and History*, 22 (1980), 639–52

Frödin, J., Skiftesväsendet och dess samband med jordbrukets nuvarande kritiska läge, *Ekonomisk Tidskrift*, 47 (1945), 295–308

Furuland, L., *Statarna i litteraturen: en studie i svensk dikt och samhällsdebatt* (Stockholm, Tiden, 1962)

Gadd, C. J., *Järn och potatis: jordbruk, teknik och social omvandling i Skaraborgs län 1750–1860* (Meddelanden från Ekonomisk–historiska Institutionen, Göteborgs Universitet, 53, Göteborg, 1983)

Galeski, B., Sociological problems of the occupation of farmers, in Shanin, *Peasants*, 1st edn, 180–201

Gårdlund, T., *Svensk industrifinansiering 1830–1913* (Stockholm, Svenska Bankföreningen, 1947)

Garfield, V. E. (ed.), *Proceedings of the 1961 annual spring meeting of the American Ethnological Society* (Seattle, American Ethnological Society, 1961)

Gaunt, D., Natural resources – population – local society: the case of pre-industrial Sweden, *Peasant Studies*, 6 (1977), 137–41

Pre-industrial economy and population structure: elements of variance in early modern Sweden, *Scandinavian Journal of History*, 2 (1977), 183–210

Geertz, C., *Agricultural involution: the process of ecological change in Indonesia* (Berkeley, Calif., University of California Press, 1963)

Gerard, R. W., Entitation, animorgs and other systems, in Mesarovic, *Views*, 119–24

Gerger, C., *Där nöden var som störst* (Meddelanden från Kulturgeografiska Institutionen, Stockholms Universitet, 82, Stockholm, 1992)

Gerger, T. and Hoppe, G. *Education and society: the geographer's view* (Acta Universitatis Stockholmiensis: Stockholm Studies in Human Geography, 1, Stockholm, Almqvist & Wiksell International, 1980)

Gibbon, P., Arensberg and Kimball revisited, *Economic Sociology*, 2 (1973), 479–98

Giddens, A., Time, space and regionalisation, in A. Giddens, *The constitution of society* (Cambridge, Polity Press, 1984), reprinted in Gregory and Urry, *Social relations*, 265–95

Gilbert, A., The new regional geography in English- and French-speaking countries, *Progress in Human Geography*, 12 (1988), 208–28

Glass, D. V. and Eversley, D. E. C. (eds), *Essays in historical demography* (London, Edward Arnold, 1965)

Glennie, P., In search of agrarian capitalism: manorial land markets and the acquisition of land in the Lea valley, *c.* 1450–*c.* 1560, *Continuity and Change*, 3 (1988), 11–40

Industry and towns 1500–1730, in Dodgshon and Butlin, *Historical geography*, 2nd edn, 199–222

Goodman, D. and Redclift, M., *From peasant to proletarian: capitalist development and agrarian transition* (Oxford, Blackwell, 1981)

Goody, J., Inheritance, property and women: some comparative considerations, in Goody, Thirsk and Thompson, *Family*, 10–36

Introduction, in Goody, Thirsk and Thompson, *Family*, 1–9

The development of the family and marriage in Europe (Cambridge, Cambridge University Press, 1983)

Goody, J., Thirsk, J. and Thompson, E. P. (eds.), *Family and inheritance: rural society in Western Europe 1200–1800* (Cambridge, Cambridge University Press, 1976)

Göransson, A. *Från familj till fabrik* (Arkiv avh. serie, 28, Lund, 1988)

Göransson, S., De regelbundna strukturerna i östsveriges bebyggelsegeografi, *Geografiska Regionstudier*, 15 (Uppsala, 1985), 65–82

Grant, R., *The royal forests of England* (Stroud, Alan Sutton, 1991)

Gregory, D., *Regional transformation and industrial revolution: a geography of the Yorkshire woollen industry* (London, Macmillan, 1982)

Time-geography, in Johnston, *Dictionary*, 484–7

Suspended animation: the stasis of diffusion theory, in Gregory and Urry, *Social relations*, 296–336

'A new and different face in many places': three geographies of industrialization, in Dodgshon and Butlin, *Historical geography*, 2nd edn, 351–99

Gregory, D. and Urry, J. (eds.), *Social relations and spatial structures* (London, Macmillan, 1985)

Gregson, N., On duality and dualism: the case of structuration theory and time geography, *Progress in Human Geography*, 2 (1986), 184–205

Guillemot, A., *Rask Resolut Trogen* (Acta Universitatis Umensis: Umeå Studies in the Humanities, 74, Umeå, 1986)

Gullickson, G. L., *The spinners and weavers of Auffray*, (Cambridge, Cambridge University Press, 1986)

Guinchard, J. (ed.), *Sweden: historical and statistical handbook*, 2 vols. (Stockholm, P. A. Norstedt and Söner, 1914)

Gustavson, G. G., *The small giant: Sweden enters the industrial era* (Athens, Ohio, Ohio University Press, 1986)

Guteland, G., *et al.*, *The biography of a people* (Stockholm, Royal Ministry of Foreign Affairs, 1974)

Hägerstrand, T., En landbygdsbefolknings flyttningsrörelser, *Svensk Geografisk Årsbok*, 23 (1957), 114–42

  Migration and area, in D. Hannerberg, T. Hägerstrand and B. Odeving (eds.), *Migration in Sweden: a symposium* (Lund Studies in Geography, 13B, Lund, 1957), 27–158

  Tidsanvändning och omgivningsstruktur, in Appendix 4, *Statens Offentliga Utredningar*, 14 (Allmänna Förlaget, Stockholm, 1970), 5–37

  What about people in regional science?, *Papers of the Regional Science Association*, 24 (1970), 7–21

  The domain of human geography, in R. J. Chorley, (ed.), *Directions in Geography* (London, Methuen, 1973), 67–87

  Tidsgeografisk beskrivning, syfte och postulat, *Svensk Geografisk Årsbok*, 30 (1974), 86–94

  Ecology under one perspective, in E. Bylund *et al.* (eds.), *Ecological problems of the circumpolar area* (Luleå, Norbottens Museum, 1974), 271–6

  Survival and arena: on the life-history of individuals in relation to their geographical environment, *The Monadnock*, Clark Univ. Geogr. Soc., 49 (1975), 9–29

  Space, time and human conditions, in A. Karlqvist *et al.* (eds), *Dynamic allocation of urban space* (Farnborough, Saxon House, 1975), 3–14

  Geography and the study of interaction between nature and society, *Geoforum*, 7 (1976), 329–34

  Diorama, paths and projects, *Tijdschrift voor Economische en Sociale Geografie*, 73 (1982), 323–39

  Tidsgeografi, in Carlestam and Sollbe, *Om tidens*, 133–42

Haggett, P. and Chorley, R. J., *Frontiers in geographical teaching* (London, Methuen, 1965)

Hajnal, J., European marriage patterns in perspective, in Glass and Eversley, *Essays*, 101–43

Hammel, E. A., The Zadruga as process, in Laslett and Wall, *Household*, 335–73

Hamshere, J. D., Data sources in historical geography, in Pacione, *Historical geography*, 46–69

Hannerberg, D., *Närkes landsbygd 1600–1820: Folkmängd och befolkningsrörelse, åkerbruk och spannmålsproduktion* (Medd. Göteborgs Högskolas Geogr. Inst., Göteborg, 1941)

Den äldre pappersbruksbygden i Sverige, in J. Malmros (ed.), *En bok om papper* (Klippan, Klippans finpappersbruk, 1944), 210–28

Svenskt agrarsamhälle under 1200 år: gård och åker, skörd och boskap (Stockholm, Läromedelsförlagen–Scandinavian University Books, 1971)

Kumlaboken IV: By, gård och samhälle (Kumla, Kumla Bokhandel, 1977)

Hanssen, B., *Österlen: en studie över socialantropologiska sammanhang under 1600– och 1700–talet i sydöstra Skåne* (Ystad, LT:s Förlag, 1952)

Commonfolk and gentlefolk, Ethnologia Scandinavica (1973), 67–100

Haraldsson, K., *Tradition, regional specialisering och industriell utveckling, Geografiska Regionstudier*, 21 (Uppsala, 1989)

Hardin, G., The tragedy of the commons, *Science*, 162 (1968), 1,243–8

Harley, J. B., Historical geography and its evidence: reflections on modelling sources, in Baker and Billinge, *Period and place*, 261–73

Harriss, J., *Capitalism and peasant farming: agrarian structure and ideology in northern Tamil Nadu* (Oxford, Oxford University Press, 1982)

(ed.), *Rural development: theories of peasant economy and agrarian change* (London, Hutchinson, 1982)

Hart, C., *The commoners of Dean Forest* (Gloucester, British Publishing Co., 1951)

Harvey, D., Money, time, space and the city, in D. Harvey, *Consciousness and the urban experience: studies in the history and theory of capitalist urbanization* (Oxford, Blackwell, 1985), 1–35

The condition of post-modernity: an enquiry into the origins of social change (Oxford, Blackwell, 1989)

Havinden, M., Agricultural progress in open field Oxfordshire, *Agricultural History Review*, 9 (1971), 73–83

Heckscher, E. F., *Industrialismen. Den ekonomiska utvecklingen sedan 1750* (Stockholm, Norstedts, 1931)

The place of Sweden in modern economic history, *Economic History Review*, 4 (1932), 1–22

Svenskt arbete och liv (Stockholm, Bonniers, 1941)

Det svenska jordbrukets äldre ekonomiska historia, en skiss, *Ekonomisk Tidskrift*, 47 (1945), 197–224

Sveriges ekonomiska historia från Gustaf Vasa, vol. III (Stockholm, Bonniers, 1949)

An economic history of Sweden (Cambridge, Mass., Harvard University Press, 1954)

Hefner, R., The culture problem in human ecology: a review article, *Comparative Studies of Society and History*, 25 (1983), 547–56

Hellspong, M. and Löfgren, O., *Land och stad: Svenska samhällstyper och livsformer från medeltid till nutid* (Lund, Liber Läromedel, 1974)

Helmfrid, B., *Tre sekler med Häfla bruk* (Linköping, Ytforum Förlags A. B., 1989)

Helmfrid, S., The storskifte, enskifte and laga skifte in Sweden - general features, *Geografiska Annaler*, 43 (1961), 114–29

Östergötland Västanstång. Studien über die ältere Agrarlandschaft und ihre Genese, Geografiska Annaler, 44 (1962), 1–277

Herlitz, L., *Jordegendom och ränta: omfördelningen av jordbrukets merprodukt i*

434   Bibliography

Skaraborgs län under frihetstiden (Meddelanden från Ekonomisk–Historiska Institutionen, Göteborgs Universitet, 31, Göteborg, 1974)

Homespun capitalism, *Scandinavian Economic History Review*, 25 (1977), 101–3

Markegångspriser och relativ prisutveckling vid 1700–talets mitt, in *Ekonomisk-historiska studier tillägnade Artur Attman* (Meddelanden från Ekonomisk–historiska Institutionen, Göteborgs Universitet, 39, Göteborg, 1977)

Herlitz, U., *Restadtegen i världsekonomin* (Meddelanden från Ekonomisk–historiska Institutionen, Göteborgs Universitet, 58, Göteborg, 1988)

Hildebrand, K. G., *Svenskt järn* (Jernkontorets Bergshist. Skriftserie, 20, 1987)

Foreign markets for Swedish iron in the eighteenth century, *Scandinavian Economic History Review*, 6 (1958), 3–52

Hilton, R., The small town as part of peasant society, in R. Hilton, *The English peasantry in the later middle ages* (Oxford, Clarendon Press, 1975), 76–94

Introduction, in R. Hilton (ed.), The *transition from feudalism to capitalism* (London, New Left Books, 1976), 1–30

Hodges, R., *Primitive and peasant markets* (Oxford, Blackwell, 1988)

Holderness, B. A., 'Open' and 'close' parishes in England in the eighteenth and nineteenth centuries, *Agricultural History Review*, 20 (1972), 126–39

Homans, G. C., The explanation of English regional differences, *Past and Present*, 42 (1969), 18–34

Hoppe, G., *Skolutbildning och individbanans utveckling* (Stockholm University, Department of Geography, Kulturgeografiskt Seminarium, 5/77, Stockholm, 1977)

Enclosure in Sweden: Background and consequences (Stockholm University, Department of Geography, Kulturgeografiskt Seminarium, 9/79, Stockholm, 1981)

At the ventilation of the suggested redistribution, much controversy was disclosed...': enclosure in Väversunda village, Östergötland (Stockholm University, Department of Geography, Kulturgeografiskt Seminarium, 5/82, Stockholm, 1982)

Skiftesreformerna och agrarutvecklingen, *Bebyggelsehist. Tidskr.*, 5 (1983), 32–45

Review of Isacson and Magnusson, *Proto-industrialisation*, in *Journal of Historical Geography*, 14 (1988), 314–15

Från jordbruk till järnverk, *Bebyggelsehist. Tidskr.*, 16 (1988), 77–92

Hoppe, G. and Langton, J., *Countryside and town in industrialisation* (Stockholm University, Department of Geography, Kulturgeografiskt Sem., 7/79, Stockholm, 1979)

Time-geography and economic development: the changing structure of livelihood positions on arable farms in nineteenth-century Sweden, *Geografiska Annaler*, 68B (1986), 115–37

Flows of labour in the early phase of capitalist development: the time-geography of longitudinal migration paths in nineteenth-century Sweden, Historical Geography Research Series, 29, (1992)

Hörsell, A., *Borgare, smeder och änkor: ekonomi och befolkning i Eskilstuna gamla stad och fristad 1750–1850* (Stockholm, Almqvist & Wiksell International, 1983)

Houston, R. and Snell, K., Proto-industrialization? Cottage industry, social change, and industrial revolution, *Historical Journal*, 27 (1984), 473–92

Howell, C., Stability and change, 1300–1700: the socio-economic context of the self-perpetuating family farm in England, *Journal of Peasant Studies*, 2 (1975), 468–82

Hoyle, R. W., Tenure and the land market in early modern England: or a later contribution to the Brenner debate, *Economic History Review*, 2nd series, 53 (1990), 1–20

Hudson, P., Proto-industrialization: the case of the West Riding wool textile industry in the 18th and early 19th centuries, *Historical Workshop Journal*, 12 (1981), 34–61
The regional perspective, in Hudson, *Regions*, 5–38
Capital and credit in the West Riding wool textile industry *c*. 1750–1850, in Hudson, *Regions*, 69–99
Regions and industries: a perspective on the industrial revolution in Britain (Cambridge, Cambridge University Press, 1989)
From manor to mill: the West Riding in transition, in Berg, Hudson and Sonenscher, *Manufacturing*, 124–44

Hyde, C. K., *Technological change and the British iron industry, 1700–1870* (Princeton, NJ, Princeton University Press, 1977)

Hymer, S. and Resnick, S., A model of an agrarian economy with nonagricultural activities, *American Economic Review*, 59 (1969), 493–506

Ingers, E., *Bonden i svensk historia*, 2 vols. (Stockholm, Lantbruksförbundets tidskrifts AB, 1948)

Isacson, M., *Ekonomisk tillväxt och social differentiering 1680–1860* (Acta Universitatis Upsaliensis: Uppsala Studies in Economic History, 18, 1979)

Isacson, M. and Magnusson, L., *Proto-industrialisation in Scandinavia: craft skills in the industrial revolution* (Leamington Spa, Berg, 1987)

Johansson, S., *Knyppling i Östergötland: tradition och nyskapande* (Föreningen Gamla Vadstena, Småskriftserien, 28, Vadstena, 1983)

Johnson, B. L. C., The charcoal iron industry in the early eighteenth century, *Geographical Journal*, 118 (1951), 167–177

Johnson, R. E., *Peasant and proletarian: the working class of Moscow in the late nineteenth century* (Leicester, Leicester University Press, 1979)

Johnston R. J. (ed.), *The dictionary of human geography*, 2nd edn (Oxford, Blackwell 1986)

Johnston, R. J., Hauer, J. and Hoekveld, G. A. (eds), *Regional geography: current developments and future prospects* (London, Routledge, 1990)

Jones, E. L., The agricultural origins of industry, *Past and Present*, 40 (1960), 58–71
The environment and the economy, in Burke, *Cambridge History*, 15–42
The European miracle: environments, economies and geopolitics in the history of Europe and Asia (Cambridge, Cambridge University Press, 1981)

Jones, P. M., Parish, seigneurie and the community of inhabitants in southern central France during the eighteenth and nineteenth centuries, *Past and Present*, 91 (1981), 74–108

Jonsson, J. O., *Arbetsorganisitionen vid Motala verkstad 1822–1843* (Acta Universitatis Stockholmiensis, Stockholm Studies in Economic History, 13. 1990)

Jonsson, U., *Jordmagnater, landbönder och torpare i sydöstra Södermanland 1800–1880* (Acta Universitatis Stockholmiensis, Stockholm Studies in Economic History, 5, 1980)

Jörberg, L., *A history of prices in Sweden, 1732–1914*, vol. I (Lund, Gleerups, 1972)
Structural change and economic growth in nineteenth-century Sweden, in Koblik, *Sweden's development*, 92–135

436   Bibliography

Kain, R. J. P. and Prince, H., *The tithe surveys of England and Wales* (Cambridge, Cambridge University Press, 1985)

Kälvemark, A-S., The country that kept track of its population, in Sundin and Söderlund, *Time*, 221–38

Karlsson, F., *Mark och försörjning* (Meddelanden från Ekonomisk–historiska Institutionen, Göteborgs Universitet, 58, Göteborg, 1978)

Kenchington, F. E., *The commoners' New Forest* (London, Hutchinson, 1944)

Kerridge, E., *The farmers of old England* (London, Allen and Unwin, 1973)

Textiles manufacture in early modern England (Manchester, Manchester University Press, 1985)

Koblik, S. (ed.), *Sweden's development from poverty to affluence, 1750–1970* (Minneapolis, University of Minnesota Press, 1975)

Köll, A-M., *Tradition och reform i västra Södermanlands jordbruk 1810–1890* (Acta Universitatis Stockholmiensis: Stockholm Studies in Economic History, Stockholm, Almqvist & Wiksell International, 1983)

Kriedte, P., Medick, H. and Schlumbohm, J., *Industrialization before industrialization* (Cambridge, Cambridge University Press, 1981)

Proto-industrialization on test with the guild of historians: response to some critics, *Economy and Society*, 15 (1986), 254–72

Kula, W., *An economic theory of the feudal system: towards a model of the Polish economy 1500–1800* (London, New Left Books, 1979)

Kumm, E. *Indelt soldat och rotebonde* (LT:s Förlag, Stockholm, 1949)

Kussmaul, A. *A general view of the rural economy of England, 1538–1840* (Cambridge, Cambridge University Press, 1990)

Kuuse, J., *Från redskap till maskiner. Mekaniseringsspridning och kommersialisering inom svenskt jordbruk 1860–1910* (Meddelanden från Ekonomisk–historiska Institutionen, Göteborgs Universitet, 20, Göteborg, 1970)

Kyle, J., *Striden om hemmanen* (Meddelanden från Ekonomisk–historiska Institutionen, Göteborgs Universitet, 31, Göteborg, 1987)

Le Roy Ladurie, E., *The peasants of Languedoc* (Urbana, University of Illinois Press, 1976)

Peasants, in Burke, *Cambridge History*, 115–64

Lägnert, F., *Syd-och mellansvenska växtföljder I: De äldre brukningssystemens upplösning under 1800–talet* (Meddelanden från Lunds Universitets Geografiska Institutioner, Avhandlingar 29, Lund, 1955)

Langton, J., Potentialities and problems of adopting a systems approach to the study of change in human geography, *Progress in Geography* 4, (1972), 125–79

Geographical change and industrial revolution: coal mining in south-west Lancashire, 1590–1799 (Cambridge, Cambridge University Press, 1979)

Landowners and the development of coalmining in south-west Lancashire 1590–1799, in Fox and Butlin, *Change*, 123–44

Industry and towns 1500–1730, in Dodgshon and Butlin, *Historical geography*, 1st edn, 182–3

The industrial revolution and the regional geography of England, *Transactions, Institute of British Geographers*, Series 2, 9 (1984), 145–67

Habitat, economy and society revisited: peasant ecotypes and economic development in Sweden, *Cambria*, 13 (1986), 5–24

The two traditions of geography, historical geography and the study of landscapes, *Geografiska Annaler*, 70B (1988), 17–25

The end of peasantry in nineteenth-century Sweden?, in Clarke and Langton, *Peasantry and progress*, 13–23

Langton, J. and Hoppe, G., *Town and country in the development of early modern Western Europe*, Historical Geography Research Series, 11 (1983)

Urbanization, social structure and population circulation in pre-industrial times: flows of people through Vadstena (Sweden) in the mid-nineteenth century, in Corfield and Keene, *Work*, 138–63

Patterns of migration and regional identity: economic development, social change and the lifepaths of individuals in nineteenth-century western Östergötland, in D. Postles (ed.), *Naming, society and regional identity* (Leicester, Leicester University Press, 1994)

Langton, J. and Morris, R. J., Introduction, in Langton and Morris (eds.), *Atlas of Industrializing Britain* (London, Methuen, 1986)

Larsson, A., Match industry, in Guinchard, *Sweden*, vol. II, 398–400

Metal and machine industry, in Guinchard, *Sweden*, vol. II, 438–63

Laslett, P. and Wall R. (eds.), *Household and family in past time* (Cambridge, Cambridge University Press, 1972)

Layton, I. G., *The evolution of upper Norrland's ports and loading places 1750–1976* (University of Umeå, Department of Geography, Reports, 11, 1981)

Industry and the local economy: changing contact patterns within the Luleå district during the nineteenth century, *Geografiska Annaler Series*, 70B (1988), 205–18

En handelsmans död, in *Geografi och samhällsplanering, studier tillägnade Erik Bylund* (GERUM Report, B:7, Umeå, 1982), 255–80

Lenin, V. I., *The development of capitalism in Russia* (Moscow, Foreign Languages Publishing House, 1956)

Levine, D., *Family formation in an age of nascent capitalism* (London, Academic Press, 1977)

Lext, G., *Mantalsskrivningen i Sverige före 1860* (Meddelanden från Ekonomisk-historiska Institutionen, Göteborgs Universitet, 13, Göteborg, 1968)

Ley, D. and Samuels, M. (eds.), *Humanistic geography: prospects and problems* (Chicago, Ill., Maaroufa Press, 1978)

Lindegren, J., *Utskrivning och utsugning* (Acta Universitatis Upsaliensis; Studia Historica Upsaliensa, 117, Uppsala, 1980)

Lithell, U. B., *Breast feeding and reproduction: studies in marital fertility in nineteenth-century Finland and Sweden* (Acta Universitatis Upsaliensis, Studia Historica Upsaliensia, 120, Uppsala, 1981)

Ljunggren, C. J. F., Handicrafts, in Guinchard, *Sweden*, vol. II, 483–90

Löfgren, O., Family and household among Scandinavian peasants: an exploratory essay, *Ethnologia Scandinavica*, (1974), 17–52

Peasant ecotypes: problems in the comparative study of ecological adaptation, *Ethnologia Scandinavica*, (1976), 100–115

Historical perspectives on Scandinavian peasantries, *Annual Review of Anthropology*, 9 (1980), 187–215

Lucassen, J., *Migrant labour in Europe 1600–1900* (London, Croom Helm, 1987)

Lundahl, M. and Svensson, T. (eds.), *Agrarian society in history: essays in honour of*

Magnus Mörner (London, Routledge, 1990), 48–67

Lundsjö, O., *Fattigdomen på den svenska landsbygden under 1800–talet* (Acta Universitatis Stockholmiensis: Stockholm Studies in Economic History, 1, Stockholm, 1975)

McCloskey, D., The economics of enclosure: a market analysis, in Parker and Jones, *European peasants*, 123–60

Macfarlane, A., *The origins of English individualism*, (Oxford, Blackwell, 1978)

Peasants, in A. Macfarlane, *The culture of capitalism* (Oxford, Blackwell, 1987), 1–24

McGuire, R. and Netting, R. M., Leveling peasants? the maintenance of equality in a Swiss Alpine community, *American Ethnology*, 9 (1982), 269–90

McLeod, C., *Inventing the industrial revolution: the English patent system, 1660–1800* (Cambridge, Cambridge University Press, 1988)

Magnusson, L., *Ty som ingenting angelägnare är än mina bönders conservation* (Acta Universitatis Upsaliensis, Uppsala Studies in Economic History, 20, Uppsala, 1980)

Marshall, J., *Travels thro' Holland . . . Sweden . . . and Poland in the years 1768, 1769 and 1770* (London, J. Almond, 1772)

Marshall, J. D., The study of local and regional 'communities': some problems and possibilities, *Northern History*, 27 (1981), 203–30

Martin, J. E., *Feudalism to capitalism: peasant and landlord in English agrarian development* (London, Macmillan, 1983)

Martinius, S., *Agrar kapitalbildning och finansiering 1833–1892* (Meddelanden från Ekonomisk–historiska Institutionen, Göteborgs Universitet, 17, Göteborg, 1970)

Peasant destinies: the history of 552 Swedes born 1810–12 (Stockholm, Almqvist & Wicksell, 1977)

Jordbrukets omvandling på 1700– och 1800–talen (Malmö, Liber, 1982)

Massey, D., *Spatial divisions of labour: social structures and the geography of production* (London, Macmillan, 1984)

Medick, H., The proto-industrial family economy, in Kriedte, Medick and Schlumbohm, *Industrialization*, 38–73

The structures and function of population-development under the proto-industrial system, in Kriedte, Medick and Schlumbohm, *Industrialization*, 74–93

Mendels, F. F., Proto-industrialization: the first phase of the industrialization process, *Journal of Economic History*, 32 (1972), 241–61

Agriculture and peasant industry in eighteenth-century Flanders, in Parker and Jones, *European peasants*, 179–204

Social mobility and phases of industrialization, *Journal of Interdisciplinary History*, 7 (1976), 193–216

Seasons and regions in agriculture and industry during the process of industrialisation, in Pollard, *Region*, 177–95

Mesarovic, M. D. (ed.), *Views on general systems theory* (New York, Wiley, 1964)

Miller, R. and Gerger, T., *Social change in 19th-century Swedish agrarian society* (Acta Universitatis Stockholmiensis, Stockholm Studies in Human Geography, 5, Stockholm, Almqvist & Wicksell, 1985)

Mills, D. R., *Lord and peasant in nineteenth-century Britain* (London, Croom Helm, 1980)

Mingay, G. E., *Enclosure and the small farmer in the age of the industrial revolution*,

(London, Macmillan, 1968)

The gentry: the rise and fall of a ruling class (London, Longman, 1976)

Mintz, S. W., *Pratik*: Haitian personal economic relationships, in Garfield, *Proceedings*, 54–63

Moberg, V., *The emigrants* (New York, Popular Library, 1951)

Möller, J., *Godsen och den agrara revolutionen* (Meddelanden från Lunds Universitets Geografiska Institutioner, Avhandlingar 106, Lund University Press, 1989)

Montelius, S., Recruitment and conditions of life of Swedish ironworkers during the eighteenth and nineteenth centuries, *Scandinavian Economic History Review*, 16 (1966), 1–17

Montgomery, G. A., *Industrialismens genombrott i Sverige* (Stockholm, Skoglunds, 1931)

The rise of modern industry in Sweden (London, P. S. King, 1939)

Mui, H-C. and Mui, L.H., *Shops and shopkeeping in eighteenth-century England* (London, Routledge, 1989)

Musgrove, F., *The North of England: a history from Roman times to the present* (Oxford, Blackwell, 1990)

Myers, A. R., *Parliaments and estates in Europe to 1789* (London, Thames and Hudson, 1975)

Myrdal, G., *The cost of living in Sweden* (London, P. S. King and Son, 1933)

Myrdal, J., *Medeltidens åkerbruk* (Nordiska Museet, Handlingar 105, Stockholm, 1985)

Nash, M., Markets and Indian peasant economies, in Shanin, *Peasants*, 1st edn, 161–77

Nertorp, H., Swedish artisans in the nineteenth century, *Scandinavian Economic History Review*, 5 (1967), 77–823

Netting, R. M., *Balancing on an Alp: ecological change and continuity in a Swiss mountain community*, (Cambridge, Cambridge University Press, 1981)

Nilsson, S. A., Landbor och skattebönder, in Tedebrand, *Historieforskning*, 143–68

Norddolum, H., The 'Dugnad' in the pre-industrial peasant community. An attempt at explanation, *Ethnologia Scandinavica* (1980), 102–12

Nordström, O., *Relationer mellan bruk och omland i östra Småland 1750–1900* (Meddelanden från Lunds Universitets Geografiska Institutioner, Avhandlingar, 22, Lund, 1952)

Norman, H. *Från Bergslagen till Nordamerika* (Acta Universitatis Upsaliensis, Studia Historica Upsaliensa, 62, Uppsala, 1974)

Nygren, I. *Svensk kreditmarknad 1820–35: översikt av det institutionella kreditväsendets utveckling* (Meddelanden från Ekonomisk–historiska Institutionen, Göteborgs Universitet, 47, Göteborg, 1981)

North, D. C. and Thomas, R. P., *The rise of the Western World: a new economic history* (Cambridge, Cambridge University Press, 1973)

Odén, S., Domestic industries, in Guinchard, *Sweden*, vol. II, 491–496

Olai, B., *Storskiftet i Ekebyborna* (Acta Universitatis Upsaliensis, Studia Historica Upsaliensa, 130, Uppsala, 1983)

'... *till vinnande av ett redigt Storskifte ...*', (Acta Universitatis Upsaliensis, Studia Historica Upsaliensa, 145, Uppsala, 1987)

Orlove, B. S., Ecological anthropology, *Annual Review of Anthropology*, 9 (1980), 235–73

Österberg, E., Den gamla goda tiden, *Scandia*, 47 (1982), 31–61

Østerud, Ø., *Agrarian structure and peasant politics in Scandinavia: a comparative study of rural response to economic change* (Oslo, Universitetsforlaget, 1978)

Overton, M., Agricultural revolution? Development of the agrarian economy in early modern England, in Baker and Gregory, *Explorations*, 118–39

Pacione, M. (ed.), *Historical geography: progress and prospect* (London, Croom Helm, 1987)

van Paassen, C., Human geography in terms of existential anthropology, *Tijdschrift voor Economische en Sociale Geografie*, 67 (1976), 324–41

Parker, D., Sovereignty, absolutism and the function of the law in seventeenth-century France, *Past and Present*, 122 (1989), 36–74

Parker, W. N. and Jones, E. L., *European peasants and their markets: essays in agrarian economic history* (Princeton, NJ, Princeton University Press, 1975)

Parkes, D. N. and Thrift, N., *Time, spaces and places: a chronogeographical perspective* (New York, Wiley, 1980)

Persson, C., Freehold farmers and landownership in a parish of south-eastern Sweden during the nineteenth century, *Journal of Historical Geography*, 14 (1988), 245–59 *Jorden, bonden och hans familj: en studie av bondejordbruket i en socken i norra Småland under 1800–talet* ... (Meddelanden serie B 79, Kulturgeografiska Institutionen, Stockholms Universitet, Stockholm, 1992)

Petersson, G., *Jordbrukets omvandling i västra Östergötland, 1810–1890* (Acta Universitatis Stockholmiensis, Stockholm Studies in Economic History, 13, Almqvist & Wicksell, Stockholm, 1989)

Pettersson, R., *Laga skifte i Hallands län 1827–1876. Förändring mellan regeltvång och handlingsfrihet* (Acta Universitatis Stockholmiensis, Stockholm Studies in Economic History, 6, Stockholm, Almqvist & Wicksell, 1983)

Pfeifer, G., The quality of peasant living in central Europe, in W. L. Thomas (ed.), *Man's role in changing the face of the Earth* (Chicago, Ill., University of Chicago Press, 1956), 240–77

Pfister, U., Appendix: a note on the administrative structure and social stratification in the countryside of Zurich during the Ancien Régime, in Braun, *Industrialisation*

Polanyi, K., *The great transformation: the political and economic origin of our time* (Boston, Beacon Press, 1957)

Pollard, S., *Peaceful conquest: the industrialization of Europe 1760–1970* (Oxford, Oxford University Press, 1981)

(ed.), *Region und Industrialisierung* (Göttingen, Vandenhoek und Ruprecht, 1980)

Poor Law Commissioners' Report; Appendix F, Foreign communications: Sweden, *English Parliamentary Papers*, 1834 (xxxix), 343–89

Popkin, S., *The rational peasant: the political economy of rural society in Vietnam* (Berkeley, Calif., University of California Press, 1979)

Postan, M. M., A plague of economists? on some current myths, errors and fallacies, in M. M. Postan, *Fact and relevance: essays on historical method* (Cambridge, Cambridge University Press, 1971), 80–91

Pounds, J. N. G., *An historical geography of Europe, 1500–1840* (Cambridge, Cambridge University Press, 1976)

Pred, A. R., Of paths and projects: individual behaviour and its societal context, in Cox and Golledge, *Behavioral problems*, 231–55

The choreography of existence: some comments on Hägerstrand's time-geography and its effectiveness, *Economic Geography*, 53 (1977), 207–21

Social reproduction and the time-geography of everyday life, *Geografiska Annaler*, 63B (1981), 5–22

Place, practice and structure: social and spatial transformation in southern Sweden: 1750–1850 (Oxford, Polity Press, 1986)

Making histories and constructing human geographies: The local transformation of practice, power relations and consciousness (Oxford, Westview Press, 1990)

(ed.), *Space and time in geography* (Lund, Gleerups, 1981)

Pressnell, L. S. (ed.), *Studies in the industrial revolution* (London, Athlone Press, 1960)

Price, R., *The modernization of rural France: communication networks and agricultural market structures in nineteenth-century France* (London, Hutchinson, 1983)

Prince, H., Regional contrasts in agrarian structures, in H. Clout (ed.), *Themes in the historical geography of France* (London, Academic Press, 1977), 129–84

Pudup, M. B., Arguments within regional geography, *Progress in Human Geography*, 12 (1988), 369–90

Qvarsell, R., Locked up or put to bed: psychiatry and the treatment of the mentally ill in Sweden, 1800–1920, in Bynum, Porter and Shepherd, *Anatomy*, 86–97

Redfield, R., *The little community and peasant society and culture* (Chicago, Ill., University of Chicago Press, 1966)

Reed, M., The peasantry of nineteenth-century England: a neglected case?, *Historical Workshop Journal*, 18 (1984), 53–76

Nineteenth-century rural England: a case for peasant studies?, *Journal of Peasant Studies*, 14 (1986), 78–99

Revera, M., *Gods och gård 1650–1680: Magnus Gabriel De la Gardies godsbildning och godsdrift i Västergötland* (Acta Universitatis Upsaliensis, Studia Historica Upsaliensa, 70, Uppsala, 1975)

Hur bönders hemman blev säterier, in Revera and Torstendahl, *Bördor*, 73–114

Revera, M. and Torstendahl, R. (eds.), *Bördor, bönder, börd i 1600–talets Sverige* (Lund, Studentlitteratur, 1979)

Resnick, S., The decline of rural industry under export expansion: a comparison among Burma, Philippines and Thailand, 1870–1938, *Journal of Economic History*, 30 (1970), 51–73

Rickman, A. F., *Swedish iron ores*, (London, Faber and Faber, 1939)

Ridgeway, H., Medieval mentalités, *Journal of Historical Sociology*, 4 (1991), 175–81

Rydeberg, G., *Skatteköpen i Örebro län* (Acta Universitatis Upsaliensis, Studia Historica Upsaliensia, 141, 1986)

Salaman, R., *The history and social influence of the potato*, 2nd edn (Cambridge, Cambridge University Press, 1985)

Samuelson, P. A., Dynamic process analysis, in Ellis, *Survey*, 352–87

Samuelsson, K., *From great power to welfare state: 300 years of Swedish social history* (London, George Allen and Unwin, 1968)

Sayer, A., Explanation in human geography: abstraction versus generalization, *Progress in Human Geography*, 6 (1982), 68–88

Schmal H. (ed.), *Patterns of European urbanization since 1500* (London, Croom Helm, 1981)

Schön, L., *Från hantverk till fabriksindustri. Svensk textiltillverkning 1820–1870* (Arkiv

Avhandlingsserie, 9, Kristianstad, 1979)

Scott, F. D., *Sweden: the nation's history* (Minneapolis, University of Minnesota Press, 1977)

Scott, J. C., *The moral economy of the peasant: rebellion and subsistence in Southeast Asia* (New Haven, Yale University Press, 1976)

Scoville, W. C., Minority migrations and the diffusion of technology, *Journal of Economic History*, 11 (1951), 347–60

Seavoy, R. E., *Famine in peasant societies* (New York, Greenwood Press, (1986)

Seebass, F., *Bergslagen* (Brunswick, Verlag Georg Westermann, 1928)

Segalen, M., *Love and power in the peasant family: rural France in the nineteenth century* (Oxford, Blackwell, 1983)

Sellergren, G., Textile and clothing industries, in Guinchard, *Sweden*, vol. II, 358–89

Shammas, C., *The pre-industrial consumer in England and America* (Oxford, Clarendon Press, 1990)

Shanin, T., *The awkward class* (Oxford, Clarendon Press, 1972)

Defining peasants: essays concerning rural societies, expolary economies, and learning from them in the contemporary world (Oxford, Blackwell, 1990)

(ed.), *Peasants and peasant societies*, 1st edn (Penguin, Harmondsworth, 1971); 2nd edn (Blackwell, Oxford, 1986)

Sheridan, T. E., *Where the dove calls: the political ecology of a peasant corporate community in northwestern Mexico* (Tucson, University of Arizona Press, 1988)

Slicher van Bath, B. H., *The agrarian history of western Europe AD 500–1850*, (London, Edward Arnold, 1963)

Smith, C. T., Historical geography: current trends and prospects, in Haggett and Chorley, *Frontiers*, 118–43

Snell, K. D. M., *Annals of the labouring poor: social change and agrarian England 1660–1900* (Cambridge, Cambridge University Press, 1985)

Söderbäck, P., *Rågöborna* (Stockholm, Nordiska Museet, 1940)

Söderberg, J., *Agrar fattigdom i Sydsverige under 1800–talet* (Acta Universitatis Stockholmiensis, Stockholm Studies in Economic History, 4, Stockholm, 1978)

Söderlund, E. F., The impact of the British industrial revolution on the Swedish iron industry, in Pressnell, *Studies*, 52–65

Soja, E. J., The spatiality of social life: towards a transformative retheorisation, in Gregory and Urry, *Social relations*, 90–127

Postmodern geographies: the reassertion of space in critical social theory (London, Verso, 1989)

Spufford, M., *Contrasting communities: English villagers in the sixteenth and seventeenth centuries* (Cambridge, Cambridge University Press, 1974)

Stadin, K., *Småstäder, småborgare och stora samhällsförändringar: borgarnas sociala struktur i Arboga, Enköping och Västervik under perioden efter 1680* (Acta Universitatis Upsaliensis, Studia Historica Upsaliensia, 105, Uppsala, 1979)

Stoklund, B., Ecological succession: reflections on the relations between Man and Environment in pre-industrial Denmark, *Ethnologia Scandinavica* (1976), 84–99

Stone, L., *The crisis of the aristocracy 1558–1641* (Oxford, Clarendon Press, 1965)

Sträng, G., *Vadstena under 700 År* (Vadstena, Vadstena Affärstryck, 1983)

Stridsberg, E. and Mattsson, L., *Skogen genom tiderna* (Stockholm, LT:s Förlag, 1980)

Sundbärg, G., *Emigrationsutredningen* (Betänkanden i utvandringsfrågan, Bilaga 5, Bygdestatistik, Stockholm, 1910) (ed.), *Sweden, its people and its industry: historical and statistical handbook* (Stockholm, P. A., Norstedt & Söner, 1904)

Sundin, J. and Söderlund, E. (eds.), *Time, space and man: essays on microdemography* (Stockholm, Almqvist & Wiksell, 1979)

Swain, J. T., *Industry before the industrial revolution: northeast Lancashire* c.*1500–1640* (Manchester, Chetham Society, 1986)

Tawney, R. H., *Religion and the rise of capitalism* (London, John Murray, 1926)

Tedebrand, L. G. (ed.), *Historieforskning på nya vägar* (Lund, Studentlitteratur, 1977)

Thirsk, J., Industries in the countryside, in Fisher, *Essays*, 70–88

The farming regions of England, in J. Thirsk, *The agrarian history of England and Wales*, vol. IV (Cambridge, Cambridge University Press, 1967), 11–112

Seventeenth-century agriculture and social change, in J. Thirsk (ed.), *Land, church and people* (Reading, Museum of Rural Life, 1970), 148–77

The European debate on customs of inheritance, 1500–1700, in Goody, Thirsk and Thompson, *Family*, 177–91

Economic policy and projects: the development of a consumer society in early modern England (Oxford, Oxford University Press, 1978)

Horn and thorn in Staffordshire: the economy of a pastoral county, in Thirsk, *Rural economy*, 163–82

The rural economy of England (London, Hambledon Press, 1984)

England's agricultural regions and agrarian history, 1500–1750 (London, Macmillan, 1987)

Thomas, R. P. and McCloskey, D. N., Overseas trade and empire 1700–1800, in Floud and McCloskey, *Economic history*, 87–102

Thompson, E. P., *The making of the English working class* (Harmondsworth, Penguin, 1968)

The grid of inheritance: a comment, in Goody, Thirsk and Thompson, *Family*, 328–60

The poverty of theory (London, Merlin Press, 1978)

Cultures in common (London, Merlin Press, 1991)

Thrift, N. and Pred, A., Time-geography: a new beginning, *Progress in Human Geography*, 5 (1981), 277–86

Thulin, E. J., The temperance question, in Guinchard, *Sweden*, vol. I, 740–50

Tidander, L. G., Inland commerce, in Sundbärg, *Sweden*, 934–6

Tillhagen, C. H., *Järnet och människorna* (Stockholm, LT:s Förlag, 1981)

Todd, E., *The making of modern France: politics, ideology and culture* (Oxford, Blackwell, 1991)

Topelius, A-S., *Damastduktyg, och verksamheten vid Vadstena fabrik 1753–1843* (Nordiska Museets Handlingar, 104, Stockholm, 1985)

Torbrand, D., *Johannishus fideikommiss intill 1735* (Meddelanden Geografiska Institutionen Stockholms Universitet, 147, Stockholm, 1963)

Törnqvist, G., On arenas and systems, in Pred, *Space and time*, 109–20

Tribe, K. Enclosure: reorganisation of landscape and labour, in K. Tribe, *Genealogies of capitalism* (London, Macmillan, 1981), 50–70

Tucker, J., *Instructions for Travellers* [1758] in R. L. Schuyler, *Josiah Tucker: a selection from his economic and political writings* (New York, Columbia University Press, 1931)

Tupling, G. H., *The economic history of Rossendale* (Chetham Society, NS, 86, 1927)

Turnbull, G., Canals, coal and regional growth during the industrial revolution, *Economic History Review*, NS, 40 (1987), 537–560

Turner, M. E., Parliamentary enclosure and landownership change in Buckinghamshire, *Economic History Review*, NS, 28 (1975), 565–76

Enclosures in Britain 1750–1830 (London, Macmillan, 1984)

Ullmann, W., *The individual and society in the Middle Ages* (London, Methuen, 1967)

Underdown, D., *Revel, riot and rebellion: popular politics and culture in England 1603–1660* (Oxford, Clarendon Press, 1985)

Unwin, G., *Industrial organization in the 16th and 17th centuries* (1904; reprint London, Frank Cass, 1957)

Utterström, G., *Jordbrukets arbetare*, vol. I (Stockholm, Tiden, 1957)

Verhulst, A., The origins of towns in the Low Countries and the Pirenne thesis, *Past and Present*, 122 (1989), 3–35

de Vries, J., Peasant demand patterns and economic development: Friesland 1550–1750, in Parker and Jones, *European peasants*, 205–66

The European economy in an age of crisis, 1600–1750 (Cambridge, Cambridge University Press, 1976)

Patterns of urbanization in pre-industrial Europe, 1500–1800, in Schmal, *Patterns*, 77–109

European urbanization 1500–1800 (London, Methuen, 1984)

Wallén, A., Hydrology, in Guinchard, *Sweden*, vol. I, 20–31

Wallerstein, I., *The modern World-System: capitalist agriculture and the origins of the European world economy in the sixteenth century* (London, Academic Press, 1974)

Historical capitalism (London, Verso, 1983)

Wannerdt, A., *Den svenska folkbokföringens historia under tre sekler* (Stockholm, Riksskatteverket, 1982)

Weber, E., *Peasants into Frenchmen: the modernization of rural France 1870–1914* (London, Chatto and Windus, 1977)

Weber, M., *The Protestant ethic and the spirit of capitalism* (1904–5; London, George Allen and Unwin, 1930)

Wennberg, A., *Lantbebyggelsen i nordöstra Östergötland 1600–1875* (Meddelanden från Lunds Universitets Geografiska Institutioner, Avhandlingar 13, Gleerups, Lund, 1947)

Wik, H., *Norra Sveriges sågverksindustri*, *Geographica*, 21 (Uppsala, 1950)

Wilkinson, J. C., Traditional concepts of territory in south east Arabia, *Geography Journal*, 148 (1983), 301–15

Wilmot, S., *'The business of improvement': agriculture and scientific culture in Britain, c.1700–c.1800* (Historical Geography Research Series, 24, 1990)

Winberg, C., *Folkökning och proletarisering* (Meddelanden från Historiska Institutionen, Göteborgs Universitet, 10, Göteborg, 1975)

Grenverket (Rättshistoriskt Bibliotek, 38, Stockholm, 1985)

Another route to modern society: the advancement of the Swedish peasantry, in M. Lundahl and T. Svensson (eds.), *Agrarian society in history: essays in honour of Magnus Mörner* (London, Routledge, 1990), 48–67

Öst och väst i svensk samhällsutveckling. En hypotetisk skiss, in *Lokalt, regionalt, centralt – analysnivåer i historisk forskning* (Studier i stads- och kommunhistoria, 3, Stads- och kommunhistoriska institutet, Stockholm, 1988)

Wohlin, N., *Den jordbruksidkande befolkningen i Sverige 1751–1900* (Emigrations-utredningen, App. IX, Stockholm, 1909)

Den svenska jordstyckningspolitiken i de 18:e och 19:e århundradena (Stockholm, Norstedts, 1912)

Wolf, E., *Peasants* (Englewood Cliffs, Prentice-Hall, 1966)

Europe and the people without history, (Berkeley, Calif., University of California Press, 1982)

Wrigley, E. A., *Population and history* (London, Weidenfeld and Nicolson, 1969)

Men on the land and men in the countryside: employment in agriculture in early nineteenth-century England, in L. Bonfield, R. M. Smith and K. Wrightson (eds.), *The world we have gained. Histories of population and social structure* (Oxford, Blackwell, 1986), 295–336

Continuity, chance and change: the character of the industrial revolution in England (Cambridge, Cambridge University Press, 1988)

Some reflections on corn yields and prices in pre-industrial economies, in Wrigley, *People*, 92–130

Urban growth and agricultural change: England and the Continent in the early modern period, in Wrigley, *People*, 157–93

People, cities and wealth: the transformation of traditional society (Oxford, Blackwell, 1987)

Yelling, J. A., *Common field and enclosure in England 1450–1850* (London, Macmillan, 1977)

Index

Capital
 Farm, 191, 194
 Flows of, 46, 47, 49, 265–6, 311–12,
  316–17, 329–32
 Investment: Rates of interest, 24, 316, 322,
  331, 421; Rural, 2, 5, 24, 26, 33–4, 47,
  194, 311, 317, 323, 324, 326; Urban, 219,
  237, 238, 249, 266, 279, 289
 See also Money
Carlshult estate, 79, 93
Carlsson, Peter of Västra Ny, 200
Carnegie, David, industrialist, 219
Carlsund, O. E., Manager of Motala
 Engineering Works, 291
Charlottenborg estate, 93
Chayanov, A. V., 84
Church, 26–7
 Clergy, 11, 12, 96, 240–1, 265
 Farming by, 26, 52, 96
 Land of, 330, 401
 Loans, 316, 320, 321, 326
 Protestant, 26–7, 30, 396
 Roman Catholic, 26
 Tithes, 11, 12, 36
 See also Farm, Land
Class *see* Social differentiation
Climate, 4, 14, 130, 186
Common property, 9–10, 12, 14–15, 22, 24,
 54, 56, 115, 127, 139, 398, 403
Communal farming, 8–10, 14, 24, 54, 126
 See also Farming system – Common field
Corvée duties *see* Labour services
Cottages and cottagers, 45, 53, 66, 115
 By-employments of, 140, 201, 202, 288, 304
 Crops of, 131, 132
 on Estates, 116, 147, 149, 160
 Family size of, 165, 167
 Mobility of, 344
 Numbers of, 106, 115–16, 117, 119,
  148–50, 153
 on Peasant Farms, 118, 119, 149, 150
Crafts
 Rural, 3, 7, 12, 31–3, 46, 223, 284, 388;
  Brewing, 195, 413; Dairying, 7, 30, 32,
  138, 183, 195, 413; Distilling, 13, 191–5,
  312, 377, 413, 424; Leather, 195, 201;
  Metal, 33, 141, 202, 286; Textiles, 7,
  32–3, 141–4, 183–4, 196–201, 288, 392,
  414, 424; Women and, 7, 32–3, 196, 198,
  199–201, 212, 220; Wood, 142, 196,
  201–2
 Urban, 27, 28, 213–15, 221; Building, 244,
  272, 274, 276; Clothing, 215, 243, 244,
  245, 272, 274, 276; Leather, 243, 244,
  273, 274; Metal, 215, 243, 244, 245, 272,
  273, 274; Ropemaking, 244; Textiles,
  246, 248, 250, 273, 276–7, 279–80, 392;

 Victualling, 243, 244, 270, 272, 274;
  Women and, 33, 248, 250, 276, 280;
  Wood, 243, 244, 272, 274, 276
 See also Craftsmen, Industry
Craftsmen
 Itinerant, 142
 Rural, 17, 31–3, 284; Farming by, 143, 245,
  284, 288
 Urban, 18, 29–30, 214, 221, 243, 244–7,
  253–5, 266, 472–6, 278, 281–3, 298–300,
  302, 303; Apprentices, 214, 221, 253–5,
  303, (Mobility of) 353–6, 363, 364, 365;
  Families of, 253–5, 281–3, 300–1, 304,
  306; Farming by, 239, 240, 245, 269, 298,
  299–300, 417; Income of, 238, 240, 244,
  264, 269, 272, 276, 299, 300;
  Journeymen, 214, 221, 252, 254, 255,
  277, 281–3, 300–3, 304, 306, (Mobility
  of), 353–6, 362, 363, 364; Mobility of,
  353, 356, 362–4; Real estate of, 239,
  244–5, 263–4, 269, 272–3, 295, 298,
  299–300
Crofts and crofters, 52–3
 By-employments of, 52, 140, 143, 196, 197,
  198, 201, 288, 304
 Colonization of land by, 61, 66, 118–19,
  174
 Crops of, 130, 131, 132, 178
 Debts of, 323, 328
 Duties of, 52, 61, 116
 on Estates, 52, 61, 116, 120, 149, 159–61
 Family size of, 165, 166, 167
 Implements of, 136, 179–80
 Livestock of, 137, 182, 184
 Mobility of, 335, 336, 344, 345, 349, 350,
  351–2
 Moneylending of, 203
 Numbers of, 103, 108, 115, 116, 117–19,
  148–50, 153, 344, 352
 on Peasant farms, 117–19, 120, 147–51;
  Förpantning, 118, 119, 328, 409
 Peddling by, 199
 Size of, 52, 61, 174, 407–8
 See also Soldiers
Crops, 175–8
 Barley, 129–32, 176–7, 186
 Flax, 178, 197, 245, 246, 398
 Fodder, 13, 14, 25, 130, 133, 135, 176, 178,
  188, 191
 Harvesting of, 134–6
 Innovation of, 13–14, 54, 130–1, 178, 410
 Lentils, 132, 176
 Manuring of, 128, 129, 171–2, 173, 181,
  185
 Mixed grain, 131, 176–7, 186, 193
 Oats, 13, 130–2, 176, 177, 186, 187, 189
 Peas, 130, 131, 132, 176, 177, 178, 187, 189

452   *Index*

Cambridge Studies in Historical Geography

Titles marked with an asterisk* are available in paperback